Cell Therapy for Lung Disease

Cell Therapy for Lung Disease

editor Julia Polak
Imperial College London, UK

ICP
Imperial College Press

Published by

Imperial College Press
57 Shelton Street
Covent Garden
London WC2H 9HE

Distributed by

World Scientific Publishing Co. Pte. Ltd.
5 Toh Tuck Link, Singapore 596224
USA office: 27 Warren Street, Suite 401-402, Hackensack, NJ 07601
UK office: 57 Shelton Street, Covent Garden, London WC2H 9HE

British Library Cataloguing-in-Publication Data
A catalogue record for this book is available from the British Library.

CELL THERAPY FOR LUNG DISEASE

ISBN-13 978-1-84816-439-0
ISBN-10 1-84816-439-4

Typeset by Stallion Press
Email: enquiries@stallionpress.com

Printed in Singapore by Mainland Press Pte Ltd

Contents

List of Contributors

Malcolm R. Alison
Centre for Diabetes and Metabolic Medicine
ICMS, 4 Newark Street
London E1 2AT, UK
E-mail: m.alison@qmul.ac.uk

Alberto Bellini
Avail Biomedical Research Institute
P.O. Box 110,
CH-4003, Basel, Switzerland

Timothy J. Cole
Department of Biochemistry and Molecular Biology
Strip II, Level 3
Building 76, Faculty of Medicine
Monash University
Wellington Rd
Clayton 3800, Australia

Christine M. Finck
Division of Pediatric Surgery
Connecticut Children's Medical Center
Hartford, CT 06103, USA

Alexandra Firsova
Department of Biochemistry and Molecular Biology
Strip II, Level 3
Building 76, Faculty of Medicine
Monash University
Wellington Rd
Clayton 3800, Australia

Peter I. Lelkes
Calhoun Chair Professor of Cellular Tissue Engineering
School of Biomedical Engineering, Science and Health Systems
Drexel University
Bossone Research Enterprise, Room 707
3141 Chestnut Street
Philadelphia, PA 19104, USA
E-mail: pilelkes@drexel.edu

Michael R. Loebinger
Centre of Respiratory Research
Rayne Building
University College London
5 University Street
London WC1E 6JJ, UK
E-mail: mloebinger@doctors.org.uk

Anastasios Mantalaris
Biological Systems Engineering Laboratory
Department of Chemical Engineering
Imperial College London
South Kensington Campus
London SW7 2AZ, UK
E-mail: a.mantalaris@imperial.ac.uk

Sabrina Mattoli
Avail Biomedical Research Institute
P.O. Box 110
CH-4003 Basel, Switzerland
E-mail: smattoli@avail-research.com

Shirley H. J. Mei
Ottawa Hospital Research Institute
725 Parkdale Ave
Ottawa, ON K1Y 4E9, Canada

Richard Mollard
Department of Biochemistry and Molecular Biology
Strip II, Level 3
Building 76, Faculty of Medicine
Monash University
Wellington Rd
Clayton 3800, Australia
E-mail: mollard@med.monash.edu.au

Rohit Moudgil
Internal Medicine
University of Ottawa
550 Cumberland St
Ottawa, ON K1N 6N5, Canada

Richard Newton
Division of Surgery,
Oncology, Reproductive Biology & Anaesthetics (SORA)
Biosurgery and Surgical Technology
Imperial College
St. Mary's Campus
London W2 1PG, UK

Julia M. Polak
Biological Systems Engineering Laboratory
Department of Chemical Engineering
Imperial College London
South Kensington Campus
London SW7 2AZ, UK

Blair Roszell
Division of Pediatric Surgery
Connecticut Children's Medical Center
Hartford, CT 06103, USA

William Rust
Lonza Walkersville, Inc.
8830 Biggs Ford Rd
Walkersville, MD 21793, USA
E-mail: william.rust@lonza.com

Elizabeth K. Sage
Centre of Respiratory Research
Rayne Building
University College London
5 University Street
London WC1E 6JJ, UK
E-mail: bethsage@hotmail.co.uk

Joshua C. Snyder
Department of Medicine
Division of Pulmonary, Allergy, and Critical Care Medicine
Duke University Medical Center
Durham, NC 27710, USA

Duncan J. Stewart
Department of Medicine
Faculty of Medicine
University of Ottawa
CEO and Scientific Director, OHRI
V.P. Research, Ottawa Hospital, Civic Campus
725 Parkdale Avenue
Ottawa, Ontario, Canada K1Y 4E9
E-mail: djstewart@ohri.ca

Siân Stinchcombe
Department of Respiratory Medicine
Royal Brompton Hospital
National Heart and Lung Institute
Imperial College London
Fulham Road
London SW3 6NP, UK

Barry R. Stripp
Duke University Medical Center
2075 MSRBII
106 Research Drive
DUMC Box 103000
Durham, NC 27710, USA
E-mail: Barry.Stripp@Duke.Edu

Viranuj Sueblinvong
Division of Pulmonary and Critical Care
Emory University
Atlanta, GA, USA

Benjamin T. Suratt
Vermont Lung Center
University of Vermont College of Medicine
149 Beaumont Avenue, HSRF 230
Burlington, Vermont 05405, USA
E-mail: Benjamin.Suratt@uvm.edu

Roxana M. Teisanu
Department of Medicine
Division of Pulmonary, Allergy, and Critical Care Medicine
Duke University Medical Center
Durham, NC 27710, USA

Daniel J. Weiss
Associate Professor of Medicine
Division of Pulmonary and Critical Care
The University of Vermont and Fletcher Allen Health Care
Burlington, VT 05405, USA

Guang-Zhong Yang
Department of Computing
Visual Information Group
Imperial College London
South Kensington Campus
London, SW7 2AZ, UK

Robert Zweigerdt
Institute of Medical Biology (IMB)
8A Biomedical Grove
06-06 Immunos, Level 5, Room # 5.04
Singapore 138648
E-mail: robert.zweigerdt@imb.a-star.edu.sg

Foreword

Regenerative medicine is a multidisciplinary field aimed at enhancing the body's own reparative capacity in order to provide safe and effective therapies for a host of incurable diseases. This innovative way of treating diseases uses either exogenous sources of cells, mostly stem cells, or stimulates the further production and release of endogenous stem cells, commonly known as "reparative" cells.

The use of stem cells to help with lung regeneration and repair is a novel therapy which could help phase out the need for conventional surgical or pharmacological approaches currently employed to treat diseases of the lung or other organs. The present book explores all avenues of this new form of medical care, moving swiftly, but in depth, from the basic science of lung development, to the analyses of different stem cells types available for regeneration and on to the application of this knowledge base in initial clinical trials.

I would like to congratulate Professor Dame Julia Polak and the impressive team of contributors for giving all of us an up-to-date insight into what is currently known on stem cell therapies for lung disorders. This work is enlightening and to the point. My warmest compliments to the team for highlighting the great promise for the future of regenerative medicine.

Professor Sir Roy Anderson
Imperial College London

Preface

Discovered centuries ago, regeneration is a fascinating biological phenomenon. Organ regeneration occurs throughout life in complex organisms and has puzzled investigators for centuries. In some phylogenetically lower organisms, such as hydra and the scrutinised planaria, organ regeneration is a relatively common process.

In humans, the liver has a powerful regenerative ability after partial surgical resection; the remaining liver cells respond by growing and dividing to expand the liver to its original size. Lung diseases are common and costly. They present a serious burden to society: hospitalisation is frequent and expensive therapies are mostly palliative. Quality of life is poor and loss of productive working day expensive.

The ground breaking discovery of the presence of stem cells in mammals, including man, led to the realisation of their immense potential as therapeutic agents for a large number of diseases including those of the respiratory tract. The Nobel Prize for Medicine of 2008 was awarded to Professor Sir Martin Evans, one of the discoverers.

Stem cells are the engines of the body, generating an identical daughter cell and another one of a specific lineage (precursor cells). This means the body can maintain, in normal circumstances, a balanced homeostasis.

Respiratory diseases account for more than 845,000 hospital admissions each year, behind only injury and poisoning as a cause of emergency admissions. Of the 580,000 deaths each year in the UK, one in five is due to respiratory diseases with 35,000 deaths being attributed to lung cancer followed by pneumonia and chronic obstructive pulmonary disease (COPD). Many chronic lung diseases, such as COPD, remain without cure and are only treatable with lung transplantation. However, the demand for organ transplants is high, and the shortage of donor organs severely limits this clinical approach.

An increasing number of studies suggest that cell therapy approaches may be powerful tools both for repair of injured or diseased lungs as well as for understanding mechanisms involved in both lung development and lung repair. This rapidly progressing field encompasses a number of disciplines and conceptual approaches including study of endogenous stem and progenitor cells resident in the lung and investigations utilising exogenously administered cells for repair of injured lung.

In this volume a stellar group of researchers converge, from different angles, to help towards clarifying the basic mechanisms of lung repair. These range from basic concepts of regeneration and lung development, the analyses of a variety of cell types that may be involved in lung repair, to ways of creating complex lung structures, including artificial and bioartificial lungs. The book offers an insight into repair mechanisms of the diseased lung, the role of specific lung niches and provides information on initial clinical trials as well as the use of stem cells as vehicles for gene therapy. Ingenious technological aspects of assessing stem cell engraftment of stem cell bioprocessing are also included in this volume.

Professor Dame Julia Polak
Imperial College London

Chapter 1
A General Review of the Current Knowledge of Stem Cell Therapy for Lung Disorders

Daniel J. Weiss

An increasing number of studies suggest that cell therapy approaches may be powerful tools for repair of injured or diseased lungs as well as for understanding mechanisms involved in both lung development and lung repair. This rapidly progressing field encompasses a number of disciplines and conceptual approaches including the study of endogenous stem and progenitor cells resident in the lung, and investigations utilizing exogenously administered cells for the repair of injured lung. Moreover, the field has undergone several conceptual shifts over recent years. For example, the initial focus on engraftment of exogenously administered cells as airway or alveolar epithelium has been shifted to the current emphases on immunomodulation of inflammatory and immune pathways in the lung by stem cells, and on bioengineering approaches to grow functional lung tissue *ex vivo* for subsequent use in *in vivo* implantation for destructive lung diseases, such as emphysema. Furthermore, it has become apparent that the variety of candidate stem and progenitor cell types can have different actions in the lung. Each of these areas is the focus of a comprehensive chapter in this book. The goal of this introductory chapter is to provide an overview of the field to date.

Keywords: Lung; stem cell; cell therapy; mesenchymal stem cell; endothelial progenitor cell; fibrocyte.

Outline

1. Introduction and Terminology

Prior to embarking on a review of the field, it is essential to establish the terminology to be utilized. This is all the more important as there is no uniform agreement on a number of definitions utilized in the published literature, and the lack of consistency in terminology between different publications has confounded comparisons between findings of different investigations. The issue of terminology affects both general basic definitions of "stem" and "progenitor" cells as well as more lung-focused definitions such as "bronchiolar stem cell" and "bronchioalveolar stem cell". Furthermore, precise definitions and characterisations of specific bone marrow-derived cell populations, notably mesenchymal stem cells and endothelial progenitor cells, are not agreed upon.[1–3]

This issue has been addressed in several forums, most recently at the conference, "Stem Cells and Cell Therapies in Lung Biology and Diseases", held in 2007 at the University of Vermont. A glossary of relevant working definitions applicable to the lung was developed at that conference and is depicted in Table 1.[1] This glossary does not necessarily reflect an overall consensus for the definition of each term and will undergo continuing revision as overall understanding of the cell types and mechanisms involved in lung repair continue to be elucidated. Nonetheless, it is a useful framework.

Table 1: Glossary and definitions of terminology.

Potency	Sum of developmental options available to cell.
Totipotent	Ability to (re)generate an organism in total. In mammals only the zygote and the first cleavage blastomeres are totipotent.
Pluripotent	Ability to form all lineages of body. Example: embryonic stem cells.
Multipotent	Ability of adult stem cells to form multiple cell types of one lineage. Example: haematopoietic stem cells.
Unipotent	Cells form one cell type. Example: spermatogonial stem cells (can only generate sperm).
Reprogramming	Change in epigenetics that can lead to an increase in potency, dedifferentiation. Can be induced by nuclear transfer, cell fusion, genetic manipulation.
Transdifferentiation	The capacity of a differentiated somatic cell to acquire the phenotype of a differentiated cell of the same or different lineage. An example is epithelial–mesenchymal transition (EMT), a process whereby fully differentiated epithelial cells undergo transition to a mesenchymal phenotype giving rise to fibroblasts and myofibroblasts.
Plasticity	Hypothesis that somatic stem cells have broadened potency and can generate cells of other lineages, a concept that is controversial in mammals.
Embryonic Stem Cell	Cells isolated from the inner mass of early developing blastocysts. ES cells have the capacity for self-renewal and are pluripotent, having the ability to differentiate into cells of all embryologic lineages and all adult cell types. However, ES cells cannot form extraembryonic tissue such as trophectoderm.
Adult Stem Cell	Cells isolated from adult tissues including bone marrow, adipose tissue, nervous tissue, skin, umbilical cord blood, and placenta that have the capacity for self-renewal. In general, adult stem cells are multipotent, having the capacity to differentiate into mature cell types of the parent tissue. Some populations of adult stem cells, such as MSCs, exhibit a range of lineage differentiation that is not limited to a single tissue type. Whether adult stem cells exhibit plasticity and can differentiate into a wider variety of differentiated cells and tissues remains controversial.
Adult Tissue-Specific Stem Cell	Same as adult stem cells but with defined tissue specificity. A relatively undifferentiated cell within a given tissue that has the capacity for self-renewal through stable maintenance within a stem cell niche. Adult tissue-specific (endogenous) stem cells have a differentiation potential

(*Continued*)

Table 1: (*Continued*)

	equivalent to the cellular diversity of the tissue in which they reside. The haematopoietic stem cell is a prototypical adult tissue stem cell.
Induced Pluripotent Stem Cell	Reprogrammed adult somatic cells that have undergone dedifferentiation, such as dermal fibroblasts, reprogrammed by retroviral transduction to express four transcription factors: Oct 3/4, Sox2, c-Myc, and Klf4. iPS cells are similar to ES cells in morphology, proliferation, gene expression, and ability to form teratomas. *In vivo* implantation of iPS cells results in formation of tissues from all three embryonic germ layers. iPS cells have been generated from both mouse and human cells.
Progenitor Cell	A collective term used to describe any proliferative cell that has the capacity to differentiate into different cell lineages within a given tissue. Unlike stem cells, progenitor cells have limited or no self-renewal capacity. The term progenitor cell is commonly used to indicate a cell can expand rapidly, but undergoes senescence after multiple cell doublings. Terminology that takes into account the functional distinctions among progenitor cells is suggested below.
Transit-Amplifying Cell	The progeny of endogenous tissue stem cells that retain relatively undifferentiated character, although more differentiated than the parent stem cell, and have a finite capacity for proliferation. The sole function of transit-amplifying cells is generation of a sufficient number of specialised progeny for tissue maintenance.
Obligate Progenitor Cell	A cell that loses its ability to proliferate once it commits to a differentiation pathway. Intestinal transit-amplifying cells are obligate progenitor cells.
Facultative Progenitor Cell	A cell that exhibits differentiated features when in the quiescent state yet has the capacity to proliferate for normal tissue maintenance and in response to injury. Bronchiolar Clara cells are an example of this cell type.
Classical Stem Cell Hierarchy	A stem cell hierarchy in which the adult tissue stem cell actively participates in normal tissue maintenance and gives rise to a transit-amplifying cell. Within this type of hierarchy, renewal potential resides in cells at the top of the hierarchy, i.e. the stem and transit-amplifying cell, and cells at each successive stage of proliferation become progressively more differentiated.

(*Continued*)

Table 1: *(Continued)*

Non-Classical Stem Cell Hierarchy	A stem cell hierarchy in which the adult tissue stem cell does not typically participate in normal tissue maintenance but can be activated to participate in repair following progenitor cell depletion.
Rapidly Renewing Tissue	Tissue in which homeostasis is dependent on maintenance of an active mitotic compartment. Rapid turnover of differentiated cell types requires continuous proliferation of stem and/or transit-amplifying cells. A prototypical rapidly renewing tissue is the intestinal epithelium.
Slowly Renewing Tissue	Tissues in which the steady state mitotic index is low. Specialised cell types are broadly distributed, long-lived, and a subset of these cells, the facultative progenitor cell, retains the ability to enter the cell cycle. The relative stability of the differentiated cell pool is paralleled by infrequent proliferation of stem and/or transit amplifying cells. The lung is an example of a slowly renewing tissue.
Haematopoietic Stem Cell	Cell that has the capacity for self-renewal and ability to differentiate into mature leukocytes, erythrocytes, and platelets. Whether HSCs exhibit plasticity and can differentiate into mature cells of other lineages remains controversial.
Endothelial Progenitor Cell	Circulating cells that have the potential to proliferate and differentiate into mature endothelial cells. Studies of EPCs have been complicated by the use of the same terminology to define at least two different cell populations that have different cell surface markers, different cell sources, and different abilities to differentiate into mature endothelial cells *in vitro* and *in vivo*. There is a critical need to develop a consensus definition of EPCs with particular emphasis on the functional capabilities of these cells.
Mesenchymal Stromal (Stem) Cell	Cells of stromal origin that can self-renew and have the ability to differentiate into a variety of cell lineages. Initially described in a population of bone marrow stromal cells, they were first described as fibroblastic colony forming units, subsequently as marrow stromal cells, then as mesenchymal stem cells, and most recently as multipotent mesenchymal stromal cells or MSCs. MSCs have now been isolated from a wide variety of tissues including umbilical cord blood, Wharton's jelly, placenta, adipose tissue, and lung. The Mesenchymal and Tissue Stem Cell Committee of the International Society for

(Continued)

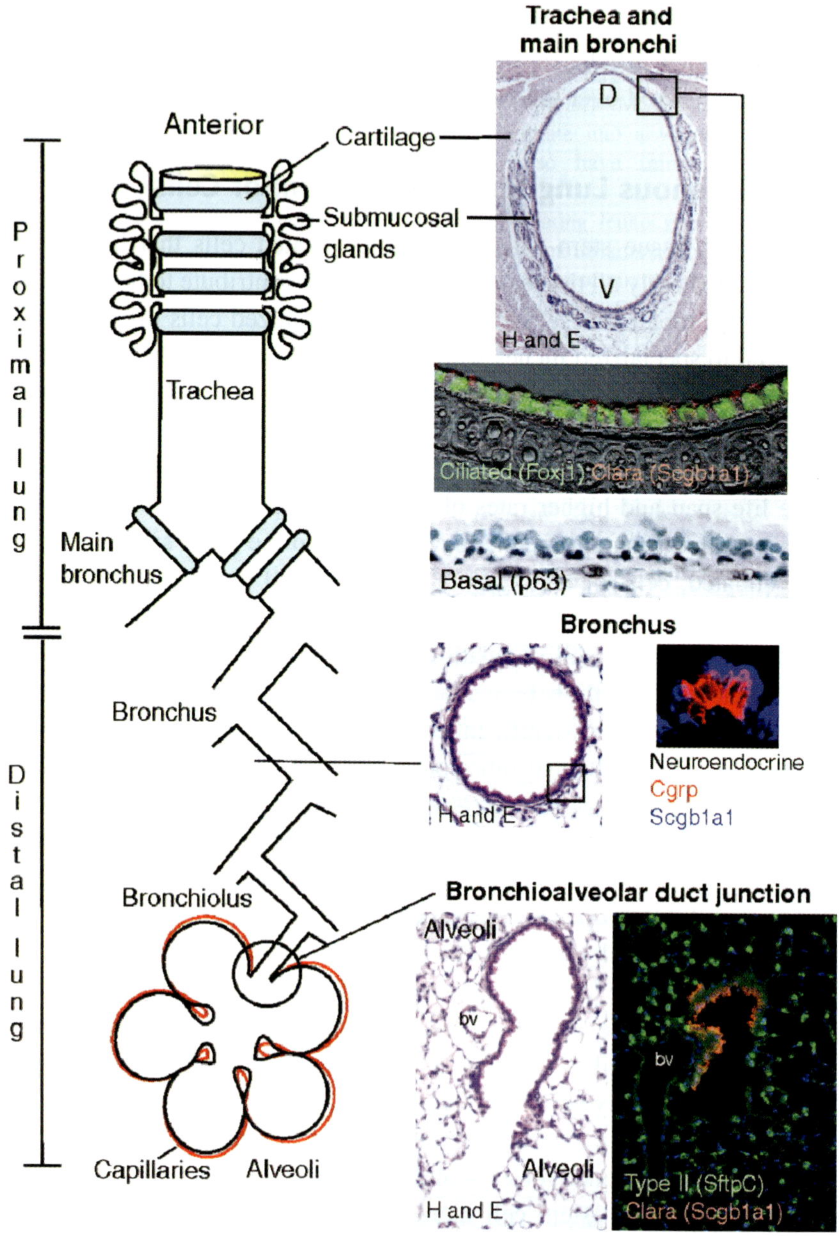

Figure 1: Schematic of endogenous progenitor cells. Adapted with permission from Rawlins and Hogan (2006).[10]

cytokeratins 5 and 14 have been implicated.[36–38] Aquaporin-3 has also been suggested as a marker of human foetal airway progenitor cells.[39] In lower airways in mice, Clara cells exhibit characteristics of transit-amplifying cells following injury to ciliated airway epithelial cells. However, unlike transit-amplifying cells in tissues with higher rates of epithelial turnover, such as intestine, Clara cells exhibit a low proliferative frequency in the steady-state, are broadly distributed throughout the bronchiolar epithelium, and contribute to the specialized tissue function. In more distal airways, toxin (i.e. naphthalene)-resistant variant Clara cells have been identified as having stem cell functions and have been termed as bronchiolar stem cells.[40–42] Naphthalene-resistant cells are also located within discrete microenvironments within bronchioles that include the neuroepithelial body (NEB) and bronchoalveolar duct junction (BADJ).[41,42]

At the bronchioalveolar duct junction in mouse lungs, a population of naphthalene-resistant cells that stain *in vivo* both for CCSP and for pro-SPC have been described.[17,18,43] Although rare, these cells were found to proliferate in response to naphthalene injury. Further, when pro-SPC/CCSP dual positive cells were isolated using methods developed for enrichment of type 2 alveolar epithelial cells, some of the dual labelled cells exhibited a unique cell surface phenotype, $Sca1^{pos}$/$CD34^{pos}$/$CD45^{neg}$/$CD31^{neg}$. These isolated cells were found to self-renew in culture and give rise to progeny expressing CCSP, pro-SPC, or aquaporin 5. As such, these cells have been termed bronchioalveolar stem cells (BASCs). However, whether the properties of these cells *in vitro* also apply *in vivo* is unknown. Moreover, the precise identity of the putative BASCs has come under question and it is possible that toxin-resistant Clara cells, BASCs, and other cells may represent different interpretations of the same cell population(s). This highlights the need both for rigorous methods of lineage tracing as well as further underscoring the importance of the *in vivo* microenvironment on cell behaviour. Most recently, another population of putative progenitor cells expressing CCSP, stem cell antigen (SCA-1), stage specific embryonic antigen 1 (SSEA-1), and the embryonic stem cell marker Oct-4 have been identified in neonatal mice.[44,45] These cells were able to form epithelial colonies and differentiate into both type 1 and type 2 alveolar epithelial cells. Interestingly, these cells were susceptible to infection with the SARS (severe acute respiratory syndrome) virus, raising the possibility that endogenous lung progenitor cells may be specific disease targets. The possibility remains that other endogenous stem or progenitor populations exist in mouse lungs and there is much room for additional information on comparable populations in human lungs.

Recent investigations have begun clarifying cell signalling and other mechanisms regulating putative lung progenitor populations. For example, tumorigenic insults, including deletion of MAPK, p18 deletion, and p27 oncogenic mutation, have been shown to induce an expansion of CCSP/pro-SPC dual-labelled BADJ cells number and also to enhance tumorigenesis in mouse lungs.[46–48] Most recently, it has been demonstrated that conditional potentiation of beta-catenin signalling in the embryonic mouse lung results in amplification of airway stem cells through attenuated differentiation rather than augmented proliferation.[49] Gata6, operating through canonical Wnt signalling pathways, has also been recently demonstrated to regulate the temporal appearance and number of bronchioalveolar stem cells (BASCs) in mouse lungs.[50] Absence of Gata6 resulted in pronounced increase of canonical Wnt signalling in lung epithelium, leading to the precocious appearance of BASCs and concurrent loss in epithelial differentiation. However, the precise role of these and other pathways in regulation of endogenous progenitor cells remains to be determined.

Endogenous progenitor cells may also be attractive candidates for targeting with gene transfer vectors that provide sustained expression. Using adult transgenic Rosa26-Flox/LacZ reporter mice, Liu and colleagues have recently demonstrated that airway-based administration of Cre-expressing recombinant adeno-associated virus vectors (rAAV1Cre and rAAV5Cre) appeared to result in preferential transduction of naphthalene-resistant variant Clara cells.[51] These findings support the possibility of selectively or preferentially transducing stem/progenitor cell populations in lung. While further investigations need to be done, this provides a new potential therapeutic approach for diseases affecting airway and alveolar epithelium.

Less information is available about the differences in endogenous stem and progenitor cells in different human lung disorders. Airway epithelium in cystic fibrosis patients contains primitive cuboidal cells that express primitive cell markers, including thyroid transcription factor and cytokeratin 7.[52] Neuroepithelial cells in mouse airway epithelium also express the cystic fibrosis transmembrane conductance regulator (CFTR), which appears to play a role in neuropeptide secretion.[53,54] CFTR knockout mice also contain fewer pulmonary neuroendocrine cells during embryonic development but increased numbers of these cells postnatally.[55] This suggests that endogenous progenitor cell pathways in cystic fibrosis lungs may be altered, but this has not been extensively investigated.

In addition to the role of endogenous lung stem and progenitor cells in repair of lung injury, increasing information suggests that mature differentiated lung cells may transdifferentiate and change phenotype. Best described for epithelial–mesenchymal transition, recent investigations describe a wider range of reversible phenotypes in epithelial and mucus cells.[56–61] These mechanisms may also play significant roles in injury or repair from injury. Overall, there remains much to be learned about endogenous stem and lung progenitor cells, including clarification of human counterparts to the cells identified in mouse models.

3. Endogenous Lung Progenitor Cells as Lung Cancer Stem Cells

There is substantial interest in the connections between endogenous stem or progenitor cells and cancer stem cells. Cancer stem cells have been defined in transplantation assays as the critical, generally chemo- or radiation therapy resistant, cells from tumours that are capable of propagating disease and are hypothesized to be the cells that maintain tumour progression and treatment resistance.[62,63] Although cancer stem cells have now been described in a variety of solid organ cancers, the existence of a lung cancer stem cell or cells is less well established.[15] A recent study has shown that CD133+ cells from cultured lung cancer cells and primary lung tumours exhibit the ability to propagate lung cancer when injected subcutaneously in mice.[64] Oct-4 expression was associated with CD133 expression and these may represent dual markers of lung cancer stem cells.[65] Side population CD45- Hoechst-effluxing cells have also recently been identified in several human lung cancer cell lines and exhibit tumorigenic properties when subcutaneously implanted into NOD SCID mice.[66] Side population cells have also been identified in lung cancer specimens.[67] However, despite suggestive data from these two reports, further work is needed to clarify the connections between endogenous lung stem cells, the cells of origin of cancer, and cancer stem cells, and also to determine their potential role as therapeutic targets. BASCs have been suggested to have a role in the development of lung cancer in mice but their role in tumour maintenance has not been established, nor has the human correlation been demonstrated.[43] Recent studies have begun elucidating cell signalling and gene expression pathways including Wnt, hedgehog, and others that may play roles in the transformation of endogenous progenitor cells into cancer cells.[68–72] Nonetheless, there is much room for additional information about the identity of lung cancer stem cells and their connections with endogenous lung stem and progenitor cells, particularly in human lungs.

4. Structural Engraftment of Circulating or Exogenously Administered Stem or Progenitor Cells

A number of publications over the past ten years have suggested that a variety of adult bone marrow-derived cells including haematopoeitic stem cells (HSCs), mesenchymal stem cells (MSCs), multipotent adult progenitor cells (MAPCs), and other populations, could structurally engraft as mature differentiated airway and alveolar epithelial cells or as vascular endothelial or interstitial lung cells. This literature has been predominantly based on studies in mice utilizing techniques that evaluated histologic demonstration of tagged donor-derived marrow cells in recipient lungs following systemic or more recently intratracheal administration of tagged donor marrow cells. Tagging techniques have included use of donor marrow cells from transgenic green fluorescent protein (GFP) expressing mice, or use of sex-mismatched transplantation ie. male donor cells into female recipients and subsequent identification of donor cells utilizing fluorescent *in situ* hybridisation (FISH) to detect the Y chromosome.[1] Lethal irradiation and myeloablation of the recipient mouse bone marrow prior to cell administration was frequently, but not always, utilized. Notably, prior lung injury, including radiation-induced lung injury resulting from myeloablative total body irradiation, was usually required to observe engraftment, although lung injury did not always result in apparent increase of engraftment.[73–75] A smaller body of literature in human bone marrow and lung transplantation also appears to demonstrate varying degrees of apparent chimerism in lungs of the transplant recipients.[1,76–83]

However, several reports have called into question whether epithelial engraftment does in fact occur.[84,85] Several technical issues contributed to misinterpretation of results in the initial reports, including inadequate microscopic techniques in which donor-derived cells superimposed on resident airway or alveolar epithelial cells were not effectively discriminated. Exquisite care and sophisticated microscopic approaches, including confocal and deconvolution techniques, must be utilized to effectively demonstrate potential engraftment.[1,4,84–87] Further, a variety of leukocytes, notably airway and alveolar macrophages, reside in the lung. Many of the early reports did not utilise antibodies directed against CD45 or other leukocyte markers, to exclude the possibility that cells of donor origin detected in airway or alveolar epithelium were donor-derived leukocytes rather than epithelial cells. Other tools, such as detection of GFP by direct fluorescence as a marker of donor-derived marrow cells, can be subject to error in the presence of autofluorescent cells resident in the lung, notably alveolar macrophages.[86]

Nonetheless, recent studies with more rigorous techniques continue to demonstrate that engraftment of airway and alveolar epithelium, as well as of pulmonary vascular endothelium and of lung interstitium, with adult bone marrow or cord blood-derived stem cells, although rare, can occur[88–97] (Fig. 2). Notably, engraftment of pulmonary vascular endothelium and stimulation of neo-angiogenesis by exogenously administered endothelial progenitor cells (EPCs) has fostered recent clinical trials for treatment of pulmonary hypertension (see contribution in this book).[98–100] Further, fusion of marrow-derived cells with resident organ cells, rather than phenotypic conversion of the marrow cells, has been demonstrated in several organs, notably liver and skeletal muscle, but also in lung.[101–103] The extent of fusion in the lung following sex-mismatched transplantation may be underestimated, as the Y chromosome may be lost from the resulting heterokaryon cells resulting from fusion of donor-derived marrow cells with type 2 alveolar epithelial cells in mouse lungs.[103]

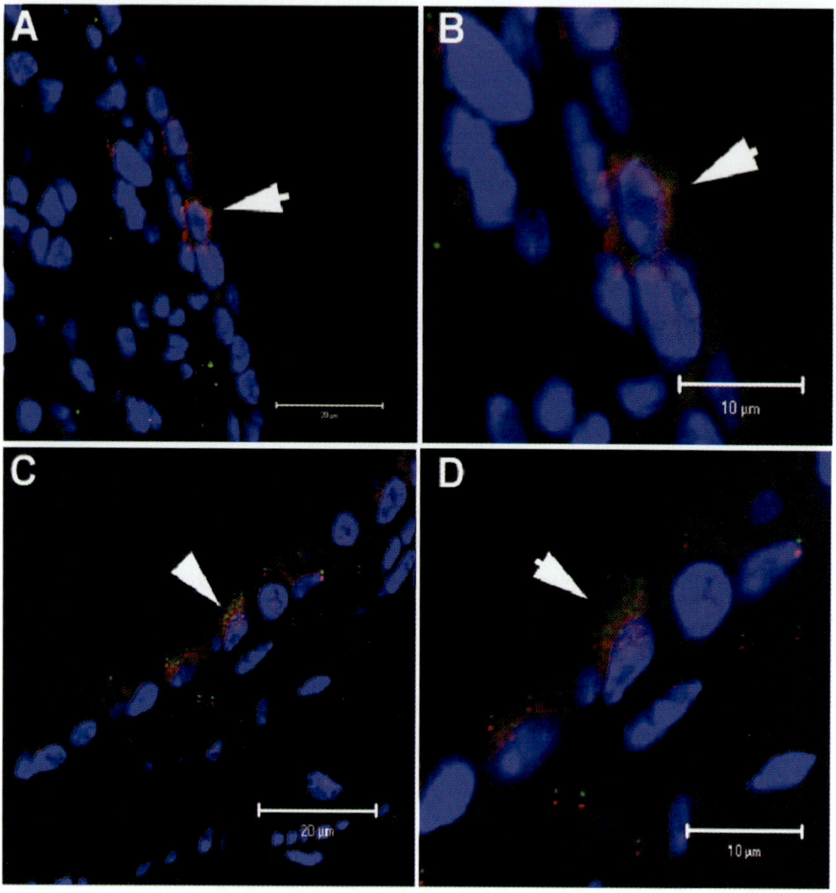

Figure 2: Human b2-microglobulin and cystic fibrosis transmembrane conductance regulator (CFTR)-positive cells can be detected in NOD-SCID mice airways after systemic administration of cord blood mesenchymal stem cells (CB-MSCs). (A and B) NOD-SCID lung sections two weeks and one month after CB-MSC administration, respectively. (C and D) Corresponding higher power images of (A) and (B). Representative photomicrographs from one of four mice per experimental condition. Original magnification 40×, 200×. Figure adapted with permission from Sueblinvong *et al.* (2008).[88] © American Thoracic Society.

These reports suggest that engraftment of lung tissues with circulating or donor-derived cells can occur under certain conditions, usually following previous perturbation through induction of lung injury. However, there are many variables still to be explored that may increase epithelial, interstitial, or pulmonary vascular engraftment with circulating or donor-derived cells. Interestingly, several recent reports suggest that chronic or progressive lung injury may result in more substantial engraftment of type 2 alveolar epithelial cells and of interstitial and pulmonary vascular cells with donor-derived cells in mouse or rat models.[92,97] However, not all chronic lung injury models resulted in more substantial engraftment.[104] The effect of age of either donor cells or of recipients is also less well explored, although one report demonstrated that transplantation of whole marrow into one day-old mouse pups, using a variety of conditioning regimens, did not increase the number of bone marrow-derived cells over that observed following total marrow administration to adult mice.[96] Furthermore, the route of administration of donor-derived cells is less well characterised, as most studies have investigated engraftment following systemic administration of donor cells. Direct intratracheal administration of MSCs may enhance retention of donor-derived cells in the lung and possible epithelial engraftment.[94] However, this area is controversial and remains under-explored.[105]

The types of marrow-derived, cord blood-derived, or fully differentiated nonpulmonary cells that might engraft as lung epithelium, interstitium, or pulmonary vasculature remain to be fully explored. In addition to existing studies of HSCs, MSCs (of bone marrow and cord blood origin), EPCs, fibrocytes, and MAPCs, the possibility remains that there may be other cell populations that could be recruited to the lung or localised to the lung following systemic or other routes of administration.[106,107] A population of circulating bone marrow-derived CD45+/CXCR4+/cytokeratin+ cells has been described to participate in re-epitheliasation of denuded tracheal xenografts.[108] Most recently, a population of CCSP-expressing marrow cells has been suggested to engraft as airway epithelium.[109] Other sources of stem or progenitor cells, such as adipose tissues, also have not yet been extensively characterised for ability to engraft as lung tissue.[109–112]

Moreover, there has not been a good demonstration that either mouse or human embryonic stem cells can engraft as lung epithelium or other lung cell type following either systemic or intratracheal administration. Both mouse and human embryonic stem cells can be induced *in vitro* to express pro-surfactants B and C and other markers of alveolar epithelial phenotype.[113–116] Further, exposure of mouse ES cells to dissociated foetal lung cells induces pseudoglandular formation and surfactant protein C expression.[117,118] In one notable study, mouse embryonic stem cells cultured at air–liquid interface formed a pseudostratified epithelium resembling mouse tracheal epithelium.[119] These studies demonstrate the capacity of ES cells to acquire lung epithelial phenotype *in vitro*, yet it is somewhat surprising that epithelial or other lung cell engraftment has not yet been demonstrated *in vivo*. Comparably, recent descriptions of induction of pluripotency in adult skin fibroblasts and other adult cell types offers another possibility of generating lung cells from autologous somatic cells.[120–123] In particular, it might be possible to generate induced pluripotent (iPS) cell lines from lung patients as models of disease. Human deltaF 508 embryonic stem cell lines have been established in England and Belgium.[124,125] These cells offer the possibility of investigating disease-specific cell therapy approaches for lung diseases and are areas ripe for continued investigation.

The ability to structurally engraft in adult lung may not solely be a property of stem or progenitor cells. Intratracheal administration of neonatal mouse lung fibroblasts resulted in apparent alveolar and interstitial engraftment, and engraftment was higher in areas of elastase-induced lung injury.[126] More recently, it has been demonstrated that intratracheal administration of type 2 alveolar epithelial cells results in rare engraftment in areas of injured lung following bleomycin administration to rats.[127] Notably bleomycin-injured rats that received the type 2 cells had less histologic injury and decreased hydroxyproline content. In another report, systemically administered skin fibroblasts, transduced *ex vivo* to express angiopoietin-1, protected against lung injury produced by intratracheal endotoxin administration in rats.[128] These results suggest that lung injuries might be amenable to cell therapy approaches utilizing a variety of cell types.

The mechanisms by which circulating or systemically administered stem or progenitor cells might be recruited to lung remain poorly understood. Following systemic (i.e. venous) administration, many cells initially localise in lung and recent studies continue to confirm that lung injury results in increased localisation and/or retention of marrow-derived cells in lung.[129–131] The timing of cell administration after lung injury can also influence recruitment and phenotypic conversion. Systemic administration of MSCs four hours after lung irradiation resulted in apparent engraftment of cells as epithelial and vascular endothelial cells.[132] However, MSCs administered at later time points appeared to engraft as interstitial cells and participate in development of fibrosis.[129,132] Recipient immune responses also play significant yet poorly characterised roles in retention of cells in lung.[133] Commonly utilised approaches of sex-mismatched transplantation or cell administration may also result in clearance of cells.[103] The range and identity of chemotactic soluble mediators released by injured lung cells, and the role of upregulation of adhesion molecules with which circulating cells might interact remains poorly understood.[1,4,134–136] As with engraftment, a number of factors including age of donor or recipient, type of cell administered, route of administration, amongst others might affect recruitment to lung.

Comparably, the mechanisms by which stem or progenitor cells might be induced to acquire phenotype of lung epithelial, interstitial, or vascular endothelial cells remain poorly understood. *In vitro* studies continue to demonstrate that soluble factors released from lung epithelial cells or from injured lung homogenates can induce expression of lung epithelial markers in several types of marrow-derived cells, possibly through activation of β-catenin signalling pathways. One novel mechanism of inducing phenotypic change might involve release of membrane-derived microvesicles, a recently appreciated means of inter-cellular communication that involves horizontal transfer of mRNA and proteins between cells.[140,141]

Overall, at present, engraftment of any type of circulating or endogenously administered stem or progenitor cell as alveolar or airway epithelium appears to be a rare phenomenon of uncertain physiologic or therapeutic significance. Engraftment of EPCs as pulmonary vascular endothelium coupled with EPC-stimulated neo-angiogenesis appears to be more robust and is the basis of current clinical trials of autologous EPCs for pulmonary hypertension (see other contributions in this book). Notably, the use of cell therapy as a gene or drug delivery device offers further potential promise. For example, transduction of EPCs to express nitric oxide synthetase is the basis of augmenting the potential therapeutic effect of autologous EPC administration in the current Canadian trial for pulmonary hypertension. However, engraftment of circulating cells in lung may not always be beneficial. Circulating fibrocytes have been implicated in the pathogenesis of lung fibrosis in several mouse models including irradiation and bleomycin-induced lung injury.[13,16,21,142–148]

Circulating fibrocytes have also been implicated in the sub-epithelial fibrosis that can develop in severe asthma[149–152] and in clinical bronchiolitis obliterans in lung transplant patients.[153] Moreover, bone marrow-derived or circulating MSCs, HSC, EPCs, and fibrocytes may contribute to development of lung and other malignancies in mouse models, in part by providing a supportive stroma for the cancers and/or by participating in tumour vascularisation.[154–166] However, MSCs and EPCs have been demonstrated to home to areas of tumour development and engineered EPCs, and MSCs, as well as HSCs, have been utilised to suppress tumour growth in mouse tumour models.[161,167–176] Cell-based therapies may thus be useful in lung cancer therapeutics.

5. Immunomodulation of Lung Inflammatory and Immune Pathways

There is an increasing number of studies demonstrating a functional role of adult marrow-derived stem cells in rodent models of acute lung inflammation and fibrosis. This occurs in the absence of significant lung engraftment, has been described in models of acute lung inflammation, emphysema, asthma, and fibrosis, and has been predominatly observed with either systemic or intratracheal administration of MSCs. For example, systemic administration of MSCs immediately after intratracheal bleomycin administration decreased subsequent lung collagen accumulation, fibrosis, and levels of matrix metalloproteinases.[177] Only minimal putative engraftment of the MSCs as lung epithelial cells was observed, and secretion of IL-1 receptor antagonist by the MSCs has been hypothesised to account for at least some of these effects.[178] Intratracheal administration of MSCs four hours after intratracheal endotoxin administration decreased mortality, tissue inflammation, and concentration of pro-inflammatory mediators, such as TNFα and MIP-1β, in bronchoalveolar lavage fluid compared to endotoxin-only treated mice.[105] Systemic MSC administration also decreased lung inflammation following endotoxin administration in mice, and co-culture of MSC with lung cells obtained from LPS-treated mice resulted in decreased pro-inflammatory cytokine release from the lung cells.[179,180] More recent data suggests that release of angiopoietin-1 or of keratinocyte growth factor by the MSCs and resulting stabilisation of alveolar-capillary permeability and endothelial fluid leak in the setting of endotoxin effects on the alveolar capillary barrier may be a relevant mechanism.[179–181] Notably, transduction of the MSCs to express high levels of angiopoietin-1 further decreased endotoxin-mediated lung injury.[180] However, systemic administration of skin fibroblasts transduced to express angiopoietin-1 also decreased acute endotoxin-induced lung injury, suggesting that a variety of cell types might be utilised for cell therapy approaches to acute lung injury.[128] In other lung injury models, intratracheal administration of bone marrow-derived MSCs decreased pulmonary hypertension and other manifestations of monocrotaline-induced pulmonary vascular injury.[182] Systemic administration of a heterogenous population of autologous adipose-derived stromal cells decreased manifestations of elastase-induced emphysema in rats.[183] Hepatocyte growth factor secreted by the stromal cells was postulated as a potential mechanism of injury repair in this report.

These results demonstrate that MSCs can have significant immunomodulatory effects in the lung in the absence of significant engraftment although the mechanisms by which this occurs are largely unknown. However, comparisons between studies and attempts to elucidate common potential mechanisms are somewhat complicated by using potentially different cell populations designated as MSCs. The Mesenchymal and Tissue Stem Cell Committee of the International Society for Cellular Therapy (ISCT) has recently updated the minimum criteria for defining (human) MSCs.[184] It is hoped that rigorous adherence to these criteria will help to focus comparative investigations of their potential utility in lung diseases. Nonetheless, consistencies between cells is further complicated as MSC characteristics can change with culture conditions and microenvironment and there is growing evidence that MSCs are heterogeneous and that different MSC subtypes exist.[185–187] Additionally, MSCs have now been isolated from a wide variety of tissues including umbilical cord blood, Wharton's jelly, placenta, adipose tissue, and most recently from adult mouse lungs and from lungs of both neonates and lung transplant recipients.[88,110–112,188–198] MSCs isolated from each of these sources generally express comparable cell surface markers and differentiate along recognised lineage pathways. However, differences in gene expression, lineage tendencies, and

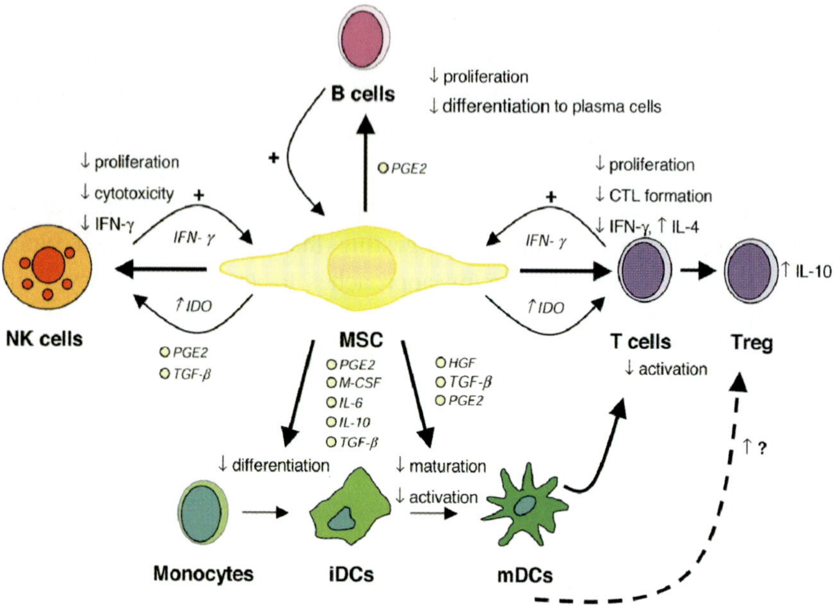

Figure 3: Schematic illustrating the range of *in vitro* immune-modulating effects described for MSCs. Figure adapted with permission from Nauta and Fibbe (2007).[210] © American Society of Hematology.

other properties have been described between MSCs isolated from different sources.[199–206]

Despite these issues, growing information suggests several possible anti-inflammatory actions of MSCs in lung. MSCs produce a wide variety of soluble mediators and can be influenced by specific microenvironments to release different patterns of mediators[207–219] (Fig. 3). For example, MSCs in bone marrow secrete cytokines and growth factors supportive of haematopoeitic cell proliferation and development, including granulocyte colony stimulating factor (G-CSF), stem cell factor (SCF), leukaemia inhibitory factor (LIF), inteleukin 6 (IL-6) and interleukin 11 (IL-11).[214] Stimulation with interleukin 1α (IL-1α), a pro-inflammatory cytokine that enhances bone marrow haematopoeisis, increases release of G-CSF, IL-6, and IL-11 by the MSCs. It is likely that release of inflammatory mediators from MSCs is influenced by the type of inflammatory environment found in different conditions of lung injury. For example,

MSCs express a wide variety of chemokine and cytokine receptors, including those for tumour necrosis factor α (TNFα), interleukin 4 (IL-4), interleukin 17 (IL-17), and interferon (IFNγ), as well as several toll-like receptors including the endotoxin receptor TLR4.[212] The IL-17_1 receptor in particular is expressed in high abundance[212] and IL-17 has been described as a proliferative stimulus for MSCs.[220] Nonetheless, the effects on MSC secretion of soluble mediators by these and other cytokines and chemokines found in different lung injury conditions have been less well explored.

MSCs can also inhibit the proliferation and function of a broad range of immune cells, including T cells, B cells, natural killer (NK) cells and dendritic cells (DCs)[207–210] (Fig. 3). Notably, MSCs inhibit T lymphocyte proliferation, activation, and cytokine release in response to either alloantigens or to mitogenic stimuli through a dose-dependent direct suppressive effect on proliferation. The mechanisms of MSC inhibition of T cell proliferation and function *in vitro* are only partly understood, and both direct cell–cell contact as well as release of soluble mediators has been proposed.[207–210] Further, MSCs constitutively express low levels of HLA class I molecules and do not express HLA class II molecules or the co-stimulatory molecules CD40, CD80, and CD86, which are essential for activation of T-lymphocyte-mediated immune responses.[207–210] As such, these properties render MSCs generally non-immunogenic and have been the basis of several clinical reports and, more recently, clinical trials utilizing both autologous and allogeneic MSCs for immune-mediated diseases, such as graft vs host disease and Crohn's disease.[211,221–228] These results suggest that administration of MSCs may be a safe and feasible clinical approach for inflammatory and immune-mediated lung diseases.

6. Lung Tissue Bioengineering

The use of three dimensional matrices or other artificial scaffolding for growth of functional tissues has been increasingly successful in regeneration of skin, vasculature, cartilage, bone, and other tissues.[229–231] Given the complex three-dimensional architecture of the lung, this is a daunting task; nonetheless, there has been significant progress in several areas.[232]

Adipose-derived MSCs, cultured *ex vivo* in sheets of polyglycolic acid and then applied to wound edges following lung volume reduction surgery in rats, accelerated alveolar and vascular regeneration.[183] Little other information exists at present evaluating the development of lung tissue from either adult or embryonic stem cells cultured in three-dimensional or other matrices.

In addition to genetic regulatory programs, mechanical forces also play a role in how differentiating tissues respond to gene instructions. During breathing, lung cells undergo complex mechanical loading that includes compression due to pressurisation and also stretch due to the expansion of the lung tissue during inhalation. *In utero*, the mechanical forces are, in part, generated by foetal breathing-like movements produced by rhythmic contractions of the respiratory muscles with varying frequency and amplitude.[248–251] The normal movement is essential in the development of foetal lung, differentiation of alveolar type 2 cells, and synthesis of surfactant protein.[252–256] The absence of foetal breathing-like movements will lead to pulmonary hypoplasia and is associated with poor infant outcome.[249–251] Several *in vitro* studies have demonstrated that application of cyclic biomechanical stretch to cultures of mixed foetal rat lung cells increases SPC mRNA expression compared to dexamethasone stimulation or control (no forces applied).[252–256] However, the mechanisms by which cyclic mechanical forces promote differentiation of foetal lung cells or influence repair of injured adult lung cells are poorly understood.[257–261]

These studies demonstrate that the application of bioengineering approaches to *ex vivo* lung tissue generation offers a powerful new approach for generating functional lung tissue that might be utilised for *in vivo* implantation to repair damaged lung in destructive diseases such as emphysema.

7. Clinical Trials in Lung Cell Therapy

While a relatively large number of clinical trials in cell therapy approaches have occurred in other disciplines, for example cardiovascular diseases, only a small number of trials have been initiated for pulmonary diseases. In part this reflects the complexity of the lung and relatively poor understanding of mechanisms by which exogenously administered stem or progenitor cells might have therapeutic effects in lung. Nonetheless, given the overall safety record of the cardiovascular trials as well as the safety and efficacy record of recent trials evaluating systemic MSC administration to patients with either graft vs host disease or with Crohn's disease, there is more enthusiasm to push ahead with clinical trials in pulmonary diseases. Three recent trials have focused on the use of systemic administration of autologous EPCs for primary pulmonary hypertension (see contribution in this book). Most recently, a multicentre double-blinded placebo control Phase 2 trial of allogeneic MSC infusions (PROCHYMAL™, Osiris Therapeutics Inc., Columbia MD) for patients with moderate–severe chronic obstructive pulmonary disease (COPD) (FEV1/FVC < 0.70, $30\% \leq$ FEV1 $\leq 70\%$) was initiated in May 2008.[262] This trial parallels comparable trials utilizing allogeneic MSC infusion for graft vs host disease and for Crohn's disease, and is based on the hypothesis that anti-inflammatory actions of MSCs will decrease pulmonary and perhaps systemic inflammation associated with COPD and improve lung function, dyspnea, and quality of life. Engraftment and/or regeneration of destroyed lung tissue is not hypothesised to be a significant potential mechanism of MSC action in this trial. Primary safety and efficacy endpoint assessments include pulmonary function testing, Borg dyspnea scale, and health-related quality of life assessments. Safety assessments further include the monitoring of blood counts, electrolytes, liver function tests, urinalyses, physician global assessments, time to hospitalisation and hospitalisation rates, time to COPD exacerbation and COPD exacerbation rates, use of rescue inhalers, and assessment of pulmonary hypertension by echocardiography. Safety endpoints also include the monitoring of adverse events, infusional toxicity, and overall survival and survival time.

These are groundbreaking and exciting trials of cell therapy approaches for lung diseases and will hopefully provide a firm basis of safety and potential efficacy for future trials in other lung diseases. Notably, MSCs can home to tumours, through as yet unclear mechanisms, and serve as vehicles for delivery of chemotherapeutic and other anti-tumour agents.[167–172] This has recently been described in mouse lung tumour models and may provide a viable therapy for lung cancers.[173–176] Administration of MSCs to patients with acute respiratory distress syndrome (ARDS) may also be another viable area for potential clinical investigations and trials.

8. Conclusions

The study of endogenous lung stem cells and investigation of cell therapy approaches for lung diseases are rapidly evolving fields that hold promise for both increasing understanding of the mechanisms of lung development and lung repair, and providing potential new therapeutic avenues for the treatment of lung diseases. As further explored in subsequent chapters, there are a variety of approaches being pursued, each of which may result in significant advances in both basic science and translational aspects of lung biology and diseases.

References

1. Weiss, D.J., Kolls, J.K., Ortiz, L.A., Panoskaltsis-Mortari, A., and Prockop, D.J. Stem cells and cell therapies in lung biology and lung diseases. *Proc Am Thorac Soc* **5**(5), 637–667 (2008).
2. Stripp, B.R. and Reynolds, S.D. Maintenance and repair of the bronchiolar epithelium. *Proc Am Thorac Soc* **5**(3), 328–333 (2008).
3. Jaenisch, R. and Young, R. Stem cells, the molecular circuitry of pluripotency and nuclear reprogramming. *Cell* **132**(4), 567–582 (2008).
4. Weiss, D.J., Berberich, M.A., Borok, Z., Gail, D.B., Kolls, J.K., Penland, C., and Prockop, D.J. Adult stem cells, lung biology, and lung disease. NHLBI/Cystic Fibrosis Foundation Workshop. *Proc Am Thorac Soc* **3**(3), 193–207 (2006).
5. Brown, J.K., Hogan, B.L.M., Randell. S.H., Stripp, B., and Weiss, D.J. Human embryonic stem cell research. *Am J Respir Crit Care Med* **173**(9), 1043–1045 (2006).
6. Fisher, K.A. and Summer, R.S. Stem and progenitor cells in the formation of the pulmonary vasculature. *Curr Top Dev Biol* **74**, 117–131 (2006).
7. Herzog, E.L. and Krause, D.S. Engraftment of marrow-derived epithelial cells: the role of fusion. *Proc Am Thorac Soc* **3**(8), 691–695 (2006).
8. Liu, X., Driskell, R.R., and Engelhardt, J.F. Stem cells in the lung. *Methods Enzymol* **419**, 285–321 (2006).
9. Neuringer, I.P. and Randell, S.H. Lung stem cell update: promise and controversy. *Monaldi Arch Chest Dis* **65**(1), 47–51 (2006).

10. Rawlins, E.L. and Hogan, B.L. Epithelial stem cells of the lung: privileged few or opportunities for many? *Development* **133**(13), 2455–2465 (2006).
11. van Haaften, T. and Thebaud, B. Adult bone marrow-derived stem cells for the lung: implications for pediatric lung diseases. *Pediatr Res* **59**(4 Pt 2), 94R–99R (2006).
12. Yen, C.C., Yang, S.H., Lin, C.Y., and Chen, C.M. Stem cells in the lung parenchyma and prospects for lung injury therapy. *Eur J Clin Invest* **36**(5), 310–319 (2006).
13. Bellini, A. and Mattoli, S. The role of the fibrocyte, a bone marrow-derived mesenchymal progenitor, in reactive and reparative fibroses. *Lab Invest* **87**(9), 858–870 (2007).
14. Varanou, A., Page, C.P., and Minger, S.L. Human embryonic stem cells and lung regeneration. *Br J Pharmacol* **155**(3), 316–325 (2008).
15. Giangreco, A., Groot, K.R., and Janes, S.M. Lung cancer and lung stem cells: strange bedfellows? *Am J Respir Crit Care Med* **175**(6), 547–553 (2007).
16. Gomperts, B.N. and Strieter, R.M. Fibrocytes in lung disease. *J Leukoc Biol* **82**(3), 449–456 (2007).
17. Kim, C.B. Advancing the field of lung stem cell biology. *Front Biosci* **12**, 3117–3124 (2007).
18. Kim, C.F. Paving the road for lung stem cell biology: bronchioalveolar stem cells and other putative distal lung stem cells. *Am J Physiol Lung Cell Mol Physiol* **293**(5), L1092–L1098 (2007).
19. Kotton, D.N. and Fine, A. Lung stem cells. *Cell Tissue Res* **331**(1), 145–156 (2008).
20. Loebinger, M.R. and Janes, S.M. Stem cells for lung disease. *Chest* **132**(1), 279–285 (2007).
21. Gharaee-Kermani, M., Gyetko, M.R., Hu, B., and Phan, S.H. New insights into the pathogenesis and treatment of idiopathic pulmonary fibrosis: a potential role for stem cells in the lung parenchyma and implications for therapy. *Pharm Res* **24**(5), 819–841 (2007).
22. Phinney, D.G. and Prockop, D.J. Concise review: mesenchymal stem/multipotent stromal cells: the state of transdifferentiation and modes of tissue repair — current views. *Stem Cells* **25**(11), 2896–2902 (2007).
23. Prockop, D.J. "Stemness" does not explain the repair of many tissues by mesenchymal stem/multipotent stromal cells (MSCs). *Clin Pharmacol Ther* **82**(3), 241–243 (2007).

24. Sueblinvong, V., Suratt, B.T., and Weiss, D.J. Novel therapies for the treatment of cystic fibrosis: new developments in gene and stem cell therapy. *Clin Chest Med* **28**(2), 361–379 (2007).
25. Reynolds, S.D., Shen, H., Reynolds, P.R., Betsuyaku, T., Pilewski, J.M., Gambelli, F., Di Giuseppe, M., Ortiz, L.A., and Stripp, B.R. Molecular and functional properties of lung SP cells. *Am J Physiol Lung Cell Mol Physiol* **292**(4), L972–L983 (2007).
26. Lama, V.N. and Phan, S.H. The extrapulmonary origin of fibroblasts: stem/progenitor cells and beyond. *Proc Am Thorac Soc* **3**(4), 373–376 (2006).
27. Olsson, F., Denham, M., Cole, T.J., Hooper, S.B., and Mollard, R. Deriving respiratory cell types from stem cells. *Curr Stem Cell Res Ther* **2**(3), 197–208 (2007).
28. Kannan, S. and Wu, M. Respiratory stem cells and progenitors: overview, derivation, differentiation, carcinogenesis, regeneration and therapeutic application. *Curr Stem Cell Res Ther* **1**(1), 37–46 (2006).
29. Murphy, J., Summer, R., and Fine, A. Stem cells in airway smooth muscle: state of the art. *Proc Am Thorac Soc* **5**(1), 11–14 (2008).
30. Krause, D.S. Bone marrow-derived cells and stem cells in lung repair. *Proc Am Thorac Soc* **5**(3), 323–327 (2008).
31. Martin, U. Methods for studying stem cells: adult stem cells for lung repair. *Methods* **45**(2), 121–132 (2008).
32. Loebinger, M.R., Sage, E.K., and Janes, S.M. Mesenchymal stem cells as vectors for lung disease. *Proc Am Thorac Soc* **5**(6), 711–716 (2008).
33. Warburton, D., Perin, L., Defilippo, R., Bellusci, S., Shi, W., and Driscoll, B. Stem/progenitor cells in lung development, injury repair, and regeneration. *Proc Am Thorac Soc* **5**(6), 703–706 (2008).
34. Rawlins, E.L. Lung epithelial progenitor cells: lessons from development. *Proc Am Thorac Soc* **5**(6), 675–681 (2008).
35. Van Vranken, B.E., Rippon, H.J., Samadikuchaksaraei, A., Trounson, A.O., and Bishop, A.E. The differentiation of distal lung epithelium from embryonic stem cells. *Curr Protoc Stem Cell Biol,* Chapter 1, Unit 1G 1 (2007).
36. Borthwick, D.W., Shahbazian, M., Krantz, Q.T., Dorin, J.R., and Randell, S.H. Evidence for stem-cell niches in the tracheal epithelium. *Am J Respir Cell Mol Biol* **24**(6), 662–670 (2001).
37. Schoch, K.G., Lori, A., Burns, K.A., Eldred, T., Olsen, J.C., and Randell, S.H. A subset of mouse tracheal epithelial basal cells generates large colonies *in vitro*. *Am J Physiol Lung Cell Mol Physiol* **286**(4), L631–L642 (2004).

38. Engelhardt, J.F., Schlossberg, H., Yankaskas, J.R., and Dudus, L. Progenitor cells of the adult human airway involved in submucosal gland development. *Development* **121**(7), 2031–2046 (1995).
39. Avril-Delplanque, A., Casal, I., Castillon, N., Hinnrasky, J., Puchelle, E., and Peault, B. Aquaporin-3 expression in human fetal airway epithelial progenitor cells. *Stem Cells* **23**(7), 992–1001 (2005).
40. Hong, K.U., Reynolds, S.D., Giangreco, A., Hurley, C.M., and Stripp, B.R. Clara cell secretory protein-expressing cells of the airway neuroepithelial body microenvironment include a label-retaining subset and are critical for epithelial renewal after progenitor cell depletion. *Am J Respir Cell Mol Biol* **24**(6), 671–681 (2001).
41. Giangreco, A., Reynolds, S.D., and Stripp, B.R. Terminal bronchioles harbor a unique airway stem cell population that localizes to the bronchoalveolar duct junction. *Am J Pathol* **161**(1), 173–182 (2002).
42. Reynolds, S.D., Giangreco, A., Power, J.H., and Stripp, B.R. Neuroepithelial bodies of pulmonary airways serve as a reservoir of progenitor cells capable of epithelial regeneration. *Am J Pathol* **156**(1), 269–278 (2000).
43. Kim, C.F., Jackson, E.L., Woolfenden, A.E., Lawrence, S., Babar, I., Vogel, S., Crowley, D., Bronson, R.T., and Jacks, T. Identification of bronchioalveolar stem cells in normal lung and lung cancer. *Cell* **121**(6), 823–835 (2005).
44. Ling, T.Y., Kuo, M.D., Li, C.L., Yu, A.L., Huang, Y.H., Wu, T.J., Lin, Y.C., Chen, S.H., and Yu, J. Identification of pulmonary Oct-4+ stem/progenitor cells and demonstration of their susceptibility to SARS coronavirus (SARS-CoV) infection *in vitro*. *Proc Natl Acad Sci USA* **103**(25), 9530–9535 (2006).
45. Ratajczak, M.Z., Machalinski, B., Wojakowski, W., Ratajczak, J., and Kucia, M. A hypothesis for an embryonic origin of pluripotent Oct-4(+) stem cells in adult bone marrow and other tissues. *Leukemia* **21**(5), 860–867 (2007).
46. Besson, A., Hwang, H.C., Cicero, S., Donovan, S.L., Gurian-West, M., Johnson, D., Clurman, B.E., Dyer, M.A., and Roberts, J.M. Discovery of an oncogenic activity in p27Kip1 that causes stem cell expansion and a multiple tumor phenotype. *Genes Dev* **21**(14), 1731–1746 (2007).
47. Ventura, J.J., Tenbaum, S., Perdiguero, E., Huth, M., Guerra, C., Barbacid, M., Pasparakis, M., and Nebreda, A.R. p38alpha MAP kinase is essential in lung stem and progenitor cell proliferation and differentiation. *Nat Genet* **39**(6), 750–758 (2007).

48. Yanagi, S., Kishimoto, H., Kawahara, K., Sasaki, T., Sasaki, M., Nishio, M., Yajima, N., Hamada, K., Horie, Y., Kubo, H., Whitsett, J.A., Mak, T.W., Nakano, T., Nakazato, M., and Suzuki, A. Pten controls lung morphogenesis, bronchioalveolar stem cells, and onset of lung adenocarcinomas in mice. *J Clin Invest* **117**(10), 2929–2940 (2007).
49. Reynolds, S.D., Zemke, A.C., Giangreco, A., Brockway, B.L., Teisanu, R.M., Drake, J.A., Mariani, T., Di, P.Y., Taketo, M.M., and Stripp, B.R. Conditional stabilization of beta-catenin expands the pool of lung stem cells. *Stem Cells* **26**(5), 1337–1346 (2008).
50. Zhang, Y., Goss, A.M., Cohen, E.D., Kadzik, R., Lepore, J.J., Muthukumaraswamy, K., Yang, J., DeMayo, F.J., Whitsett, J.A., Parmacek, M.S., and Morrisey, E.E. A Gata6-Wnt pathway required for epithelial stem cell development and airway regeneration. *Nat Genet* **40**(7), 862–870 (2008).
51. Liu, X., Luo, M., Guo, C., Yan, Z., Wang, Y., Lei-Butters, D.C., and Engelhardt, J.F. Analysis of adeno-associated virus progenitor cell transduction in mouse lung. *Mol Ther* online (November 2008).
52. Hollande, E., Cantet, S., Ratovo, G., Daste, G., Bremont, F., and Fanjul, M. Growth of putative progenitors of type II pneumocytes in culture of human cystic fibrosis alveoli. *Biol Cell* **96**(6), 429–441 (2004).
53. Pan, J., Bear, C., Farragher, S., Cutz, E., and Yeger, H. Cystic fibrosis transmembrane conductance regulator modulates neurosecretory function in pulmonary neuroendocrine cell-related tumor cell line models. *Am J Respir Cell Mol Biol* **27**(5), 553–560 (2002).
54. Yeger, H., Pan, J., Fu, X.W., Bear, C., and Cutz, E. Expression of CFTR and Cl(−) conductances in cells of pulmonary neuroepithelial bodies. *Am J Physiol Lung Cell Mol Physiol* **281**(3), L713–L721 (2001).
55. Pan, J., Luk, C., Kent, G., Cutz, E., and Yeger, H. Pulmonary neuroendocrine cells, airway innervation, and smooth muscle are altered in Cftr null mice. *Am J Respir Cell Mol Biol* **35**(3), 320–326 (2006).
56. Willis, B.C. and Borok, Z. TGF-beta-induced EMT: mechanisms and implications for fibrotic lung disease. *Am J Physiol Lung Cell Mol Physiol* **293**(3), L525–L534 (2007).
57. Willis, B.C., duBois, R.M., and Borok, Z. Epithelial origin of myofibroblasts during fibrosis in the lung. *Proc Am Thorac Soc* **3**(4), 377–382 (2006).
58. Demayo, F., Minoo, P., Plopper, C.G., Schuger, L., Shannon, J., and Torday, J.S. Mesenchymal-epithelial interactions in lung development and repair: Are

modeling and remodeling the same process? *Am J Physiol Lung Cell Mol Physiol* **283**(3), L510–L517 (2002).

59. Park, K.S., Wells, J.M., Zorn, A.M., Wert, S.E., Laubach, V.E., Fernandez, L.G., and Whitsett, J.A. Transdifferentiation of ciliated cells during repair of the respiratory epithelium. *Am J Respir Cell Mol Biol* **34**(2), 151–157 (2006).
60. Wu, R., Harper, R., Kao, C.Y., Thai, P., Wu, D., Chen, Y., Chang, M.M.J. New insights into airway mucous cell differentiation. *Journal of Organ Dysfunction* **2**(1), 30–36 (2006).
61. Wang, J., Edeen, K., Manzer, R., Chang, Y., Wang, S., Chen, X., Funk, C.J., Cosgrove, G.P., Fang, X., and Mason, R.J. Differentiated human alveolar epithelial cells and reversibility of their phenotype *in vitro*. *Am J Respir Cell Mol Biol* **36**(6), 661–668 (2007).
62. Morrison, S.J. Cancer stem cells. *Clin Adv Hematol Oncol* **3**(3), 171–172 (2005).
63. Reya, T., Morrison, S.J., Clarke, M.F., and Weissman, I.L. Stem cells, cancer, and cancer stem cells. *Nature* **414**(6859), 105–111 (2001).
64. Eramo, A., Lotti, F., Sette, G., Pilozzi, E., Biffoni, M., Di Virgilio, A., Conticello, C., Ruco, L., Peschle, C., and De Maria, R. Identification and expansion of the tumorigenic lung cancer stem cell population. *Cell Death Differ* **15**(3), 504–514 (2008).
65. Chen, Y.C., Hsu, H.S., Chen, Y.W., Tsai, T.H., How, C.K., Wang, C.Y., Hung, S.C., Chang, Y.L., Tsai, M.L., Lee, Y.Y., Ku, H.H., and Chiou, S.H. Oct-4 expression maintained cancer stem-like properties in lung cancer-derived CD133-positive cells. *PLoS ONE* **3**(7), e2637 (2008).
66. Ho, M.M., Ng, A.V., Lam, S., and Hung, J.Y. Side population in human lung cancer cell lines and tumors is enriched with stem-like cancer cells. *Cancer Res* **67**(10), 4827–4833 (2007).
67. Tian, J., Mahmood, R., Hnasko, R., and Locker, J. Loss of Nkx2.8 deregulates progenitor cells in the large airways and leads to dysplasia. *Cancer Res* **66**(21), 10399–10407 (2006).
68. Sicinski, P., Zacharek, S., and Kim, C. Duality of p27Kip1 function in tumorigenesis. *Genes Dev* **21**(14), 1703–1706 (2007).
69. Watkins, D.N., Berman, D.M., Burkholder, S.G., Wang, B., Beachy, P.A., and Baylin, S.B. Hedgehog signalling within airway epithelial progenitors and in small-cell lung cancer. *Nature* **422**(6929), 313–317 (2003).

70. Pei, X.H., Bai, F., Smith, M.D., and Xiong, Y. p18Ink4c collaborates with Men1 to constrain lung stem cell expansion and suppress non-small-cell lung cancers. *Cancer Res* **67**(7), 3162–3170 (2007).
71. Seo, D.C., Sung, J.M., Cho, H.J., Yi, H., Seo, K.H., Choi, I.S., Kim, D.K., Kim, J.S., El-Aty, A.A., and Shin, H.C. Gene expression profiling of cancer stem cell in human lung adenocarcinoma A549 cells. *Mol Cancer* **6**, 75 (2007).
72. Zhang, X., Zhao, P., Kennedy, C., Chen, K., Wiegand, J., Washington, G., Marrero, L., and Cui, Y. Treatment of pulmonary metastatic tumors in mice using lentiviral vector-engineered stem cells. *Cancer Gene Ther* **15**(2), 73–84 (2008).
73. Beckett, T., Loi, R., Prenovitz, R., Poynter, M., Goncz, K.K., Suratt, B.T., and Weiss, D.J. Acute lung injury with endotoxin or NO_2 does not enhance development of airway epithelium from bone marrow. *Mol Ther* **12**(4), 680–686 (2005).
74. Theise, N.D., Henegariu, O., Grove, J., Jagirdar, J., Kao, P.N., Crawford, J.M., Badve, S., Saxena, R., and Krause, D.S. Radiation pneumonitis in mice: a severe injury model for pneumocyte engraftment from bone marrow. *Exp Hematol* **30**(11), 1333–1338 (2002).
75. Herzog, E.L., Van Arnam, J., Hu, B., and Krause, D.S. Threshold of lung injury required for the appearance of marrow-derived lung epithelia. *Stem Cells* **24**(8), 1986–1992 (2006).
76. Suratt, B.T., Cool, C.D., Serls, A.E., Chen, L., Varella-Garcia, M., Shpall, E.J., Brown, K.K., and Worthen, G.S. Human pulmonary chimerism after hematopoietic stem cell transplantation. *Am J Respir Crit Care Med* **168**(3), 318–322 (2003).
77. Mattsson, J., Jansson, M., Wernerson, A., and Hassan, M. Lung epithelial cells and type II pneumocytes of donor origin after allogeneic hematopoietic stem cell transplantation. *Transplantation* **78**(1), 154–157 (2004).
78. Kleeberger, W., Versmold, A., Rothamel, T., Glockner, S., Bredt, M., Haverich, A., Lehmann, U., and Kreipe, H. Increased chimerism of bronchial and alveolar epithelium in human lung allografts undergoing chronic injury. *Am J Pathol* **162**(5), 1487–1494 (2003).
79. Spencer, H., Rampling, D., Aurora, P., Bonnet, D., Hart, S.L., and Jaffe, A. Transbronchial biopsies provide longitudinal evidence for epithelial chimerism in children following sex mismatched lung transplantation. *Thorax* **60**(1), 602 (2005).

80. Bittmann, I., Dose, T., Baretton, G.B., Muller, C., Schwaiblmair, M., Kur, F., and Lohrs, U. Cellular chimerism of the lung after transplantation. An interphase cytogenetic study. *Am J Clin Pathol* **115**(4), 525–533 (2001).
81. Davies, J.C., Potter, M., Bush, A., Rosenthal, M., Geddes, D.M., and Alton, E.W. Bone marrow stem cells do not repopulate the healthy upper respiratory tract. *Pediatr Pulmonol* **34**(4), 251–256 (2002).
82. Zander, D.S., Baz, M.A., Cogle, C.R., Visner, G.A., Theise, N.D., and Crawford, J.M. Bone marrow-derived stem-cell repopulation contributes minimally to the Type II pneumocyte pool in transplanted human lungs. *Transplantation* **80**(2), 206–212 (2005).
83. Albera, C., Polak, J.M., Janes, S., Griffiths, M.J., Alison, M.R., Wright, N.A., Navaratnarasah, S., Poulsom, R., Jeffery, R., Fisher, C., Burke, M., and Bishop, A.E. Repopulation of human pulmonary epithelium by bone marrow cells: a potential means to promote repair. *Tissue Eng* **11**(7–8), 1115–1121 (2005).
84. Kotton, D.N., Fabian, A.J., and Mulligan, R.C. Failure of bone marrow to reconstitute lung epithelium. *Am J Respir Cell Mol Biol* **33**(4), 328–334 (2005).
85. Chang, J.C., Summer, R., Sun, X., Fitzsimmons, K., and Fine, A. Evidence that bone marrow cells do not contribute to the alveolar epithelium. *Am J Respir Cell Mol Biol* **33**(4), 335–342 (2005).
86. Swenson, E.S., Price, J.G., Brazelton, T., and Krause, D.S. Limitations of green fluorescent protein as a cell lineage marker. *Stem Cells* **25**(10), 2593–2600 (2007).
87. Trotman, W., Beckett, T., Goncz, K.K., Beatty, B.G., and Weiss, D.J. Dual Y chromosome painting and *in situ* cell-specific immunofluorescence staining in lung tissue: an improved method of identifying donor marrow cells in lung following bone marrow transplantation. *Histochem Cell Biol* **121**(1), 73–79 (2004).
88. Sueblinvong, V., Loi, R., Eisenhauer, P.L., Bernstein, I.M., Suratt, B.T., Spees, J.L., and Weiss, D.J. Derivation of lung epithelium from human cord blood-derived mesenchymal stem cells. *Am J Respir Crit Care Med* **177**(7), 701–711 (2008).
89. Macpherson, H., Keir, P., Webb, S., Samuel, K., Boyle, S., Bickmore, W., Forrester, L., and Dorin, J. Bone marrow-derived SP cells can contribute to the respiratory tract of mice *in vivo*. *J Cell Sci* **118**(Pt 11), 2441–2450 (2005).

90. MacPherson, H., Keir, P.A., Edwards, C.J., Webb, S., and Dorin, J.R. Following damage, the majority of bone marrow-derived airway cells express an epithelial marker. *Respir Res* **7**, 145 (2006).
91. Loi, R., Beckett, T., Goncz, K.K., Suratt, B.T., and Weiss, D.J. Limited restoration of cystic fibrosis lung epithelium *in vivo* with adult bone marrow-derived cells. *Am J Respir Crit Care Med* **173**(2), 171–179 (2006).
92. Yan, C., Lian, X., Dai, Y., Wang, X., Qu, P., White, A., Qin, Y., and Du, H. Gene delivery by the hSP-B promoter to lung alveolar type II epithelial cells in LAL-knockout mice through bone marrow mesenchymal stem cells. *Gene Ther* **14**(20), 1461–1470 (2007).
93. Serikov, V.B., Popov, B., Mikhailov, V.M., Gupta, N., and Matthay, M.A. Evidence of temporary airway epithelial repopulation and rare clonal formation by BM-derived cells following naphthalene injury in mice. *Anat Rec (Hoboken)* **290**(9), 1033–1045 (2007).
94. Wong, A.P., Dutly, A.E., Sacher, A., Lee, H., Hwang, D.M., Liu, M., Keshavjee, S., Hu, J., and Waddell, T.K. Targeted cell replacement with bone marrow cells for airway epithelial regeneration. *Am J Physiol Lung Cell Mol Physiol* **293**(3), L740–L752 (2007).
95. Bruscia, E.M., Price, J.E., Cheng, E.C., Weiner, S., Caputo, C., Ferreira, E.C., Egan, M.E., and Krause, D.S. Assessment of cystic fibrosis transmembrane conductance regulator (CFTR). activity in CFTR-null mice after bone marrow transplantation. *Proc Natl Acad Sci USA* **103**(8), 2965–2970 (2006).
96. Bruscia, E.M., Ziegler, E.C., Price, J.E., Weiner, S., Egan, M.E., and Krause, D.S. Engraftment of donor-derived epithelial cells in multiple organs following bone marrow transplantation into newborn mice. *Stem Cells* **24**(10), 2299–2308 (2006).
97. Spees, J.L., Pociask, D.A., Sullivan, D.E., Whitney, M.J., Lasky, J.A., Prockop, D.J., and Brody, A.R. Engraftment of bone marrow progenitor cells in a rat model of asbestos-induced pulmonary fibrosis. *Am J Respir Crit Care Med* **176**(4), 385–394 (2007).
98. Zhao, Y.D., Courtman, D.W., Deng, Y., Kugathasan, L., Zhang, Q., and Stewart, D.J. Rescue of monocrotaline-induced pulmonary arterial hypertension using bone marrow-derived endothelial-like progenitor cells: efficacy of combined cell and eNOS gene therapy in established disease. *Circ Res* **96**(4), 442–450 (2005).

99. Raoul, W., Wagner-Ballon, O., Saber, G., Hulin, A., Marcos, E., Giraudier, S., Vainchenker, W., Adnot, S., Eddahibi, S., and Maitre, B. Effects of bone marrow-derived cells on monocrotaline- and hypoxia-induced pulmonary hypertension in mice. *Respir Res* **8**, 8 (2007).
100. Zhu, J.H., Wang, X.X., Zhang, F.R., Shang, Y.P., Tao, Q.M., and Chen, J.Z. Safety and efficacy of autologous endothelial progenitor cells transplantation in children with idiopathic pulmonary arterial hypertension: open-label pilot study. *Pediatr Transplant* **12**(6), 650–655 (2008).
101. Spees, J.L., Olson, S.D., Ylostalo, J., Lynch, P.J., Smith, J., Perry, A., Peister, A., Wang, M.Y., and Prockop, D.J. Differentiation, cell fusion, and nuclear fusion during *ex vivo* repair of epithelium by human adult stem cells from bone marrow stroma. *Proc Natl Acad Sci USA* **100**(5), 2397–2402 (2003).
102. Harris, R.G., Herzog, E.L., Bruscia, E.M., Grove, J.E., Van Arnam, J.S., and Krause, D.S. Lack of a fusion requirement for development of bone marrow-derived epithelia. *Science* **305**(5680), 90–93 (2004).
103. Herzog, E.L., Van Arnam, J., Hu, B., Zhang, J., Chen, Q., Haberman, A.M., and Krause, D.S. Lung-specific nuclear reprogramming is accompanied by heterokaryon formation and Y chromosome loss following bone marrow transplantation and secondary inflammation. *FASEB J* **21**(10), 2592–2601 (2007).
104. Levis, J., Loi, R., Butnor, K.J., Vacek, P., Steele, C., Mossman, B.T., and Weiss, D.J. Decreased asbestos-induced lung inflammation and fibrosis after radiation and bone marrow transplant. *Am J Respir Cell Mol Biol* **38**(1), 16–25 (2008).
105. Gupta, N., Su, X., Popov, B., Lee, J.W., Serikov, V., and Matthay, M.A. Intrapulmonary delivery of bone marrow-derived mesenchymal stem cells improves survival and attenuates endotoxin-induced acute lung injury in mice. *J Immunol* **179**(3), 1855–1863 (2007).
106. Kucia, M., Wysoczynski, M., Ratajczak, J., and Ratajczak, M.Z. Identification of very small embryonic like (VSEL) stem cells in bone marrow. *Cell Tissue Res* **331**(1), 125–134 (2008).
107. Kucia, M., Zuba-Surma, E.K., Wysoczynski, M., Wu, W., Ratajczak, J., Machalinski, B., and Ratajczak, M.Z. Adult marrow-derived very small embryonic-like stem cells and tissue engineering. *Expert Opin Biol Ther* **7**(10), 1499–1514 (2007).

108. Gomperts, B.N., Belperio, J.A., Rao, P.N., Randell, S.H., Fishbein, M.C., Burdick, M.D., and Strieter, R.M. Circulating progenitor epithelial cells traffic via CXCR4/CXCL12 in response to airway injury. *J Immunol* **176**(3), 1916–1927 (2006).
109. Wong, A.P., Keating, A., Lu, W.-Y., Duschesneau, P., Wang, X., Sacher, A., Hu, J., and Waddell, T.K. Identification of a bone marrow-derived epithelial-like population capable of repopulating injured mouse airway epithelium. *J Clin Invest* **119**, 336–348 (2009).
110. Berger, M.J., Adams, S.D., Tigges, B.M., Sprague, S.L., Wang, X.J., Collins, D.P., and McKenna, D.H. Differentiation of umbilical cord blood-derived multilineage progenitor cells into respiratory epithelial cells. *Cytotherapy* **8**(5), 480–487 (2006).
111. Miao, Z., Jin, J., Chen, L., Zhu, J., Huang, W., Zhao, J., Qian, H., and Zhang, X. Isolation of mesenchymal stem cells from human placenta: comparison with human bone marrow mesenchymal stem cells. *Cell Biol Int* **30**(9), 681–687 (2006).
112. Traktuev, D.O., Merfeld-Clauss, S., Li, J., Kolonin, M., Arap, W., Pasqualini, R., Johnstone, B.H., and March, K.L. A population of multipotent CD34-positive adipose stromal cells share pericyte and mesenchymal surface markers, reside in a periendothelial location, and stabilize endothelial networks. *Circ Res* **102**(1), 77–85 (2008).
113. Samadikuchaksaraei, A. and Bishop, A.E. Derivation and characterization of alveolar epithelial cells from murine embryonic stem cells *in vitro*. *Methods Mol Biol* **330**, 233–248 (2006).
114. Rippon, H.J., Polak, J.M., Qin, M., and Bishop, A.E. Derivation of distal lung epithelial progenitors from murine embryonic stem cells using a novel three-step differentiation protocol. *Stem Cells* **24**(5), 1389–1398 (2006).
115. Wang, D., Haviland, D.L., Burns, A.R., Zsigmond, E., and Wetsel, R.A. A pure population of lung alveolar epithelial type II cells derived from human embryonic stem cells. *Proc Natl Acad Sci USA* **104**(11), 4449–4454 (2007).
116. Samadikuchaksaraei, A., Cohen, S., Isaac, K., Rippon, H.J., Polak, J.M., Bielby, R.C., and Bishop, A.E. Derivation of distal airway epithelium from human embryonic stem cells. *Tissue Eng* **12**(4), 867–875 (2006).
117. Denham, M., Conley, B.J., Olsson, F., Gulluyan, L., Cole, T.J., and Mollard, R. A murine respiratory-inducing niche displays variable efficiency across

human and mouse embryonic stem cell species. *Am J Physiol Lung Cell Mol Physiol* **292**(5), L1241–L1247 (2007).

118. Denham, M., Cole, T.J., and Mollard, R. Embryonic stem cells form glandular structures and express surfactant protein C following culture with dissociated fetal respiratory tissue. *Am J Physiol Lung Cell Mol Physiol* **290**(6), L1210–L1215 (2006).
119. Coraux, C., Nawrocki-Raby, B., Hinnrasky, J., Kileztky, C., Gaillard, D., Dani, C., and Puchelle, E. Embryonic stem cells generate airway epithelial tissue. *Am J Respir Cell Mol Biol* **32**(2), 87–92 (2005).
120. Takahashi, K. and Yamanaka, S. Induction of pluripotent stem cells from mouse embryonic and adult fibroblast cultures by defined factors. *Cell* **126**(4), 663–676 (2006).
121. Takahashi, K., Tanabe, K., Ohnuki, M., Narita, M., Ichisaka, T., Tomoda, K., and Yamanaka, S. Induction of pluripotent stem cells from adult human fibroblasts by defined factors. *Cell* **131**(5), 861–872 (2007).
122. Yu, J., Vodyanik, M.A., Smuga-Otto, K., Antosiewicz-Bourget, J., Frane, J.L., Tian, S., Nie, J., Jonsdottir, G.A., Ruotti, V., Stewart, R., Slukvin, II, and Thomson, J.A. Induced pluripotent stem cell lines derived from human somatic cells. *Science* **318**(5858), 1917–1920 (2007).
123. Nakagawa, M., Koyanagi, M., Tanabe, K., Takahashi, K., Ichisaka, T., Aoi, T., Okita, K., Mochiduki, Y., Takizawa, N., and Yamanaka, S. Generation of induced pluripotent stem cells without Myc from mouse and human fibroblasts. *Nat Biotechnol* **26**(1), 101–106 (2008).
124. Pickering, S.J., Minger, S.L., Patel, M., Taylor, H., Black, C., Burns, C.J., Ekonomou, A., and Braude, P.R. Generation of a human embryonic stem cell line encoding the cystic fibrosis mutation deltaF508, using preimplantation genetic diagnosis. *Reprod Biomed Online* **10**(3), 390–397 (2005).
125. Mateizel, I., De Temmerman, N., Ullmann, U., Cauffman, G., Sermon, K., Van de Velde, H., De Rycke, M., Degreef, E., Devroey, P., Liebaers, I., and Van Steirteghem, A. Derivation of human embryonic stem cell lines from embryos obtained after IVF and after PGD for monogenic disorders. *Hum Reprod* **21**(2), 503–511 (2006).
126. Kuang, P.P., Lucey, E., Rishikof, D.C., Humphries, D.E., Bronsnick, D., and Goldstein, R.H. Engraftment of neonatal lung fibroblasts into the normal and elastase-injured lung. *Am J Respir Cell Mol Biol* **33**(4), 371–377 (2005).

127. Serrano-Mollar, A., Nacher, M., Gay-Jordi, G., Closa, D., Xaubet, A., and Bulbena, O. Intratracheal transplantation of alveolar type II cells reverses bleomycin-induced lung fibrosis. *Am J Respir Crit Care Med* **176**(12), 1261–1268 (2007).
128. McCarter, S.D., Mei, S.H., Lai, P.F., Zhang, Q.W., Parker, C.H., Suen, R.S., Hood, R.D., Zhao, Y.D., Deng, Y., Han, R.N., Dumont, D.J., and Stewart, D.J. Cell-based angiopoietin-1 gene therapy for acute lung injury. *Am J Respir Crit Care Med* **175**(10), 1014–1026 (2007).
129. Yan, X., Liu, Y., Han, Q., Jia, M., Liao, L., Qi, M., and Zhao, R.C. Injured microenvironment directly guides the differentiation of engrafted Flk-1(+) mesenchymal stem cell in lung. *Exp Hematol* **35**(9), 1466–1475 (2007).
130. Kahler, C.M., Wechselberger, J., Hilbe, W., Gschwendtner, A., Colleselli, D., Niederegger, H., Boneberg, E.M., Spizzo, G., Wendel, A., Gunsilius, E., Patsch, J.R., and Hamacher, J. Peripheral infusion of rat bone marrow derived endothelial progenitor cells leads to homing in acute lung injury. *Respir Res* **8**, 50 (2007).
131. Suzuki, H., Hogg, J.C., and van Eeden, S.F. Sequestration and homing of bone marrow-derived lineage negative progenitor cells in the lung during pneumococcal pneumonia. *Respir Res* **9**, 25 (2008).
132. Epperly, M.W., Guo, H., Gretton, J.E., and Greenberger, J.S. Bone marrow origin of myofibroblasts in irradiation pulmonary fibrosis. *Am J Respir Cell Mol Biol* **29**(2), 213–224 (2003).
133. Tolar, J., O'Shaughnessy M, J., Panoskaltsis-Mortari, A., McElmurry, R.T., Bell, S., Riddle, M., McIvor, R.S., Yant, S.R., Kay, M.A., Krause, D., Verfaillie, C.M., and Blazar, B.R. Host factors that impact the biodistribution and persistence of multipotent adult progenitor cells. *Blood* **107**(10), 4182–4188 (2006).
134. Viswanathan, A., Painter, R.G., Lanson, N.A., Jr., and Wang, G. Functional expression of N-formyl peptide receptors in human bone marrow-derived mesenchymal stem cells. *Stem Cells* **25**(5), 1263–1269 (2007).
135. Satoh, K., Kagaya, Y., Nakano, M., Ito, Y., Ohta, J., Tada, H., Karibe, A., Minegishi, N., Suzuki, N., Yamamoto, M., Ono, M., Watanabe, J., Shirato, K., Ishii, N., Sugamura, K., and Shimokawa, H. Important role of endogenous erythropoietin system in recruitment of endothelial progenitor cells in hypoxia-induced pulmonary hypertension in mice. *Circulation* **113**(11), 1442–1450 (2006).

136. Son, B.R., Marquez-Curtis, L.A., Kucia, M., Wysoczynski, M., Turner, A.R., Ratajczak, J., Ratajczak, M.Z., and Janowska-Wieczorek, A. Migration of bone marrow and cord blood mesenchymal stem cells *in vitro* is regulated by stromal-derived factor-1-CXCR4 and hepatocyte growth factor-c-met axes and involves matrix metalloproteinases. *Stem Cells* **24**(5), 1254–1264 (2006).
137. Popov, B.V., Serikov, V.B., Petrov, N.S., Izusova, T.V., Gupta, N., and Matthay, M.A. Lung epithelial cells induce endodermal differentiation in mouse mesenchymal bone marrow stem cells by paracrine mechanism. *Tissue Eng* **13**(10), 2441–2450 (2007).
138. Field-Corbett, C. and O'Dea, S. Soluble signals from mechanically disrupted lung tissue induce lung-related gene expression in bone marrow-derived cells *in vitro*. *Stem Cells Dev* **16**(2), 231–242 (2007).
139. Xu, J., Woods, C.R., Mora, A.L., Joodi, R., Brigham, K.L., Iyer, S., and Rojas, M. Prevention of endotoxin-induced systemic response by bone marrow-derived mesenchymal stem cells in mice. *Am J Physiol Lung Cell Mol Physiol* **293**(1), L131–L141 (2007).
140. Ratajczak, J., Miekus, K., Kucia, M., Zhang, J., Reca, R., Dvorak, P., and Ratajczak, M.Z. Embryonic stem cell-derived microvesicles reprogram hematopoietic progenitors: evidence for horizontal transfer of mRNA and protein delivery. *Leukemia* **20**(5), 847–856 (2006).
141. Aliotta, J.M., Sanchez-Guijo, F.M., Dooner, G.J., Johnson, K.W., Dooner, M.S., Greer, K.A., Greer, D., Pimentel, J., Kolankiewicz, L.M., Puente, N., Faradyan, S., Ferland, P., Bearer, E.L., Passero, M.A., Adedi, M., Colvin, G.A., and Quesenberry, P.J. Alteration of marrow cell gene expression, protein production, and engraftment into lung by lung-derived microvesicles: a novel mechanism for phenotype modulation. *Stem Cells* **25**(9), 2245–2256 (2007).
142. Phillips, R.J., Burdick, M.D., Hong, K., Lutz, M.A., Murray, L.A., Xue, Y.Y., Belperio, J.A., Keane, M.P., and Strieter, R.M. Circulating fibrocytes traffic to the lungs in response to CXCL12 and mediate fibrosis. *J Clin Invest* **114**(3), 438–446 (2004).
143. Moore, B.B., Murray, L., Das, A., Wilke, C.A., Herrygers, A.B., and Toews, G.B. The role of CCL12 in the recruitment of fibrocytes and lung fibrosis. *Am J Respir Cell Mol Biol* **35**(2), 175–181 (2006).
144. Moore, B.B., Kolodsick, J.E., Thannickal, V.J., Cooke, K., Moore, T.A., Hogaboam, C., Wilke, C.A., and Toews, G.B. CCR2-mediated recruitment

of fibrocytes to the alveolar space after fibrotic injury. *Am J Pathol* **166**(3), 675–684 (2005).

145. Ishida, Y., Kimura, A., Kondo, T., Hayashi, T., Ueno, M., Takakura, N., Matsushima, K., and Mukaida, N. Essential roles of the CC chemokine ligand 3-CC chemokine receptor 5 axis in bleomycin-induced pulmonary fibrosis through regulation of macrophage and fibrocyte infiltration. *Am J Pathol* **170**(3), 843–854 (2007).
146. Vannella, K.M., McMillan, T.R., Charbeneau, R.P., Wilke, C.A., Thomas, P.E., Toews, G.B., Peters-Golden, M., and Moore, B.B. Cysteinyl leukotrienes are autocrine and paracrine regulators of fibrocyte function. *J Immunol* **179**(11), 7883–7890 (2007).
147. Mehrad, B., Burdick, M.D., Zisman, D.A., Keane, M.P., Belperio, J.A., and Strieter, R.M. Circulating peripheral blood fibrocytes in human fibrotic interstitial lung disease. *Biochem Biophys Res Commun* **353**(1), 104–108 (2007).
148. Andersson-Sjoland, A., de Alba, C.G., Nihlberg, K., Becerril, C., Ramirez, R., Pardo, A., Westergren-Thorsson, G., and Selman, M. Fibrocytes are a potential source of lung fibroblasts in idiopathic pulmonary fibrosis. *Int J Biochem Cell Biol* **40**(10), 2129–2140 (2008).
149. Schmidt, M., Sun, G., Stacey, M.A., Mori, L., and Mattoli, S. Identification of circulating fibrocytes as precursors of bronchial myofibroblasts in asthma. *J Immunol* **171**(1), 380–689 (2003).
150. Nihlberg, K., Larsen, K., Hultgardh-Nilsson, A., Malmstrom, A., Bjermer, L., and Westergren-Thorsson, G. Tissue fibrocytes in patients with mild asthma: a possible link to thickness of reticular basement membrane? *Respir Res* **7**, 50 (2006).
151. Larsen, K., Macleod, D., Nihlberg, K., Gurcan, E., Bjermer, L., Marko-Varga, G., and Westergren-Thorsson, G. Specific haptoglobin expression in bronchoalveolar lavage during differentiation of circulating fibroblast progenitor cells in mild asthma. *J Proteome Res* **5**(6), 1479–1483 (2006).
152. Wang, C.H., Huang, C.D., Lin, H.C., Lee, K.Y., Lin, S.M., Liu, C.Y., Huang, K.H., Ko, Y.S., Chung, K.F., and Kuo, H.P. Increased circulating fibrocytes in asthma with chronic airflow obstruction. *Am J Respir Crit Care Med* **178**(6), 583–591 (2008).
153. Brocker, V., Langer, F., Fellous, T.G., Mengel, M., Brittan, M., Bredt, M., Milde, S., Welte, T., Eder, M., Haverich, A., Alison, M.R., Kreipe, H., and

Lehmann, U. Fibroblasts of recipient origin contribute to bronchiolitis obliterans in human lung transplants. *Am J Respir Crit Care Med* **173**(11), 1276–1282 (2006).

154. Hilbe, W., Dirnhofer, S., Oberwasserlechner, F., Schmid, T., Gunsilius, E., Hilbe, G., Woll, E., and Kahler, C.M. CD133 positive endothelial progenitor cells contribute to the tumour vasculature in non-small cell lung cancer. *J Clin Pathol* **57**(9), 965–969 (2004).
155. Aghi, M. and Chiocca, E.A. Contribution of bone marrow-derived cells to blood vessels in ischemic tissues and tumors. *Mol Ther* **12**(6), 994–1005 (2005).
156. Aghi, M., Cohen, K.S., Klein, R.J., Scadden, D.T., and Chiocca, E.A. Tumor stromal-derived factor-1 recruits vascular progenitors to mitotic neovasculature, where microenvironment influences their differentiated phenotypes. *Cancer Res* **66**(18), 9054–9064 (2006).
157. Oh, H.K., Ha, J.M., O, E., Lee, B.H., Lee, S.K., Shim, B.S., Hong, Y.K., and Joe, Y.A. Tumor angiogenesis promoted by *ex vivo* differentiated endothelial progenitor cells is effectively inhibited by an angiogenesis inhibitor, TK1-2. *Cancer Res* **67**(10), 4851–4859 (2007).
158. Nolan, D.J., Ciarrocchi, A., Mellick, A.S., Jaggi, J.S., Bambino, K., Gupta, S., Heikamp, E., McDevitt, M.R., Scheinberg, D.A., Benezra, R., and Mittal, V. Bone marrow-derived endothelial progenitor cells are a major determinant of nascent tumor neovascularization. *Genes Dev* **21**(12), 1546–1558 (2007).
159. Gao, D., Nolan, D.J., Mellick, A.S., Bambino, K., McDonnell, K., and Mittal, V. Endothelial progenitor cells control the angiogenic switch in mouse lung metastasis. *Science* **319**(5860), 195–198 (2008).
160. Direkze, N.C., Hodivala-Dilke, K., Jeffery, R., Hunt, T., Poulsom, R., Oukrif, D., Alison, M.R., and Wright, N.A. Bone marrow contribution to tumor-associated myofibroblasts and fibroblasts. *Cancer Res* **64**(23), 8492–8495 (2004).
161. Studeny, M., Marini, F.C., Dembinski, J.L., Zompetta, C., Cabreira-Hansen, M., Bekele, B.N., Champlin, R.E., and Andreeff, M. Mesenchymal stem cells: Potential precursors for tumor stroma and targeted-delivery vehicles for anticancer agents. *J Natl Cancer Inst* **96**(21), 1593–1603 (2004).
162. Karnoub, A.E., Dash, A.B., Vo, A.P., Sullivan, A., Brooks, M.W., Bell, G.W., Richardson, A.L., Polyak, K., Tubo, R., and Weinberg, R.A. Mesenchymal

stem cells within tumour stroma promote breast cancer metastasis. *Nature* **449**(7162), 557–563 (2007).

163. Avital, I., Moreira, A.L., Klimstra, D.S., Leversha, M., Papadopoulos, E.B., Brennan, M., and Downey, R.J. Donor-derived human bone marrow cells contribute to solid organ cancers developing after bone marrow transplantation. *Stem Cells* **25**(11), 2903–2909 (2007).
164. Djouad, F., Bony, C., Apparailly, F., Louis-Plence, P., Jorgensen, C., and Noel, D. Earlier onset of syngeneic tumors in the presence of mesenchymal stem cells. *Transplantation* **82**(8), 1060–1066 (2006).
165. Houghton, J., Stoicov, C., Nomura, S., Rogers, A.B., Carlson, J., Li, H., Cai, X., Fox, J.G., Goldenring, J.R., and Wang, T.C. Gastric cancer originating from bone marrow-derived cells. *Science* **306**(5701), 1568–1571 (2004).
166. Cogle, C.R., Theise, N.D., Fu, D., Ucar, D., Lee, S., Guthrie, S.M., Lonergan, J., Rybka, W., Krause, D.S., and Scott, E.W. Bone marrow contributes to epithelial cancers in mice and humans as developmental mimicry. *Stem Cells* **25**(8), 1881–1887 (2007).
167. Arap, W. and Pasqualini, R. Engineered embryonic endothelial progenitor cells as therapeutic Trojan horses. *Cancer Cell* **5**(5), 406–408 (2004).
168. Wei, J., Blum, S., Unger, M., Jarmy, G., Lamparter, M., Geishauser, A., Vlastos, G.A., Chan, G., Fischer, K.D., Rattat, D., Debatin, K.M., Hatzopoulos, A.K., and Beltinger, C. Embryonic endothelial progenitor cells armed with a suicide gene target hypoxic lung metastases after intravenous delivery. *Cancer Cell* **5**(5), 477–488 (2004).
169. Kim, S.M., Lim, J.Y., Park, S.I., Jeong, C.H., Oh, J.H., Jeong, M., Oh, W., Park, S.H., Sung, Y.C., and Jeun, S.S. Gene therapy using TRAIL-secreting human umbilical cord blood-derived mesenchymal stem cells against intracranial glioma. *Cancer Res* **68**(23), 9614–9623 (2008).
170. Hall, B., Dembinski, J., Sasser, A.K., Studeny, M., Andreeff, M., and Marini, F. Mesenchymal stem cells in cancer: tumor-associated fibroblasts and cell-based delivery vehicles. *Int J Hematol* **86**(1), 8–16 (2007).
171. Stoff-Khalili, M.A., Rivera, A.A., Mathis, J.M., Banerjee, N.S., Moon, A.S., Hess, A., Rocconi, R.P., Numnum, T.M., Everts, M., Chow, L.T., Douglas, J.T., Siegal, G.P., Zhu, Z.B., Bender, H.G., Dall, P., Stoff, A., Pereboeva, L., and Curiel, D.T. Mesenchymal stem cells as a vehicle for targeted delivery of CRAds to lung metastases of breast carcinoma. *Breast Cancer Res Treat* **105**(2), 157–167 (2007).

172. Miletic, H., Fischer, Y., Litwak, S., Giroglou, T., Waerzeggers, Y., Winkeler, A., Li, H., Himmelreich, U., Lange, C., Stenzel, W., Deckert, M., Neumann, H., Jacobs, A.H., and von Laer, D. Bystander killing of malignant glioma by bone marrow-derived tumor-infiltrating progenitor cells expressing a suicide gene. *Mol Ther* **15**(7), 1373–1381 (2007).
173. Xin, H., Kanehira, M., Mizuguchi, H., Hayakawa, T., Kikuchi, T., Nukiwa, T., and Saijo, Y. Targeted delivery of CX3CL1 to multiple lung tumors by mesenchymal stem cells. *Stem Cells* **25**(7), 1618–1626 (2007).
174. Rachakatla, R.S., Marini, F., Weiss, M.L., Tamura, M., and Troyer, D. Development of human umbilical cord matrix stem cell-based gene therapy for experimental lung tumors. *Cancer Gene Ther* **14**(10), 828–835 (2007).
175. Kanehira, M., Xin, H., Hoshino, K., Maemondo, M., Mizuguchi, H., Hayakawa, T., Matsumoto, K., Nakamura, T., Nukiwa, T., and Saijo, Y. Targeted delivery of NK4 to multiple lung tumors by bone marrow-derived mesenchymal stem cells. *Cancer Gene Ther* **14**(11), 894–903 (2007).
176. Hakkarainen, T., Sarkioja, M., Lehenkari, P., Miettinen, S., Ylikomi, T., Suuronen, R., Desmond, R.A., Kanerva, A., and Hemminki, A. Human mesenchymal stem cells lack tumor tropism but enhance the antitumor activity of oncolytic adenoviruses in orthotopic lung and breast tumors. *Hum Gene Ther* **18**(7), 627–641 (2007).
177. Ortiz, L.A., Gambelli, F., McBride, C., Gaupp, D., Baddoo, M., Kaminski, N., and Phinney, D.G. Mesenchymal stem cell engraftment in lung is enhanced in response to bleomycin exposure and ameliorates its fibrotic effects. *Proc Natl Acad Sci USA* **100**(14), 8407–8411 (2003).
178. Ortiz, L.A., Dutreil, M., Fattman, C., Pandey, A.C., Torres, G., Go, K., and Phinney, D.G. Interleukin 1 receptor antagonist mediates the antiinflammatory and antifibrotic effect of mesenchymal stem cells during lung injury. *Proc Natl Acad Sci USA* **104**(26), 11002–11007 (2007).
179. Mei, S.H., McCarter, S.D., Deng, Y., Parker, C.H., Liles, W.C., and Stewart, D.J. Prevention of LPS-induced acute lung injury in mice by mesenchymal stem cells overexpressing angiopoietin 1. *PLoS Med* **4**(9), e269 (2007).
180. Xu, J., Qu, J., Cao, L., Sai, Y., Chen, C., He, L., and Yu, L. Mesenchymal stem cell-based angiopoietin-1 gene therapy for acute lung injury induced by lipopolysaccharide in mice. *J Pathol* **214**(4), 472–481 (2008).
181. Lee, J.W., Fang, X., Gupta, N., Serikov, V., and Matthay, M.A. Allogeneic human mesenchymal stem cells for treatment of *E. coli* endotoxin-induced

acute lung injury in the *ex vivo* perfused human lung. *Proc Natl Acad Sci USA* **106**(26), 16357–16362 (2009).

182. Baber, S.R., Deng, W., Master, R.G., Bunnell, B.A., Taylor, B.K., Murthy, S.N., Hyman, A.L., and Kadowitz, P.J. Intratracheal mesenchymal stem cell administration attenuates monocrotaline-induced pulmonary hypertension and endothelial dysfunction. *Am J Physiol Heart Circ Physiol* **292**(2), H1120–H1128 (2007).
183. Shigemura, N., Okumura, M., Mizuno, S., Imanishi, Y., Matsuyama, A., Shiono, H., Nakamura, T., and Sawa, Y. Lung tissue engineering technique with adipose stromal cells improves surgical outcome for pulmonary emphysema. *Am J Respir Crit Care Med* **174**(11), 1199–1205 (2006).
184. Dominici, M., Le Blanc, K., Mueller, I., Slaper-Cortenbach, I., Marini, F., Krause, D., Deans, R., Keating, A., Prockop, D., and Horwitz, E. Minimal criteria for defining multipotent mesenchymal stromal cells. The International Society for Cellular Therapy position statement. *Cytotherapy* **8**(4), 315–317 (2006).
185. Lee, R.H., Hsu, S.C., Munoz, J., Jung, J.S., Lee, N.R., Pochampally, R., and Prockop, D.J. A subset of human rapidly self-renewing marrow stromal cells preferentially engraft in mice. *Blood* **107**(5), 2153–2161 (2006).
186. Crigler, L., Robey, R.C., Asawachaicharn, A., Gaupp, D., and Phinney, D.G. Human mesenchymal stem cell subpopulations express a variety of neuro-regulatory molecules and promote neuronal cell survival and neuritogenesis. *Exp Neurol* **198**(1), 54–64 (2006).
187. Phinney, D.G. Biochemical heterogeneity of mesenchymal stem cell populations: clues to their therapeutic efficacy. *Cell Cycle* **6**(23), 2884–2889 (2007).
188. Lee, O.K., Kuo, T.K., Chen, W.M., Lee, K.D., Hsieh, S.L., and Chen, T.H. Isolation of multipotent mesenchymal stem cells from umbilical cord blood. *Blood* **103**(5), 1669–1675 (2004).
189. Lee, M.W., Choi, J., Yang, M.S., Moon, Y.J., Park, J.S., Kim, H.C., and Kim, Y.J. Mesenchymal stem cells from cryopreserved human umbilical cord blood. *Biochem Biophys Res Commun* **320**(1), 273–278 (2004).
190. Bieback, K., Kern, S., Kluter, H., and Eichler, H. Critical parameters for the isolation of mesenchymal stem cells from umbilical cord blood. *Stem Cells* **22**(4), 625–634 (2004).
191. Wang, H.S., Hung, S.C., Peng, S.T., Huang, C.C., Wei, H.M., Guo, Y.J., Fu, Y.S., Lai, M.C., and Chen, C.C. Mesenchymal stem cells in the

Wharton's jelly of the human umbilical cord. *Stem Cells* **22**(7), 1330–1337 (2004).
192. Fukuchi, Y., Nakajima, H., Sugiyama, D., Hirose, I., Kitamura, T., and Tsuji, K. Human placenta-derived cells have mesenchymal stem/progenitor cell potential. *Stem Cells* **22**(5), 649–658 (2004).
193. Li, D., Wang, G.Y., Dong, B.H., Zhang, Y.C., Wang, Y.X., and Sun, B.C. Biological characteristics of human placental mesenchymal stem cells and their proliferative response to various cytokines. *Cells Tissues Organs* **186**(3), 169–179 (2007).
194. Puissant, B., Barreau, C., Bourin, P., Clavel, C., Corre, J., Bousquet, C., Taureau, C., Cousin, B., Abbal, M., Laharrague, P., Penicaud, L., Casteilla, L., and Blancher, A. Immunomodulatory effect of human adipose tissue-derived adult stem cells: comparison with bone marrow mesenchymal stem cells. *Br J Haematol* **129**(1), 118–129 (2005).
195. Yanez, R., Lamana, M.L., Garcia-Castro, J., Colmenero, I., Ramirez, M., and Bueren, J.A. Adipose tissue-derived mesenchymal stem cells have *in vivo* immunosuppressive properties applicable for the control of the graft-versus-host disease. *Stem Cells* **24**(11), 2582–2591 (2006).
196. Summer, R., Fitzsimmons, K., Dwyer, D., Murphy, J., and Fine, A. Isolation of an adult mouse lung mesenchymal progenitor cell population. *Am J Respir Cell Mol Biol* **37**(2), 152–159 (2007).
197. Hennrick, K.T., Keeton, A.G., Nanua, S., Kijek, T.G., Goldsmith, A.M., Sajjan, U.S., Bentley, J.K., Lama, V.N., Moore, B.B., Schumacher, R.E., Thannickal, V.J., and Hershenson, M.B. Lung cells from neonates show a mesenchymal stem cell phenotype. *Am J Respir Crit Care Med* **175**(11), 1158–1164 (2007).
198. Lama, V.N., Smith, L., Badri, L., Flint, A., Andrei, A.C., Murray, S., Wang, Z., Liao, H., Toews, G.B., Krebsbach, P.H., Peters-Golden, M., Pinsky, D.J., Martinez, F.J., and Thannickal, V.J. Evidence for tissue-resident mesenchymal stem cells in human adult lung from studies of transplanted allografts. *J Clin Invest* **117**(4), 989–996 (2007).
199. Kern, S., Eichler, H., Stoeve, J., Kluter, H., and Bieback, K. Comparative analysis of mesenchymal stem cells from bone marrow, umbilical cord blood, or adipose tissue. *Stem Cells* **24**(5), 1294–301 (2006).
200. Chang, Y.J., Shih, D.T., Tseng, C.P., Hsieh, T.B., Lee, D.C., and Hwang, S.M. Disparate mesenchyme-lineage tendencies in mesenchymal stem cells

from human bone marrow and umbilical cord blood. *Stem Cells* **24**(3), 679–685 (2006).

201. Tsai, M.S., Hwang, S.M., Chen, K.D., Lee, Y.S., Hsu, L.W., Chang, Y.J., Wang, C.N., Peng, H.H., Chang, Y.L., Chao, A.S., Chang, S.D., Lee, K.D., Wang, T.H., Wang, H.S., and Soong, Y.K. Functional network analysis of the transcriptomes of mesenchymal stem cells derived from amniotic fluid, amniotic membrane, cord blood, and bone marrow. *Stem Cells* **25**(10), 2511–2523 (2007).
202. Bernardo, M.E., Emons, J.A., Karperien, M., Nauta, A.J., Willemze, R., Roelofs, H., Romeo, S., Marchini, A., Rappold, G.A., Vukicevic, S., Locatelli, F., and Fibbe, W.E. Human mesenchymal stem cells derived from bone marrow display a better chondrogenic differentiation compared with other sources. *Connect Tissue Res* **48**(3), 132–140 (2007).
203. Panepucci, R.A., Siufi, J.L., Silva, W.A., Jr., Proto-Siquiera, R., Neder, L., Orellana, M., Rocha, V., Covas, D.T., and Zago, M.A. Comparison of gene expression of umbilical cord vein and bone marrow-derived mesenchymal stem cells. *Stem Cells* **22**(7), 1263–1278 (2004).
204. De Ugarte, D.A., Morizono, K., Elbarbary, A., Alfonso, Z., Zuk, P.A., Zhu, M., Dragoo, J.L., Ashjian, P., Thomas, B., Benhaim, P., Chen, I., Fraser, J., and Hedrick, M.H. Comparison of multi-lineage cells from human adipose tissue and bone marrow. *Cells Tissues Organs* **174**(3), 101–109 (2003).
205. Keyser, K.A., Beagles, K.E., and Kiem, H.P. Comparison of mesenchymal stem cells from different tissues to suppress T-cell activation. *Cell Transplant* **16**(5), 555–562 (2007).
206. Kunisaki, S.M., Fuchs, J.R., Steigman, S.A., and Fauza, D.O. A comparative analysis of cartilage engineered from different perinatal mesenchymal progenitor cells. *Tissue Eng* **13**(11), 2633–2644 (2007).
207. Stagg, J. Immune regulation by mesenchymal stem cells: two sides to the coin. *Tissue Antigens* **69**(1), 1–9 (2007).
208. Keating, A. Mesenchymal stromal cells. *Curr Opin Hematol* **13**(6), 419–425 (2006).
209. Rasmusson, I. Immune modulation by mesenchymal stem cells. *Exp Cell Res* **312**(12), 2169–2179 (2006).
210. Nauta, A.J. and Fibbe, W.E. Immunomodulatory properties of mesenchymal stromal cells. *Blood* **110**(10), 3499–3506 (2007).

211. Uccelli, A., Moretta, L., and Pistoia, V. Mesenchymal stem cells in health and disease. *Nat Rev Immunol* **8**, 726–736 (2008).
212. Phinney, D.G., Hill, K., Michelson, C., DuTreil, M., Hughes, C., Humphries, S., Wilkinson, R., Baddoo, M., and Bayly, E. Biological activities encoded by the murine mesenchymal stem cell transcriptome provide a basis for their developmental potential and broad therapeutic efficacy. *Stem Cells* **24**(1), 186–198 (2006).
213. Pevsner-Fischer, M., Morad, V., Cohen-Sfady, M., Rousso-Noori, L., Zanin-Zhorov, A., Cohen, S., Cohen, I.R., and Zipori, D. Toll-like receptors and their ligands control mesenchymal stem cell functions. *Blood* **109**(4), 1422–1432 (2007).
214. Haynesworth, S.E., Baber, M.A., and Caplan, A.I. Cytokine expression by human marrow-derived mesenchymal progenitor cells *in vitro*: Effects of dexamethasone and IL-1 alpha. *J Cell Physiol* **166**(3), 585–592 (1996).
215. Grayson, W.L., Zhao, F., Izadpanah, R., Bunnell, B., and Ma, T. Effects of hypoxia on human mesenchymal stem cell expansion and plasticity in 3D constructs. *J Cell Physiol* **207**(2), 331–339 (2006).
216. Stolzing, A. and Scutt, A. Effect of reduced culture temperature on antioxidant defences of mesenchymal stem cells. *Free Radic Biol Med* **41**(2), 326–338 (2006).
217. Lisignoli, G., Cristino, S., Piacentini, A., Cavallo, C., Caplan, A.I., and Facchini, A. Hyaluronan-based polymer scaffold modulates the expression of inflammatory and degradative factors in mesenchymal stem cells: Involvement of Cd44 and Cd54. *J Cell Physiol* **207**(2), 364–373 (2006).
218. Engler, A.J., Sen, S., Sweeney, H.L., and Discher, D.E. Matrix elasticity directs stem cell lineage specification. *Cell* **126**(4), 677–689 (2006).
219. Gregory, C.A., Ylostalo, J., and Prockop, D.J. Adult bone marrow stem/progenitor cells (MSCs) are preconditioned by microenvironmental "niches" in culture: A two-stage hypothesis for regulation of MSC fate. *Sci STKE* **2005**(294), pe37 (2005).
220. Huang, W., La Russa, V., Alzoubi, A., and Schwarzenberger, P. Interleukin-17A: A T-cell-derived growth factor for murine and human mesenchymal stem cells. *Stem Cells* **24**(6), 1512–1518 (2006).
221. Giordano, A., Galderisi, U., and Marino, I.R. From the laboratory bench to the patient's bedside: an update on clinical trials with mesenchymal stem cells. *J Cell Physiol* **211**(1), 27–35 (2007).

222. Le Blanc, K., Rasmusson, I., Sundberg, B., Gotherstrom, C., Hassan, M., Uzunel, M., and Ringden, O. Treatment of severe acute graft-versus-host disease with third party haploidentical mesenchymal stem cells. *Lancet* **363**(9419), 1439–1441 (2004).
223. Ringden, O., Uzunel, M., Rasmusson, I., Remberger, M., Sundberg, B., Lonnies, H., Marschall, H.U., Dlugosz, A., Szakos, A., Hassan, Z., Omazic, B., Aschan, J., Barkholt, L., and Le Blanc, K. Mesenchymal stem cells for treatment of therapy-resistant graft-versus-host disease. *Transplantation* **81**(10), 1390–1397 (2006).
224. Lazarus, H.M., Koc, O.N., Devine, S.M., Curtin, P., Maziarz, R.T., Holland, H.K., Shpall, E.J., McCarthy, P., Atkinson, K., Cooper, B.W., Gerson, S.L., Laughlin, M.J., Loberiza, F.R., Jr., Moseley, A.B., and Bacigalupo, A. Cotransplantation of HLA-identical sibling culture-expanded mesenchymal stem cells and hematopoietic stem cells in hematologic malignancy patients. *Biol Blood Marrow Transplant* **11**(5), 389–398 (2005).
225. Fang, B., Song, Y.P., Liao, L.M., Han, Q., and Zhao, R.C. Treatment of severe therapy-resistant acute graft-versus-host disease with human adipose tissue-derived mesenchymal stem cells. *Bone Marrow Transplant* **38**(5), 389–390 (2006).
226. Le Blanc, K., Samuelsson, H., Gustafsson, B., Remberger, M., Sundberg, B., Arvidson, J., Ljungman, P., Lonnies, H., Nava, S., and Ringden, O. Transplantation of mesenchymal stem cells to enhance engraftment of hematopoietic stem cells. *Leukemia* **21**(8), 1733–1738 (2007).
227. Ball, L.M., Bernardo, M.E., Roelofs, H., Lankester, A., Cometa, A., Egeler, R.M., Locatelli, F., and Fibbe, W.E. Cotransplantation of *ex vivo* expanded mesenchymal stem cells accelerates lymphocyte recovery and may reduce the risk of graft failure in haploidentical hematopoietic stem-cell transplantation. *Blood* **110**(7), 2764–2767 (2007).
228. Le Blanc, K., Frassoni, F., Ball, L., Locatelli, F., Roelofs, H., Lewis, I., Lanino, E., Sundberg, B., Bernardo, M.E., Remberger, M., Dini, G., Egeler, R.M., Bacigalupo, A., Fibbe, W., and Ringden, O. Mesenchymal stem cells for treatment of steroid-resistant, severe, acute graft-versus-host disease: a phase II study. *Lancet* **371**(9624), 1579–1586 (2008).
229. Bajada, S., Mazakova, I., Richardson, J.B., and Ashammakhi, N. Updates on stem cells and their applications in regenerative medicine. *J Tissue Eng Regen Med* **2**(4), 169–183 (2008).

230. Slater, B.J., Kwan, M.D., Gupta, D.M., Panetta, N.J., and Longaker, M.T. Mesenchymal cells for skeletal tissue engineering. *Expert Opin Biol Ther* **8**(7), 885–893 (2008).
231. Nichols, J.E. and Cortiella, J. Engineering of a complex organ: progress toward development of a tissue-engineered lung. *Proc Am Thorac Soc* **5**(6), 723–730 (2008).
232. Kunisaki, S.M., Armant, M., Kao, G.S., Stevenson, K., Kim, H., and Fauza, D.O. Tissue engineering from human mesenchymal amniocytes: a prelude to clinical trials. *J Pediatr Surg* **42**(6), 974–9; discussion 979–980 (2007).
233. Fuchs, J.R., Hannouche, D., Terada, S., Zand, S., Vacanti, J.P., and Fauza, D.O. Cartilage engineering from ovine umbilical cord blood mesenchymal progenitor cells. *Stem Cells* **23**(7), 958–964 (2005).
234. Kunisaki, S.M., Freedman, D.A., and Fauza, D.O. Fetal tracheal reconstruction with cartilaginous grafts engineered from mesenchymal amniocytes. *J Pediatr Surg* **41**(4), 675–82; discussion 675–682 (2006).
235. Kojima, K., Ignotz, R.A., Kushibiki, T., Tinsley, K.W., Tabata, Y., and Vacanti, C.A. Tissue-engineered trachea from sheep marrow stromal cells with transforming growth factor beta2 released from biodegradable microspheres in a nude rat recipient. *J Thorac Cardiovasc Surg* **128**(1), 147–153 (2004).
236. Kojima, K. and Vacanti, C.A. Generation of a tissue-engineered tracheal equivalent. *Biotechnol Appl Biochem* **39**(Pt 3), 257–262 (2004).
237. Macchiarini, P., Jungebluth, P., Go, T., Asnaghi, M.A., Rees, L.E., Cogan, T.A., Dodson, A., Martorell, J., Bellini, S., Parnigotto, P.P., Dickinson, S.C., Hollander, A.P., Mantero, S., Conconi, M.T., and Birchall, M.A. Clinical transplantation of a tissue-engineered airway. *Lancet* **372**(9655), 2023–2030 (2008).
238. Liu, M., Xu, J., Souza, P., Tanswell, B., Tanswell, A.K., and Post, M. The effect of mechanical strain on fetal rat lung cell proliferation: comparison of two- and three-dimensional culture systems. *In Vitro Cell Dev Biol Anim* **31**(11), 858–866 (1995).
239. Mondrinos, M.J., Koutzaki, S.H., Poblete, H.M., Crisanti, M.C., Lelkes, P.I., and Finck, C.M. *In vivo* pulmonary tissue engineering: Contribution of donor-derived endothelial cells to construct vascularization. *Tissue Eng Part A* **14**(3), 361–368 (2008).
240. Crisanti, M.C., Koutzaki, S.H., Mondrinos, M.J., Lelkes, P.I., and Finck, C.M. Novel methods for delivery of cell-based therapies. *J Surg Res* **146**(1), 3–10 (2008).

241. Hoganson, D.M., Pryor, H.I., 2nd, and Vacanti, J.P. Tissue engineering and organ structure: a vascularized approach to liver and lung. *Pediatr Res* **63**(5), 520–526 (2008).
242. Mondrinos, M.J., Koutzaki, S., Lelkes, P.I., and Finck, C.M. A tissue-engineered model of fetal distal lung tissue. *Am J Physiol Lung Cell Mol Physiol* **293**(3), L639–L650 (2007).
243. Mondrinos, M.J., Koutzaki, S., Jiwanmall, E., Li, M., Dechadarevian, J.P., Lelkes, P.I., and Finck, C.M. Engineering three-dimensional pulmonary tissue constructs. *Tissue Eng* **12**(4), 717–728 (2006).
244. Lin, Y.M., Boccaccini, A.R., Polak, J.M., Bishop, A.E., and Maquet, V. Biocompatibility of poly-DL-lactic acid (PDLLA) for lung tissue engineering. *J Biomater Appl* **21**(2), 109–118 (2006).
245. Choe, M.M., Sporn, P.H., and Swartz, M.A. Extracellular matrix remodeling by dynamic strain in a three-dimensional tissue-engineered human airway wall model. *Am J Respir Cell Mol Biol* **35**(3), 306–313 (2006).
246. Cortiella, J., Nichols, J.E., Kojima, K., Bonassar, L.J., Dargon, P., Roy, A.K., Vacant, M.P., Niles, J.A., and Vacanti, C.A. Tissue-engineered lung: an *in vivo* and *in vitro* comparison of polyglycolic acid and pluronic F-127 hydrogel/somatic lung progenitor cell constructs to support tissue growth. *Tissue Eng* **12**(5), 1213–1225 (2006).
247. Andrade, C.F., Wong, A.P., Waddell, T.K., Keshavjee, S., and Liu, M. Cell-based tissue engineering for lung regeneration. *Am J Physiol Lung Cell Mol Physiol* **292**(2), L510–L518 (2007).
248. Kitterman, J.A. The effects of mechanical forces on fetal lung growth. *Clin Perinatol* **23**(4), 727–740 (1996).
249. Inanlou, M.R., Baguma-Nibasheka, M., and Kablar, B. The role of fetal breathing-like movements in lung organogenesis. *Histol Histopathol* **20**(4), 1261–1266 (2005).
250. Blott, M., Greenough, A., and Nicolaides, K.H. Fetal breathing movements in pregnancies complicated by premature membrane rupture in the second trimester. *Early Hum Dev* **21**(1), 41–48 (1990).
251. Harding, R. Fetal pulmonary development: the role of respiratory movements. *Equine Vet J Suppl* **24**, 32–39 (1997).
252. Nakamura, T., Liu, M., Mourgeon, E., Slutsky, A., and Post, M. Mechanical strain and dexamethasone selectively increase surfactant protein C and tropoelastin gene expression. *Am J Physiol Lung Cell Mol Physiol* **278**(5), L974–L980 (2000).

253. Sanchez-Esteban, J., Cicchiello, L.A., Wang, Y., Tsai, S.W., Williams, L.K., Torday, J.S., and Rubin, L.P. Mechanical stretch promotes alveolar epithelial type II cell differentiation. *J Appl Physiol* **91**(2), 589–595 (2001).
254. Sanchez-Esteban, J., Tsai, S.W., Sang, J., Qin, J., Torday, J.S., and Rubin, L.P. Effects of mechanical forces on lung-specific gene expression. *Am J Med Sci* **316**(3), 200–204 (1998).
255. Liu, M., Skinner, S.J., Xu, J., Han, R.N., Tanswell, A.K., and Post, M. Stimulation of fetal rat lung cell proliferation *in vitro* by mechanical stretch. *Am J Physiol* **263**(3 Pt 1), L376–L383 (1992).
256. Sanchez-Esteban, J., Wang, Y., Gruppuso, P.A., and Rubin, L.P. Mechanical stretch induces fetal type II cell differentiation via an epidermal growth factor receptor-extracellular-regulated protein kinase signaling pathway. *Am J Respir Cell Mol Biol* **30**(1), 76–83 (2004).
257. Torday, J.S. and Rehan, V.K. Stretch-stimulated surfactant synthesis is coordinated by the paracrine actions of PTHrP and leptin. *Am J Physiol Lung Cell Mol Physiol* **283**(1), L130–L135 (2002).
258. Liu, M., Tanswell, A.K., and Post, M. Mechanical force induced signal transduction in lung cells. *Am J Physiol* **277**(4 Pt 1), L667–L683 (1999).
259. Liu, M. and Post, M. Invited review: mechanochemical signal transduction in the fetal lung. *J Appl Physiol* **89**(5), 2078–2084 (2000).
260. Ingber, D.E. Cellular mechanotransduction: putting all the pieces together again. *FASEB J* **20**(7), 811–827 (2006).
261. Zhao, Q., Liu, Z., Wang, Z., Yang, C., Liu, J., and Lu, J. Effect of preprocalcitonin gene-related peptide-expressing endothelial progenitor cells on pulmonary hypertension. *Ann Thorac Surg* **84**(2), 544–552 (2007).
262. http://www.osiris.com/clinical_trials_prochymal_phaseII.php.

Chapter 2
Development of the Lung: Clues for Regeneration and Repair

Sheila G. Haworth and Alison A. Hislop

Nature builds the lung from scratch and understanding the signalling cascades that control the temporal and spatial construction of the airways, parenchyma and blood vessels will provide clues about how man could learn to repair or even replace damaged lung tissue. We describe the normal development of the airways and the vasculature, then discuss interaction between these systems during development and present a brief section on functional development as the lung adapts to extra-uterine life, the most critical period of lung development. Reference is made to abnormal development because it can be viewed as a vicarious experiment. Lastly, we have summarised current information on lung regeneration and repair from a developmental perspective.

Keywords: Lung development; adaptation to extra-uterine life; pulmonary arterial hypertension; regulation of airway and arterial development.

Outline

1. Introduction

The past ten years have seen major advances in our understanding of lung development. This new information is vital to our understanding the potential of the lung to regenerate. Nature builds the lung from scratch and understanding the signalling cascades that control the temporal and spatial construction of the airways, parenchyma and blood vessels will provide clues about how man could learn to repair or even replace damaged lung tissue. Building a new "donor" lung is the ultimate goal. This chapter will review the progress made in the field of lung development, the lessons learnt and indicate some of the major deficits in our understanding which should be addressed urgently.

The lung is a complex structure composed of the airways and pulmonary vasculature, whose functions in post-natal life are obviously interdependent, but their embryonic, foetal and post-natal development is also interdependent. We will first describe the normal development of the airways and the vasculature, and then discuss the interaction between these systems during development. Structure and function are indivisible. Mitogens are contractile agonists and anti-proliferative agents have vasodilator properties. This chapter will therefore include a brief section on functional development as the lung adapts to extra-uterine life, the most critical period of lung development. Reference will also be made to abnormal development since this is informative in that it can be viewed as a vicarious experiment. Lastly, we have summarised current information on lung regeneration and repair.

2. Normal Development of the Airways and the Regulation of Development

(See also Chapter 3)

The lung is a made up of a complex and interacting branching systems of airways, arteries and veins (Fig. 1). The primary developmental objective

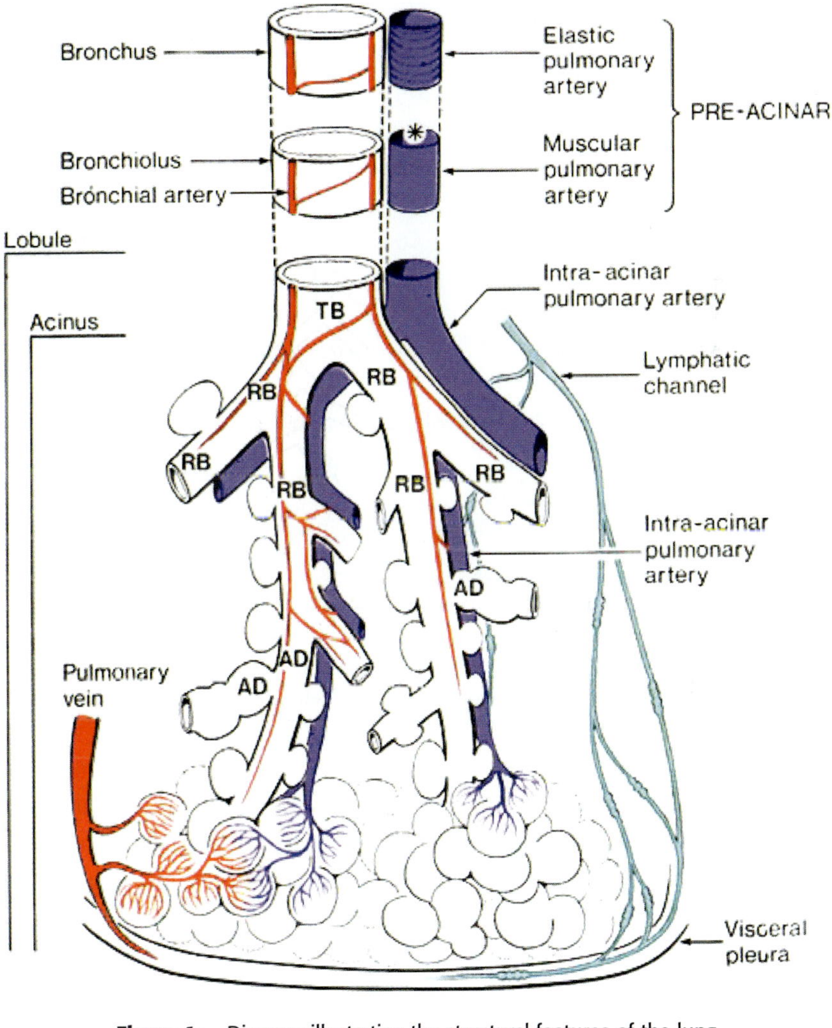

Figure 1: Diagram illustrating the structural features of the lung.

is to produce a large gas exchange area within a relatively small volume of thorax. This is achieved by branching morphogenesis of the airways and the formation of an accompanying vascular system. The heart tube is formed by the end of the third week of gestation. The lung bud appears in humans during the fourth week of gestation, and by 16 weeks all pre-acinar airways, arteries and veins have formed. Lung development has been divided into four stages: embryonic, pseudoglandular, canalicular

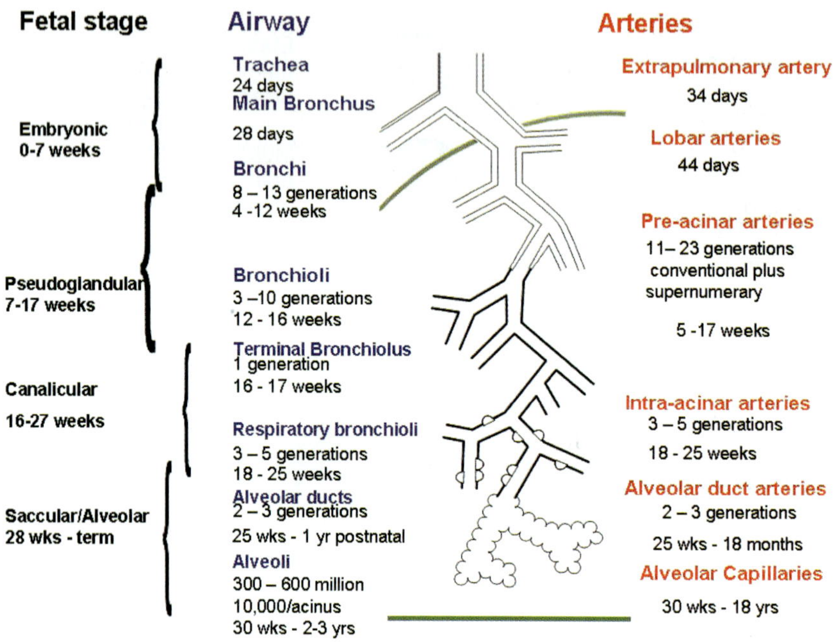

Figure 2: Diagram illustrating the timing of development of the airways and pulmonary arteries within the stages of lung development during foetal life in humans. The number of each new structure is given.

and alveolar, based on structural appearance (Fig. 2). The age of transition from stage to stage varies between individuals and species. In some species (e.g. rat and mouse) the alveolar stage is entirely post-natal.

During the embryonic stage (up to seven weeks' gestation in humans), the lung bud appears as a ventral diverticulum of the foregut and divides within the surrounding mesenchyme to form the left and right lung. The transformation of the foregut endodermal cells to form the lung bud is critically dependent on the transcription factor hepatocyte nuclear factor-3 and transforming thyroid factor-1 (TTF-1). The initial induction of lung development results from fibroblast growth factor (FGF)7 signalling in the matrix surrounding the foregut. Several reviews have addressed the involvement of this and other transcription factors.[1–4] The interaction of the surrounding mesenchyme determines the initiation and complexity of the branching pattern and it is the mesenchyme which is responsible for the final pattern of branching. An airway continues to increase in length when stripped of its surrounding mesenchyme, but does not branch, whereas mesenchyme transplanted from an area of active branching will stimulate an epithelial tube to divide.[5] The lung buds are lined by endoderm, which gives rise to the specialised epithelial cells. The other elements of the airway wall are derived from the mesoderm.

During the pseudoglandular stage, from five to 16 weeks' gestation, the airways continue to grow by dichotomous branching from the hilum towards the periphery. This process is controlled by the orchestrated development of pro- and inhibitory growth factors. (See Fig. 3.) Further branching is induced by mesenchymal FGF10 via FGF receptor 2 on the epithelium. FGF10 knockout mice do not develop lungs, though they do have a trachea. Cleft formation is also involved in airway division via

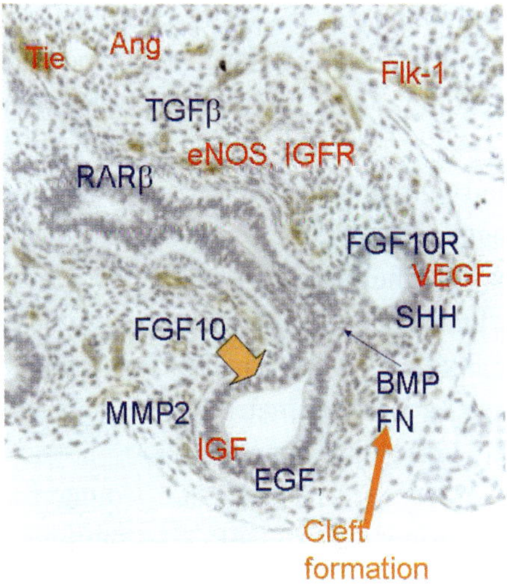

Figure 3: Orchestrated development via growth factors in the embryonic and pseudoglandular stage. Photomicrograph of an eight-week foetal lung stained with CD31 showing the location of growth factors and their receptors which are interacting during morphogenesis of the airways and blood vessels. Factors primarily involved in airway growth are in blue, and in vessel growth, in red. VEGF — vascular endothelial growth factor; IGF — insulin-like growth factor; Flk-1 — VEGF receptor; eNOS — endothelial nitric oxide synthase; IGFR — IGF receptor; Ang — angiopoietin; Tie — receptor for angiopoietin; BMP — bone morphogenic protein; FN — fibronectin; SHH — sonic hedgehog; RAR — retinoic acid receptor; EGF — epidermal growth factor; FGF — fibroblast growth factor; MMP — matrix metalloproteinase; TGFβ — transforming growth factor.

BMP-4 and fibronectin, which leads to inhibition of FGF10.[6] At the same time, stimulation from the epithelium by sonic hedgehog (SHH) leads to mesenchymal re-arrangement. Both over- and underexpression of SHH lead to hypoplasia of the lungs. Epidermal growth factor (EGF) found in epithelial cells induces proliferation and differentiation. TGFβ acts as an inhibitor of epithelial proliferation and is pro-matrix synthesis.[5,7,8] A low oxygen tension is important for normal branching morphogenesis. Mouse lung buds reduce their branching rate in 20% oxygen and there is concomitant failure of blood vessel development. In culture conditions, nitric oxide (NO) increases airway branching, but its physiological role is poorly understood.[9] However, inhaled NO restored alveolar development in hyperoxic infant rats.[10] Numerous other factors are implicated in the regulation of airway branching.[8] Retinoic acid, the active derivative of Vitamin A, influences the transcription of multiple genes. It inhibits FGF-10 and BMP-4 possibly via SHH. Both maternal overdose and deprivation of retinoic acid cause dose- and time-dependent defects of airway branching leading to lobar, unilateral, or bilateral lung agenesis or lung hypoplasia.[4]

Degradation and restructuring of the extracellular matrix is vital for branching morphogenesis, and the epithelial–mesenchymal interactions which take place within the matrix influence the differentiation and activation of the cells. It also acts as a barrier between the different structures forming within the lung bud. Matrix metalloproteinases (MMPs) regulate the formation of the extracellular matrix but also influence growth factors and therefore airway branching. Branching morphogenesis is abnormal in MMP-2 knockout mice. The MMPs are counterbalanced by tissue inhibitors of metalloproteinases (TIMPs). In the human foetal lung the expression of MMP-2 is high with a low expression of TIMP-3, while this ratio is reversed in the adult lung. The expression of MMP-14 and 20, and TIMP-1 and 2 does not alter during development. The relationship of MMPs and TIMPs is altered with exposure to hypoxia.[11,12]

By the end of the pseudoglandular period all airways down to the level of the terminal bronchioli have formed (Fig. 2). During this period the airway wall structures differentiate to form cartilage, submucosal glands, bronchial smooth muscle and the different epithelial cell types. Cartilage appears at 44 days. Smooth muscle cells are present in the trachea and large airways and are innervated by eight weeks of gestation. From 11 weeks' gestation, the epithelium differentiates into ciliated, goblet and basal cells (stem cells), with Clara cells in the peripheral airways. By 24 weeks of gestation the composition of the airway wall is similar to that in the adult.

The appearance of the lung changes in the canalicular stage (16–27 weeks' gestation). There is less undifferentiated mesenchyme between the airspaces. In the peripheral airways, branching continues to form the respiratory airways, the epithelium becomes much thinner and the airspaces enlarge. Capillaries come to lie under the epithelium and a thin blood gas barrier is formed. There is epithelial differentiation to Type I and II pneumonocytes and surfactant is detected by 24 weeks of gestation. The capillaries seem responsible for the thinning and stretching of the epithelium, which in turn leads to the differentiation of these cell types.[13] VEGF production increases and converts glycogen to surfactant in the Type II cells. This process appears to be under the control of the transcription factor HIF2α since maturation of surfactant fails in HIF2α knockout mice.[14] In transgenic mice with an excess of TGFβ1 the epithelial cells fail to differentiate[15] and VEGF is reduced in these mice. By the end of the canalicular stage the periphery of the lung is made up of thin-walled saccules, simple in outline but with a surface area sufficient to support extrauterine life (gas exchange).

The final foetal stage of lung development is the alveolar stage (27 weeks of gestation to term). The walls of the saccules contain discrete bundles of elastin and muscle which lie beneath the epithelium and form small crests subdividing the walls.[16] These crests elongate to produce primitive alveolar walls which have a double capillary supply beneath the epithelium on either side of the wall. The two layers of capillaries then coalesce to form a single sheet. Fusion is evident in the human lung up to 18 months of age. The mature cup-shaped alveoli line elongated saccules, now defined as alveolar ducts, and part of the wall of the respiratory bronchioli.[17] The number of alveoli increases with gestational age. By term, up to half the adult number of alveoli has formed in humans. In mice and rats they form entirely after birth.[18]

Postnatally, alveolar multiplication continues until at least 2–3 years of age, after which surface area is increased by enlargement of alveoli to keep pace with somatic growth. At all ages boys have more alveoli than girls and by adulthood the alveoli are of similar size in males and females.

3. Regulation of Alveolar Development

Studies by Burri and colleagues have shown that a double capillary network must be present for alveolar septa to form.[17,19] Administration of glucocorticoids leads to premature fusing of the double capillary networks and failure of alveolar development.[20] Elastin and smooth muscle are also necessary. Elastin null mice have reduced secondary crests and a reduced alveolar number.[21] Failure of maturation of airway smooth muscle cells also reduces alveolar formation.[22] Recent studies have highlighted the importance of angiogenesis. The anti-angiogenic factors fumagillin and thalidomide reduce arterial and capillary development and also reduce the number of alveoli in rats.[23] Angiogenesis during alveolarisation is also promoted by VEGF, and inhibition of the VEGF (Flk) receptor reduces the number of alveoli.[23] HIF2α is also involved in this angiogenic process and knockout mice have fewer blood vessels and a reduced alveolar number.[14] Recently a genetic basis for alveolar septation has been described in rats.[24] Rat lung fibroblasts isolated on post-natal days 2, 7 and 21 showed differential expression of mRNA for genes of interest. They found that gene expression of Hoxa5 and retinoic acid receptor, located by immunohistochemistry on the tips of growing septa, increased during the period of septation on days 3–14 and was reduced in animals in which alveolar formation had been reduced by dexamethasone or hyperoxia. While retinoic acid is inhibitory during branching morphogenesis, it is vital for the formation of alveoli and can be used to enhance alveolar development in hypoplastic and immature rodent lungs.[8,25] The retinoic acid receptors are probably even more important. Mice bearing deletions of the retinoic acid receptor gene have reduced lung elastin expression and alveolar number.[26]

Physical factors have an important influence on lung growth. Maintaining an appropriate amount of foetal lung fluid is essential for normal lung growth and development. An optimal balance is needed between the production of fluid by the distal airways and drainage through swallowing or releasing fluid orally into the amniotic space. Lung fluid movement is dependent upon spontaneous peristaltic airway contractions. Stretch is known to induce the release of mitogenic growth factors[27] and it therefore constitutes a stimulus for prenatal and post-natal growth. Foetal breathing movements are important for the proper development of the lung and respiratory muscles.[28] The frequency and amplitude of foetal breathing movements can be modified by hypercapnia, hyperglycaemia, acidosis, fever, and caffeine. Foetal breathing movements can be increased by theophylline, terbutaline and indomethacin,[28,29] and inhibited by nicotine, alcohol, several sedative and narcotic drugs, corticosteroids, hypoxia, hypoglycaemia, PGE2 and infections.[30]

Not surprisingly, the response to a growth factor depends on the stage of development at which the embryo or foetus is exposed to the agent. Adenoviral transfer of TGFβ into the embryonic rat lung leads to inhibition of airway branching, while neonatal administration leads to enlarged alveolar spaces and fibrosis in newborn rats.[31]

4. Development of the Human Pulmonary Vasculature

4.1. *Structural development*

Recent work shows that the pre-acinar arteries and post-acinar veins form by vasculogenesis from the splanchnopleural mesoderm of the lung bud.[32,33] As described above, the pre-acinar airways form in a centrifugal manner from hilum to periphery, but the vessels form by continuous coalescence of endothelial tubes alongside the newly formed airways, which appear to provide a template for pre-acinar blood vessel development[32,33] (Fig. 4). The continuous addition of endothelial tubes gradually lengthens the pulmonary arteries and veins. Angiogenesis predominates later, and is thought to be responsible for the formation of intra-acinar arteries and veins, which starts during the canalicular phase and continues after birth. All the pre-acinar arteries and airways have formed by the end of the pseudoglandular stage (approximately the 16th week of gestation). From a third to a half of the adult complement of alveoli and thus accompanying intra-acinar arteries form before birth, with additional new structures forming most rapidly within the first six months of life. The lymphatics are first seen at about 56 days; gestation,[33] originating at the

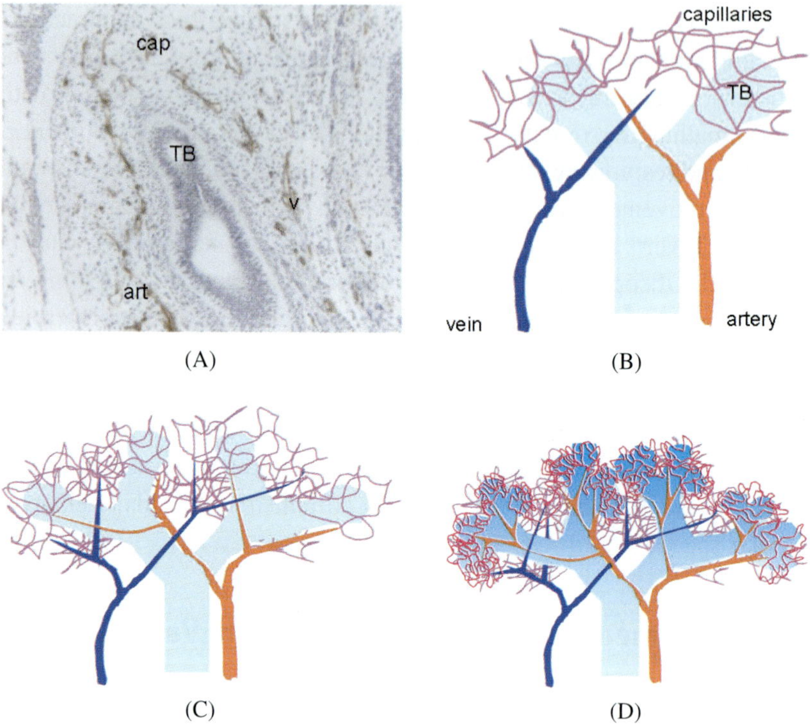

Figure 4: Photomicrograph and diagrams illustrating vasculogenesis during the embryonic and pseudoglandular stage. (A) Photomicrograph of a terminal bud (TB) in the lung of a 44-day embryo. The brown immunostaining is CD31 labelling endothelial cells. A halo of capillaries surrounds the terminal bud. (B–D) Diagrams produced by serial reconstructions through the lung buds: (B) Airways and vasculature at 44 days; (C) A new halo of capillaries and lengthening of the arteries and veins is present after further airway division; (D) Further airway division and capillary growth. There is sustained addition of newly formed endothelial tubes at the lung periphery as each airway division occurs. As the capillaries add on at the periphery so the arteries and veins increase in length. Diagram by Dr. S. M. Hall.

hilum. By the end of the pseudoglandular period, lymphatic channels have subdivided the lung periphery, positioned in the prospective connective tissue septa. The bronchial arteries develop by angiogenesis from the dorsal aorta relatively late in gestation, at about eight weeks' gestation. In the normal foetus and newborn there are many peripheral connections between the pulmonary and bronchial circulations. These connections decrease rapidly in size and number after birth unless there is a congenital cardiac abnormality causing a reduction in forward flow down the pulmonary arteries, in which case the broncho-pulmonary connections enlarge to supply the peripheral pulmonary arteries.

We have shown that the heart is connected to the two lung buds from a very early stage of embryonic life, and it is likely that they are always in continuity. By 34 days of gestation, serial reconstructions demonstrate physical continuity between the aortic sac, the pulmonary arteries, the peribronchial capillary plexus, the pulmonary veins, and the atrial component of the heart. Thus, blood can circulate through the lung earlier than had been supposed, exposing the developing structures to circulating factors generated at distant sites, including angioblasts and other cells which could become incorporated into the vessel wall.

The pulmonary arterial smooth muscle cells are derived from three sites in a temporally distinct sequence[32] (Fig. 5). The first cells to surround

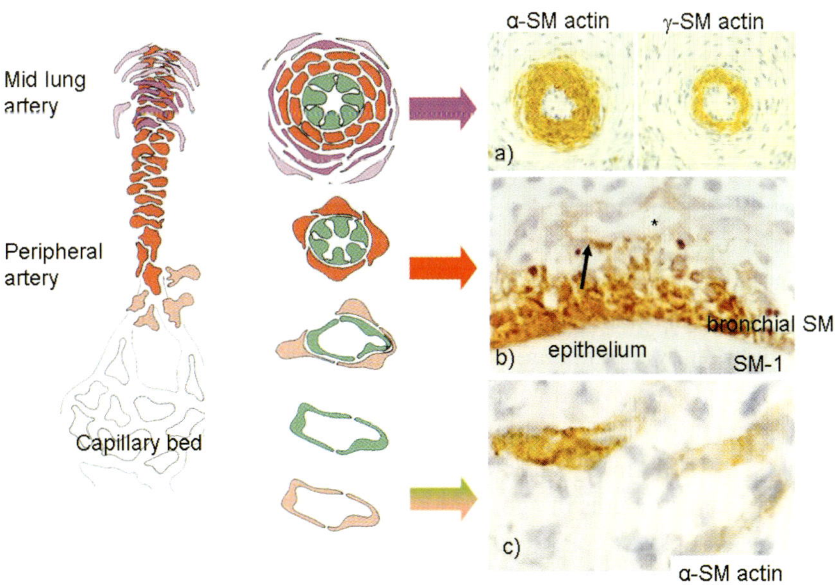

Figure 5: Illustration of the origin of pulmonary artery smooth muscle cells. a) Mid-lung artery stained for α-actin and γ-actin. The inner smooth muscle cells are more mature (γ-actin +ve) than the outer ones (α-actin +ve) that are loosely packed around the inner two layers, and have formed from mesenchymal cells. b) Bronchial smooth muscle cells in a penultimate airway are moving towards the wall of the capillary (*). c) Capillary endothelial cells showing dense (co-) expression of α-actin as they transform into smooth muscle cells. Diagram by Dr. S. M. Hall.

the endothelial tubes migrate from the bronchial smooth muscle around the neck of each terminal bud. These cells become surrounded by a second population, which differentiates from mesenchymal cells. The third source is the endothelial cells in the most peripheral arteries, which transdifferentiate into smooth muscle cells later in gestation, from 98 to 140 days. During foetal life the arterial wall is relatively thick, maintaining a high resistance. Venous smooth muscle originates from mesenchymal cells and by transdifferentiation of endothelial cells but there is no contribution from the bronchial smooth muscle.[33] Irrespective of their origin, all smooth muscle cells follow the same sequence of expression of smooth muscle-specific cytoskeletal proteins. Maturation takes place from the subendothelium to the adventitial layers, and from hilum to periphery. In the pulmonary veins, the cytoskeletal proteins appear in the same sequence as in the pulmonary arteries, but do not express caldesmon, which might help explain why pulmonary veins are so responsive to contractile agonists.

After birth, as the intrapulmonary arteries increase in size, their walls maintain a constant low relationship to the external diameter, an initial thinning having occurred immediately after birth (see below). In the arteries of the respiratory unit, smooth muscle cells differentiate from *in situ* precursor intermediate cells and pericytes, and muscle is said to have extended along the arterial pathway. The elastic laminae of the pulmonary trunk gradually become fragmented, and lose their ordered appearance, and the wall becomes thinner relative to that of the aorta.[34] The pulmonary vasculature is densely innervated at birth, and becomes more densely innervated with age.[35] In babies, as in adults, the nerves are predominantly of the sympathetic type. In pulmonary hypertensive infants, the abnormally thick-walled blood vessels are prematurely innervated by three months of age.[35]

5. Regulation of Pulmonary Vascular Development

The presence of a circulation through the pulmonary vasculature from early embryonic and foetal life means that the circulation is exposed to external, distal influences from an early stage in its development. Circulating cells could contribute to building the vessel walls. Angioblasts can stream to developing organ beds from a considerable distance[36] and incorporation of marrow-derived stem cells seems likely. Actin and calponin expressing cells are found in the pulmonary arterial intima of females following a sex-mismatched bone marrow transplantation.[37] Bone marrow-derived progenitor cells contributed to pulmonary arterial wall thickening in hypoxic calves.[38] Understanding the extent to which such cells retain their genetic memory, and the extent to which they can be modified by environmental factors, is of fundamental importance. The inner and outer medial smooth muscle cells differ with respect to cytoskeletal composition, regulation of the actin cytoskeleton by RhoGTPases, and contractility.[39,40] These phenotypic differences are stable in culture,[39,40] suggesting that they retain distinct hereditable characteristics. The innermost medial smooth muscle cells originate by migrating from bronchial smooth muscle and these cells migrate into the subendothelium in pulmonary vascular disease. It seems likely that the recruitment of cells from outside the lung bud would enhance the heterogeneity of the vessel wall, and its potential for adaptation to environmental change.[41]

The factors that determine whether an endothelial cell lying in the mesenchyme will become dedicated to either arterial or venous system are not understood. The tyrosine receptor EphB4, and its cognate ligand ephrin B2, discriminate between arteries and veins in the mouse, but less so in humans. Early angioblasts commit to either an arterial or venous fate, directed by a Notch gridlock signalling pathway.[42] It is possible, however, that pulmonary endothelial cells do not become committed until they have coalesced with an adjacent pulmonary artery or vein. It is at that time that the direction of flow, pressure, and circulating factors might influence commitment.

Experimental and gene targeting studies continue to emphasise the critical role of growth factors in lung development. Each appears to have a precise spatial and temporal expression pattern (Fig. 3). In the human lung, we found eNOS, VEGF, and the VEGF receptors Flk-1 and Flt-1 and Tie-2 expressed in the capillary plexus at 38 days.[32] eNOS stimulates endothelial proliferation, migration, formation of tubes, is a downstream mediator of growth factor-induced angiogenesis, and inhibits apoptosis. It is expressed on the endothelium of human arteries, capillaries, and veins throughout development. VEGF mediates vasculogensis and angiogenesis. VEGF-A is required for vascular development throughout the embryo, and blood vessels do not form in mice deficient in the VEGF receptor Flk-1.[43,44] Grafting of beads coated with VEGF increases the density of the capillary bed.[45] In cultured mouse lungs, VEGF is found at the branching points of peripheral airways, where the adjacent endothelial tubes coalesce and become joined to the presumptive pulmonary artery. The tyrosine receptor kinase Tie-2, and its ligand angiopoietin, is also essential in the formation of capillary networks.[46] Endothelial phosphorylation of the Tie-2 receptor by angiopoietin leads to proliferation of smooth muscle cells, perhaps indicating that it has a role in the neomuscularisation of newly formed pulmonary arteries.[47,48] Han and colleagues found that insulin-like growth factor I and II, and IGF-IR ligands and mRNA are present at four weeks' gestation.[49] Anti-IGR-IR neutralising antibody reduced the number of endothelial cells, and increased apoptosis of endothelial and mesenchymal cells in human lung explants. IGF-1 also upregulates VEGF, further emphasising the crucial role of this growth factor. A low oxygen tension, as occurs during foetal life, stimulates both vascular and airway development. This involves HIF-1α and HIF-2α, which upregulate expression of VEGF and play an important role in both vasculogenesis and angiogenesis. Studies on HIF-1α emphasise the importance of cross-talk between developing arteries and airways to ensure the normal development of both structures.[14] EGFmRNA and TGFαmRNA are expressed in the mesenchymal cells surrounding airways and alveoli in the human foetal lung from 12 to 33 weeks' gestation,[50] but the receptors for these ligands were found in the airways and were involved in differentiation. Their direct role in vascular development is uncertain. The influence of innervation on lung development is not well understood. Extracellular nucleotides help regulate mitogenic, contractile, and metabolic responses in the vasculature and the purinergic receptors, P2Y and P2X, are abundant and highly conserved. Extracellular ATP contributes to hypoxic remodelling of bovine pulmonary arteries by enhancing adventitial fibroblast proliferation.[51]

Heterogeneity of both endothelial and smooth muscle cells is evident before birth. During embryonic development, they show considerable plasticity, and undergo constant changes in protein expression to match the requirements of the developing vessel.[52,53] Plasticity of vascular cell types persists after birth. It is thought that stimuli such as hypoxia, shear stress or inflammation can lead to transdifferentiation of endothelial cells into mesenchymal fibroblasts or smooth muscle-like cells.[54] Transition between fibroblast and smooth muscle cells and *vice versa* is well described.

Bone morphogenetic proteins (BMPs) and their receptors play an important role in lung development. In the human lung the expression of BMP4 and the BMP receptors is temporally and spatially regulated during development. BMPRII expression is greatest during the late pseudoglandular and early canalicular stages when vasculogenesis is intense, and it correlates with tissue smad1 phosphorylation.[55] *In vitro*, BMPs 2 and 4 induced phosphorylation of smads 1 and 5, and pulmonary endothelial proliferation and migration, while inhibition of BMPRII inhibited smad signalling and the endothelial proliferative response to BMP4. Other investigators found evidence of BMP signalling in the later stages of murine saccular and alveolar development.[56] *In vitro* studies have shown that the BMP antagonist Drm/gremlin can modulate angiogenesis,[57] stimulating endothelial cell migration and invasion in fibrin and collagen gels, and triggering tyrosine phosphorylation of intracellular signalling molecules. It also induces neovascularisation in the chick chorioallantoic membrane. BMPII is a receptor for the TGFβ/BMP superfamily. The TGFβ receptors ALK1 and ALK5 appear to balance each other in controlling the proliferation and migration of endothelial cells during angiogenesis.[58] They show non-overlapping expression patterns during development; ALK1 in the endothelium and ALK5 in the media and adventitia of blood vessels. Cross-talk between TGFβ/BMP and the signalling pathways of mitogen-activated protein kinase, phophatidylinositol-3 kinase/Akt, Wnt, SHH, and Notch, was reviewed recently by Guo and Wang.[59]

6. Remodelling to Adapt to Extra-Uterine Life

The two most important changes taking place at birth are the reduction in pulmonary vascular resistance and the clearance of alveolar fluid. To ensure the post-natal fall in pulmonary vascular resistance, the endothelial and smooth muscle cells undergo marked cytoskeletal change to increase their surface area, become thinner and so increase lumen diameter.[60] Smooth muscle cells show a transient reduction in actin content and disassembly of actin filaments, with an increase in monomeric G-actin. The smooth muscular phenotype is regulated by the transcription factor serum response factor, which is itself negatively regulated by YY1.[61] The brief post-natal reduction in filamentous actin causes YY1 to relocate from cytoplasm to nucleus, thereby inhibiting the expression of differentiation markers, and increasing the activity of cell cycle genes. These changes do not occur in the presence of pulmonary hypertension. No single factor has yet been identified as being primarily responsible for the initiation of vasodilation at birth. It is probable that the abrupt expansion of the lungs leads to a cascade of events which facilitate the activation of vasodilatory responses, and reduce vasoconstrictor stimuli. The sudden increase in blood flow imposes a sheer stress on the endothelium to promote the release of nitric oxide and prostaglandin I_2 (Fig. 6). Endothelium-dependent relaxation is relatively poor at birth, whether receptor or non-receptor operated. The density of the muscarinic receptors responsible for acetylcholine mediated relaxation increases rapidly immediately after birth, and the subtypes change.[62] However, the vessels relax in response to exogenous nitric oxide, although this response also improves significantly during the first two to three weeks of life.[63,64] There is no lack of nitric oxide synthase at birth. The expression of both the protein and the gene increases markedly towards term, and increases further to reach a maximum at two to three days of life.[65] As in adults, the predominant enzyme present at birth is the constitutive endothelial isoform. There is no absolute or relative deficiency of the co-factor for nitric oxide synthase, BH4, at birth.[66] The efficacy of the enzyme, however, can be reduced by the action of endogenous inhibitors, primarily asymmetric dimethylarginine, which competes with the substrate L-arginine.[67] The level of the inhibitor decreases in the first days of life, metabolised to citrulline by the dimethylarginine dimethylaminohydrolase enzymes, DDAH I and II, both of which are highly expressed in the foetal lung. Each isoform is developmentally regulated, and activity of the second increases rapidly immediately after birth.[68] The pulmonary arterial smooth muscle cell being targeted by the endothelium is itself

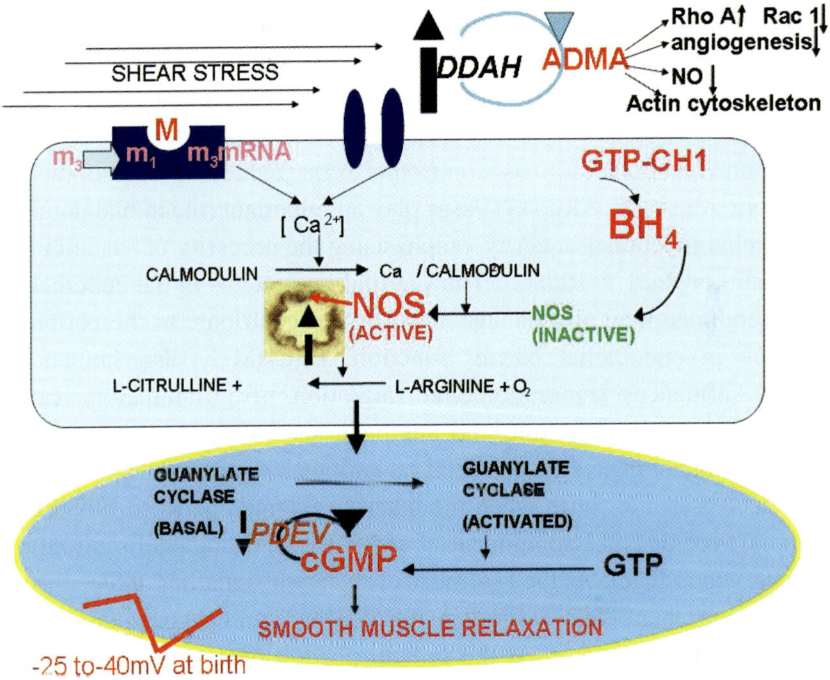

Figure 6: Diagram illustrating the pathways involved in adaptation to extra-uterine life in the pulmonary arterial endothelium and smooth muscle cells.

changing rapidly. Basal accumulation of cGMP is high at birth, but falls rapidly to a lower adult level by three days of age.[64] This post-natal fall might be explained by the high expression of phosphodiesterase (PDE) V and its hydrolytic activity found in the newborn rat lung.[69] Despite the basal accumulation and enhanced accumulation of cyclic GMP in response to nitric oxide and its donors, the relaxation response of newborn vessels is less than might be expected. This might be accounted for by the smooth muscle cell membrane being more depolarised at birth than later.[70]

The alveolar fluid is absorbed by the alveolar epithelium at birth. Foetal pulmonary endothelial intercellular junctions are complex and fenestrated, while they are tighter and less complex in older babies, indicative of improved barrier function. The higher endothelial permeability in foetal pulmonary vessels is probably due to the combined actions of hypoxia, a high level of circulating endothelin-1, VEGF and angiotensin II. Endothelin-1 induces endothelial permeability, and its receptor antagonists can prevent capillary leakage.[71,72] By contrast, an increase in nitric oxide has been shown to prevent endothelial leakage in the lung. The post-natal increase in nitric oxide, and the simultaneous reduction in endothelin, may contribute to the tightening of endothelial junctions after birth.[73] Rho GTPases play an important role in maintaining endothelial junctional integrity, emphasising the necessity of sustaining a high ratio of Rac1 to RhoA.[74] The scaffolding proteins in the endothelial cell membrane may also change at birth, and contribute to the perinatal changes in endothelial barrier function. PECAM-1, also known as CD31, influences transendothelial migration of inflammatory cells, mechanosignal transduction, and angiogenesis.[75] Considerable amounts of this agent are expressed on foetal rat endothelial junctions, and expression decreases after birth when the barrier is formed between blood and gases. Caveolin-1 is a component of caveolae, an endothelial scaffolding protein which regulates the assembly of different signalling molecules at the plasma membrane.[76] Studies on doubly homozygous caveolin-1 knockout mice indicate that the protein plays a dual regulatory role in controlling lung microvascular permeability, acting as a structural protein required for the formation of caveolae and caveolar transcytosis, and as a tonic inhibitor of eNOS.[77]

7. Interaction between Airways and Arteries during Development

The airways and pulmonary blood vessels develop in close proximity and they interact and exert reciprocal control upon each other throughout development. The terminal buds at the end of each branching pathway have a consistent appearance throughout development and those of man closely resemble those of other species, suggesting highly conserved genetic programming. Development of the vasculature appears to be controlled largely by local factors, released from the adjacent airways and mesenchyme. In the embryonic and pseudoglandular stages vasculogenesis is critically dependent on the epithelium of the terminal buds, which release VEGF into the mesenchyme. VEGF-deficient mice do not survive to term even in the presence of one allele, and show defects in vessel formation.[78] In a study of rat lung explants, at foetal day 13, presence of the epithelial cells was needed to maintain the presence of endothelial (Flk-1 +ve) cells in the distal lung mesenchyme.[79] BMP-4 acts as a stimulant for VEGF and high levels of BMP-4 are located at the tips of distal airways. Overexpression of the inhibitory vascular morphogen matrix GLA protein (MGP) in mouse lungs disrupts BMP-4 and leads to failure of peripheral arterial branching.[80] Changes in the expression of VEGF, however, alter both vascular and airway development. Overexpression of VEGF caused disrupted vascular net assembly and arrested airway branching[44] while inhibition of blood vessel development by antisense oligonucleotides to VEGF reduced airway branching in mice.[81] Other examples of the same agonist/receptor affecting both vascular and airway development include eNOS and the IGF receptor. eNOS stimulates proliferation of endothelial cells and tube formation but it also controls airway branching in the cultured rat lung.[9] Inactivation of the IGF receptor in human lung explants led to loss of endothelial cells in existing capillaries by apoptosis, which was accompanied by abnormal growth of the lung buds, suggesting that IGF receptors are involved in both airway and endothelium development.[49]

Airway branching may be more important in the organisation of the pulmonary arterial tree rather than in the formation of a capillary network. Endothelial precursor cells derived from the lung mesenchyme of foetal mice and cultured in the absence of a lung bud, formed extensive capillary-like networks but did not form larger vessels.[82] These endothelial cells expressed the RNA for VEGF receptors, the receptor tyrosine kinases Tie-1 and 2, and angiopoietins 1 and 2 (Ang-1 and 2). Endothelial progenitor cells capable of *de novo* vasculogenesis have been isolated from the microvasculature of post-natal rat lungs.[83]

In the canalicular stage, while VEGF and HIFs are considered to be instrumental in blood vessel development at this time, VEGF also causes maturation of Type II cells by converting glycogen stores to surfactant. This process is dependent upon HIF-2α and maturation does not occur in –/– mice.[14] Interaction of arteries and airways continues during alveolar development in the post-natal lung. Experimental studies have shown that as alveoli form in mice and rabbits, the Type II pneumonocytes produce VEGF mRNA.[84,85] VEGF and its receptor Flk-1 both increase in perinatal mice during this period.[86] In a study on rat alveolar development, an inhibitor of Flk-1 (Su5416) reduced the number of both arteries and alveoli, as did anti-angiogenic factors such as fumagillin and thalidomide.[23] A similar study showed that Flk-1 was essential for early alveolar and thus capillary development, but these deficiencies were reversible with time when inhibition was discontinued. By contrast, perinatal exposure to hyperoxia led to long-term structural impairment.[87] Hyperoxia reduces vascular density and alveolar development. In the rat these changes were associated with a reduction in VEGF,[88] and injecting recombinant VEGF during the recovery period led to an increase in both capillaries and alveoli.[89] Inhalation of NO also enhanced recovery from hyperoxia by increasing blood vessel development.[10] HIFs continue to play an important role in alveolar development. In a baboon model of BPD, activation of HIFs increased angiogenesis and also alveolar surface area with subsequent improvement in oxygenation and lung compliance.[90,91]

8. Can the Adult Lung Begin to Grow Again?

When an adult undergoes pneumonectomy or lobectomy the remaining lung tissue increases in volume to fill the space available. Experimental animal studies indicate that an increase in lung volume is achieved by the growth of existing structures rather than formation of new structures, that the mediators which are important in regulating normal foetal development are also important in effecting compensatory lung growth, and that the response appears to be age-dependent. In 1998 we showed that a mature adult rat lung could be stimulated to grow new, normally formed alveoli.[92] Unilateral lung transplantation with an immature donor lung led to an increase in alveolar number and size in the immature donor lung, as might be expected, but also in the mature recipient lung.

The alveolar response may be species-dependent. Studies on pneumonectomised adult foxhounds showed that compensation in the remaining lung was achieved without growth of alveolar septal tissue, but with alveolar capillary distension and an increase in Type II alveolar cells, followed by activation of other alveolar cells.[93] By contrast, alveolar number was increased by 33% in the right lungs of mice which had undergone left pneumonectomy.[94] These changes are probably initiated by abrupt expansion of the lung. Post-pneumonectomy expansion of the canine lung caused elevation of HIF-1α mRNA and protein with activation of its transcriptional activator, the serine/threonine protein kinase B, and the mRNA and protein levels of its downstream targets erythropoietin receptor and VEGF.[95] Basal HIF-1α expression was greater in immature than mature dogs and increased at least twofold after pneumonectomy.[96] Compensatory lung growth is largely attributed to Type II alveolar cells, but recent studies have shown that bronchioalveolar stem cells have the potential to contribute substantially to this process.[97] Compensatory growth of the vasculature has been less well studied than that of the airways. The remaining lung of the pneumonectomised rat showed an increase in vessel size by quantitative post-mortem angiography, more so in the upper and middle lobes than the lower lobes, accompanied by an increase in proliferating nuclear antigen.[98] VEGF accelerates compensatory growth after pneumonectomy, but βFGF, MF-1 and DC101 do not, and the authors concluded that compensatory lung growth after unilateral pneumonectomy in a murine model is, in part, angiogenesis-dependent, and can be altered with angiogenesis modulators, possibly through regulation of endothelial cell proliferation and apoptosis.[99]

9. Lung Repair and Regeneration

Lung injury can destroy the lung parenchyma, and obstruct or even obliterate either the bronchioli or pulmonary arteries. Reversing these abnormalities is not possible, despite intense research activity. The most we have achieved so far is a slowing of the disease process. In Idiopathic Pulmonary Arterial Hypertension (IPAH), for example, new therapies such as continuous intravenous prostacyclin, oral endothelin receptor antagonists and phosphodiesterase inhibitors, can slow disease progression, sometimes for many years, but are not curative.

Obliterative bronchiolitis, fibro-obliterative occlusion of the small airways, is seen in infants and young children following infection, in occupational lung injury or toxic inhalation. It is the most common cause of lung transplant failure, when it is thought to be a manifestation of chronic rejection. Recent studies suggest that while alloimmunity initiates graft rejection *de novo* autoimmunity ultimately causes progressive airway obliteration.[100] Obliterative pulmonary vascular disease is seen in IPAH and can be associated with connective tissue disease, congenital heart disease, lung disease and other disorders. It is characterised by obstruction of the terminal and respiratory bronchiolar arteries. Peripheral dilatation lesions and plexiform lesions are characteristic. Discussion of the pathobiology resonates with mention of the same signalling pathways considered important in foetal pulmonary vascular development, particularly the BMP and TGFβ signalling pathways (Fig. 7).

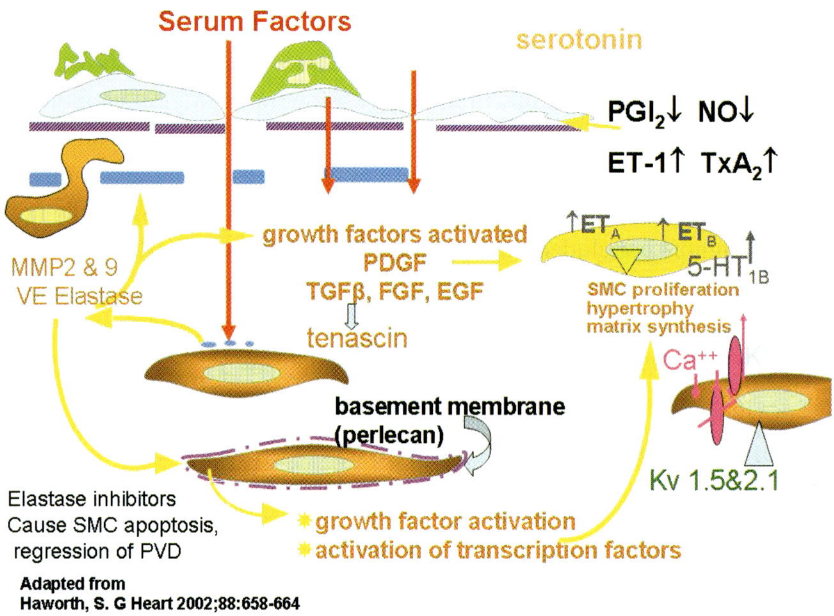

Figure 7: Summary of the factors driving the evolution of pulmonary vascular obstructive disease. Endothelial dysfunction causes adherence of activated platelets and leukocytes and loss of barrier function with leakage of serum factors into the subendothelium. This last action is thought to heighten the activity of metalloproteinases (particularly MMP2 and 9), including endogenous vascular elastase released from smooth muscle cells, to help induce structural remodelling, cause smooth muscle cell activation, disruption of the internal elastic lamina, and facilitate smooth muscle cell migration. MMPs also activate growth factors normally sequestered in the matrix in an inactive form. Increased tenascin expression is associated with cell proliferation, while its downregulation with apoptosis. Tenascin amplifies the proliferative response to epidermal growth factor and fibroblast growth factor (FGF)-2 *in vitro*. Expression of fibronectin is widespread and this glycoprotein can facilitate smooth muscle migration. The innermost smooth muscle cells cease to express many smooth muscle-specific contractile and cytoskeletal proteins before migrating through gaps in the internal elastic lamina. Changes in phenotype are widespread.

Mutations in the genes for BMPRII and activin-like kinase, which regulate specific signalling pathways for transforming growth factors, are found in patients with IPAH and in many patients with associated pulmonary hypertension. Reduced expression of the co-receptor BMPRIA is also frequently observed in patients with IPAH.[101] Smooth muscle cell replication is high in IPAH, in cells from the explanted lung. Loss of BMPRII causes proliferation of the pulmonary arterial smooth muscle cells in response to agonists such as TGF-β1 and bone morphogenetic protein, which normally inhibit proliferation[102] and induces susceptibility to apoptosis. It makes endothelial cells susceptible to apoptosis.[103] BMPRII activity induces differentiation of foetal lung fibroblasts into myofibroblasts and smooth muscle cells,[104] which could increase medial and adventitial thickness. Improving BMPRII signalling has been an effective strategy in some rodent models.[105] PPARγ is an important transcription factor downstream of BMPRII signalling, and the ApoE gene is transcribed by PPARγ. Deletion of PPARγ or apoE in mice causes pulmonary hypertension[106] and apoE can repress proliferation in systemic arterial smooth muscle cells by preventing proliferative signals through the PDGF[107] receptor. Inhibition of the PDGF receptor by imatinib can reverse monocrotaline-induced pulmonary hypertension in rats,[108] and is under investigation in patients with end-stage IPAH.[109] Such studies indicate that PPARγ may be an important therapeutic target. Other downstream targets of BMPRII include tenascin-C. Tenascin C is a glycoprotein associated with activation of growth factor receptors and survival pathways and its activity is increased in pulmonary vascular disease as is that of serine elastase. Increased activity of serine elastase leads to the release of growth factors from the extracellular matrix, and activation of matrix metalloproteinases. Elastase inhibitors prevented and reversed experimental monocrotaline-induced pulmonary hypertension by inducing apoptosis of smooth muscle cells.[110] Blocking a downstream effector of elastase, the EGF receptor, also caused regression. A dominant-negative construct of survivin[111] reversed pulmonary hypertension by smooth muscle cell apoptosis.

BMPRII signalling has been directly related to expression of the Kv potassium channels,[112] which are known to be dysfunctional in pulmonary arterial hypertension, notably the Kv1.5 and 2.1 channels. Reduced expression of these channels leads to the influx of calcium to promote vasoconstriction and cell proliferation. Elevated levels and transport of serotonin have also been implicated in the pathogenesis of pulmonary vascular disease. The fawn-hooded rat has a defect in serotonin metabolism, is readily susceptible to pulmonary hypertension and shows abnormal oxygen sensing in smooth muscle cell mitochondria, leading to reduced function of the Kv channels.[113] Serotonin also affects the proliferative response, since stimulation of the serotonin transporter and the serotonin receptors induces cyclins[114] and c-fos. Inflammation is emerging as an important factor in the pathobiology of pulmonary vascular disease, as exemplified by HIV, and demonstrated by T cell-dependent mechanisms and increased cytokine activity, both of which have been linked to BMPRII activity.

Stem cell therapy is an exciting prospect. Endothelial progenitor cells transfected with endothelial nitric oxide synthase prevent and also reverse pulmonary arterial hypertension in rats by re-establishing connections between proximal and distal pulmonary arteries.[115] This strategy is now being trialled in patients. A recent small pilot study showed that patients treated with non-engineered cells gained short-term benefit (see also Chapters 10 and 12).[116] Bone marrow-derived cells limited the remodelling seen in monocrotalline but not in hypoxia-induced pulmonary hypertension in mice.[117]

In summary, improved understanding of the genetic mechanisms, the role of chronic inflammation and autoimmunity should make it possible to manipulate the abnormal signalling pathways present in pulmonary arterial obstructive disease and understand the potential for stem cell rescue. In depth parallel studies on obstructive airways disease are mandatory.

10. Will it be Possible to Regenerate or Grow a New Lung?

At present the information we have is piecemeal. Most studies have concentrated on the development of either capillaries or alveoli, rather than the simultaneous development of both structures, and they have not focussed on the development of larger, proximal airways and blood vessels.

Alveolar regeneration is an important goal in diseases such as emphysema in adults and bronchopulmonary dysplasia in infants. Alveolar rarefaction also occurs in human starvation and in calorie-restricted mice. The expression of genes thought to be important in producing an alveolar septum from a flat alveolar wall include those related to cell replication, angiogenesis and extracellular matrix remodelling. Guided cell motion and expression of these genes was increased within three hours of allowing calorie-restricted mice food *ad libitum*.[118] Several studies emphasise the potential of endothelial progenitor cells. Endothelial precursor cells derived from the lung mesenchyme of foetal mice will form extensive capillary-like networks with a lumen when cultured in the absence of a lung bud.[82] Endothelial progenitor cells capable of *de novo* vasculogenesis have been isolated from the microvasculature of post-natal rat lungs.[83] The mature lung contains resident stem cells, the alveolar type II pneumocyte and the Clara cell,[119] but it also contains a population of primitive adult precursor cells known as the side population (SP),[120] a primitive multipotent population defined by its lack of differentiated lineage markers and the ability to differentiate into haematopoietic, epithelial and mesenchymal cells.[121] Neonatal mice recovering from hyperoxia showed a reduction in microvessel density with a simultaneous reduction in the number of side population cells and changes in CD45 analyses.[122] This study also showed that lung side population cells have endothelial potential. The authors commented that hyperoxia-induced changes in these cells may limit their ability to contribute to tissue morphogenesis during recovery. This implies that a population of progenitor cells which one might hope could aid recovery is itself susceptible to injury.

The formation of both capillaries and alveoli can be induced by recombinant human VEGF.[89] VEGF is partially controlled by HIFs. When HIF-1α is increased in preterm baboons as a result of inhibiting its degradation, there is enhancement of both arterial and alveolar growth, associated with an increase in lung compliance and oxygenation.[91] The outcome was improved in hyperoxic newborn rats by the intratracheal administration of adenoviral gene therapy with combined VEGF and angiopoietin-1, since angiopoietin-1 stabilises the wall and reduces leakage of newly formed capillaries.[123] Such strategies could be applied to the repair and regeneration of hypoplastic lungs in chronic lung disease in infancy.

Is there any evidence that we could build a new lung from scratch, *in vitro*? Primary isolates of murine lung cells from embryonic day 18 have been cultured in a three-dimensional matrigel. The isolates were a mixed population of epithelial, mesenchymal and endothelial cells. After only one day in culture, tissue resembling peripheral saccules formed, lined by epithelial cells which had the structure of alveolar type II cells producing surfactant.[124] The process continued for up to six weeks. Addition of specific growth factor FGF 10 led to the formation of a branching system of air sacs. A combination of FGF 2, 7 and 10 increased the budding of these air sacs. Endothelial cells coalesced to form an interconnecting capillary system with lumen within the matrix, lying in close proximity to the epithelial cells of the saccules.[41] Such an *in vitro* approach could become suitable for therapeutic lung augmentation in hypoplastic lungs. We do not know, however, whether these structures would function normally.

11. Conclusion and the Way Forward

We now have a good understanding of the temporal and spatial organisation of the structures forming the developing lung but a very incomplete understanding of the molecular mechanisms controlling and orchestrating these events. This information is necessary if we are to learn how best to repair the damaged lung and regenerate new lung tissue, both *in vivo* and *ex vivo*. If we are trying to copy nature then we should learn how nature itself remodels the lung during normal development.

References

1. Minoo, P. Transcriptional regulation of lung development: emergence of specificity (Review). *Respir Res* **1**, 109–115 (2000).
2. Cardoso, W.V. Molecular regulation of lung development. *Annu Rev Physiol* **63**, 471–494 (2001).
3. Warburton, D., Schwarz, M., Tefft, D., *et al.* The molecular basis of lung morphogenesis. *Mech Dev* **92**(1), 55–81 (2000).
4. Roth-Kleiner, M. and Post, M. Genetic control of lung development. *Biol Neonate* **84**(1), 83–88 (2003).

5. Shannon, J.M. and Hyatt, B.A. Epithelial-mesenchymal interactions in the developing lung. *Annu Rev Physiol* **66**, 625–645 (2004).
6. Weaver, M., Dunn, N.R., and Hogan, B.L. Bmp4 and Fgf10 play opposing roles during lung bud morphogenesis. *Development* **127**(12), 2695–2704 (2000).
7. Maeda, Y., Dave, V., and Whitsett, J.A. Transcriptional control of lung morphogenesis. *Physiol Rev* **87**(1), 219–244 (2007).
8. Roth-Kleiner, M. and Post, M. Similarities and dissimilarities of branching and septation during lung development. *Pediatr Pulmonol* **40**(2), 113–134 (2005).
9. Young, S.L., Evans, K., and Eu, J.P. Nitric oxide modulates branching morphogenesis in fetal rat lung explants. *Am J Physiol Lung Cell Mol Physiol* **282**(3), L379–L385 (2002).
10. Lin, Y.J., Markham, N.E., Balasubramaniam V., *et al.* Inhaled nitric oxide enhances distal lung growth after exposure to hyperoxia in neonatal rats. *Pediatr Res* **58**(1), 22–29 (2005).
11. Greenlee, K.J., Werb, Z., and Kheradmand, F. Matrix metalloproteinases in lung: multiple, multifarious, and multifaceted. *Physiol Rev* **87**(1), 69–98 (2007).
12. Ryu, J., Vicencio, A.G., Yeager, M.E., *et al.* Differential expression of matrix metalloproteinases and their inhibitors in human and mouse lung development. *Thromb Haemost* **94**(1), 175–183 (2005).
13. Gutierrez, J.A., Ertsey, R., Scavo, L.M., *et al.* Mechanical distention modulates alveolar epithelial cell phenotypic expression by transcriptional regulation. *Am J Respir Cell Mol Biol* **21**(2), 223–229 (1999).
14. Compernolle, V., Brusselmans, K., Acker, T., *et al.* Loss of HIF-2alpha and inhibition of VEGF impair fetal lung maturation, whereas treatment with VEGF prevents fatal respiratory distress in premature mice. *Nat Med* **8**(7), 702–710 (2002).
15. Zeng, X., Gray, M., Stahlman, M.T., *et al.* TGF-beta1 perturbs vascular development and inhibits epithelial differentiation in fetal lung *in vivo*. *Dev Dyn* **221**(3), 289–301 (2001).
16. Hislop, A.A., Wigglesworth, J.S., and Desai, R. Alveolar development in the human fetus and infant. *Early Hum Dev* **13**(1), 1–11 (1986).
17. Burri, P.H. Structural aspects of prenatal and postnatal development and growth of the lung. In: McDonald, J.A. (ed.), *Lung Growth and Development*, (Marcel Dekker, 1997).

18. Hislop, A. Developmental biology of the pulmonary circulation. *Paediatr Respir Rev* **6**(1), 35–43 (2005).
19. Zeltner, T.B. and Burri, P.H. The postnatal development and growth of the human lung. II. Morphology. *Respir Physiol* **67**(3), 269–282 (1987).
20. Roth-Kleiner, M., Berger, T.M., Tarek, M.R., *et al.* Neonatal dexamethasone induces premature microvascular maturation of the alveolar capillary network. *Dev Dyn* **233**(4), 1261–1271 (2005).
21. Wendel, D.P., Taylor, D.G., Albertine, K.H., *et al.* Impaired distal airway development in mice lacking elastin. *Am J Respir Cell Mol Biol* **23**(3), 320–326 (2000).
22. Lindahl, P., Karlsson, L., Hellstrom, M., *et al.* Alveogenesis failure in PDGF-A-deficient mice is coupled to lack of distal spreading of alveolar smooth muscle cell progenitors during lung development. *Development* **124**(20), 3943–3953 (1997).
23. Jakkula, M., Le Cras, T.D., Gebb, S., *et al.* Inhibition of angiogenesis decreases alveolarization in the developing rat lung. *Am J Physiol Lung Cell Mol Physiol* **279**(3), L600–L607 (2000).
24. Boucherat, O., Franco-Montoya, M.L., Thibault, C., *et al.* Gene expression profiling in lung fibroblasts reveals new players in alveolarization. *Physiol Genomics* **32**(1), 128–141 (2007).
25. Massaro, G.D. and Massaro, D. Retinoic acid treatment partially rescues failed septation in rats and in mice. *Am J Physiol Lung Cell Mol Physiol* **278**(5), L955–L960 (2000).
26. McGowan, S., Jackson, S.K., Jenkins-Moore, M., *et al.* Mice bearing deletions of retinoic acid receptors demonstrate reduced lung elastin and alveolar numbers. *Am J Respir Cell Mol Biol* **23**(2), 162–167 (2000).
27. Smith, P.G., Janiga, K.E., and Bruce, M.C. Strain increases airway smooth muscle cell proliferation. *Am J Respir Cell Mol Biol* **10**(1), 85–90 (1994).
28. Kotecha, S. Lung growth: implications for the newborn infant. *Arch Dis Child Fetal Neonatal Ed* **82**(1), F69–F74 (2000).
29. Mariotti, V., Marconi, A.M., and Pardi, G. Undesired effects of steroids during pregnancy. *J Matern Fetal Neonatal Med* **16**(Suppl 2), 5–7 (2004).
30. Hallak, M., Moise, K., Jr., Lira, N., *et al.* The effect of tocolytic agents (indomethacin and terbutaline) on fetal breathing and body movements: a prospective, randomized, double-blind, placebo-controlled clinical trial. *Am J Obstet Gynecol* **167**(4 Pt 1), 1059–1063 (1992).

31. Gauldie, J., Galt, T., Bonniaud, P., *et al.* Transfer of the active form of transforming growth factor-beta 1 gene to newborn rat lung induces changes consistent with bronchopulmonary dysplasia. *Am J Pathol* **163**(6), 2575–2584 (2003).
32. Hall, S.M., Hislop, A.A., Pierce, C.M., *et al.* Prenatal origins of human intrapulmonary arteries: formation and smooth muscle maturation. *Am J Respir Cell Mol Biol* **23**(2), 194–203 (2000).
33. Hall, S.M., Hislop, A.A., and Haworth, S.G. Origin, differentiation, and maturation of human pulmonary veins. *Am J Respir Cell Mol Biol* **26**(3), 333–340 (2002).
34. Heath, D., Wood, E.H., Dushane, J.W., *et al.* The structure of the pulmonary trunk at different ages and in cases of pulmonary hypertension and pulmonary stenosis. *J Pathol Bacteriol* **77**(2), 443–456 (1959).
35. Allen, K.M., Wharton, J, Polak, J.M., *et al.* A study of nerves containing peptides in the pulmonary vasculature of healthy infants and children and of those with pulmonary hypertension. *Br Heart J* **62**(5), 353–360 (1989).
36. Owens, G.K. Regulation of differentiation of vascular smooth muscle cells. *Physiol Rev* **75**(3), 487–517 (1995).
37. Suratt, B.T., Cool, C.D., Serls, A.E., *et al.* Human pulmonary chimerism after hematopoietic stem cell transplantation. *Am J Respir Crit Care Med* **168**(3), 318–322 (2003).
38. Davie, N.J., Crossno, J.T., Jr., Frid, M.G., *et al.* Hypoxia-induced pulmonary artery adventitial remodeling and neovascularization: contribution of progenitor cells. *Am J Physiol Lung Cell Mol Physiol* **286**(4), L668–L678 (2004).
39. Bailly, K., Ridley, A.J., Hall, S.M., *et al.* RhoA activation by hypoxia in pulmonary arterial smooth muscle cells is age and site specific. *Circ Res* **94**(10), 1383–1391 (2004).
40. Hall, S.M., Soueid, A., Smith, T., *et al.* Spatial differences of cellular origins and *in vivo* hypoxia modify contractile properties of pulmonary artery smooth muscle cells: lessons for arterial tissue engineering. *J Tissue Eng Regen Med* **1**(4), 287–295 (2007).
41. Mondrinos, M.J., Koutzaki, S.H., Poblete, H.M., *et al.* *In vivo* pulmonary tissue engineering: contribution of donor-derived endothelial cells to construct vascularization. *Tissue Eng Part A* **14**(3), 361–368 (2008).
42. Zhong, T.P., Childs, S., Leu, J.P., *et al.* Gridlock signalling pathway fashions the first embryonic artery. *Nature* **414**(6860), 216–220 (2001).

43. Shalaby, F., Rossant, J., Yamaguchi, T.P., *et al.* Failure of blood-island formation and vasculogenesis in Flk-1-deficient mice. *Nature* **376**(6535), 62–66 (1995).
44. Akeson, A.L., Greenberg, J.M., Cameron, J.E., *et al.* Temporal and spatial regulation of VEGF-A controls vascular patterning in the embryonic lung. *Dev Biol* **264**(2), 443–455 (2003).
45. Healy, A.M., Morgenthau, L., Zhu, X., *et al.* VEGF is deposited in the subepithelial matrix at the leading edge of branching airways and stimulates neovascularization in the murine embryonic lung. *Dev Dyn* **219**(3), 341–352 (2000).
46. Sato, T.N., Tozawa, Y., Deutsch, U., *et al.* Distinct roles of the receptor tyrosine kinases Tie-1 and Tie-2 in blood vessel formation. *Nature* **376**(6535), 70–74 (1995).
47. Sullivan, C.C., Du, L., Chu, D., *et al.* Induction of pulmonary hypertension by an angiopoietin 1/TIE2/serotonin pathway. *Proc Natl Acad Sci USA* **100**(21), 12331–12336 (2003).
48. Chu, D., Sullivan, C.C., Du, L., *et al.* A new animal model for pulmonary hypertension based on the overexpression of a single gene, angiopoietin-1. *Ann Thorac Surg* **77**(2), 449–456 (2004).
49. Han, R.N., Post, M., Tanswell, A.K., *et al.* Insulin-like growth factor-I receptor-mediated vasculogenesis/angiogenesis in human lung development. *Am J Respir Cell Mol Biol* **28**(2), 159–169 (2003).
50. Ruocco, S., Lallemand, A., Tournier, J.M., *et al.* Expression and localization of epidermal growth factor, transforming growth factor-alpha, and localization of their common receptor in fetal human lung development. *Pediatr Res* **39**(3), 448–455 (1996).
51. Gerasimovskaya, E.V., Tucker, D.A., Weiser-Evans, M., *et al.* Extracellular ATP-induced proliferation of adventitial fibroblasts requires phosphoinositide 3-kinase, Akt, mammalian target of rapamycin, and p70 S6 kinase signaling pathways. *J Biol Chem* **280**(3), 1838–1848 (2005).
52. Parker, T.A., Le Cras, T.D., Kinsella, J.P., *et al.* Developmental changes in endothelial nitric oxide synthase expression and activity in ovine fetal lung. *Am J Physiol Lung Cell Mol Physiol* **278**(1), L202–L208 (2000).
53. Gebb, S., Stevens, T. On lung endothelial cell heterogeneity. *Microvasc Res* **68**(1), 1–12 (2004).

54. Arciniegas, E., Frid, M.G., Douglas, I.S., *et al.* Perspectives on endothelial-to-mesenchymal transition: potential contribution to vascular remodeling in chronic pulmonary hypertension. *Am J Physiol Lung Cell Mol Physiol* **293**(1), L1–L8 (2007).
55. Southwood, M., Jeffery, T.K., Yang, X., *et al.* Regulation of bone morphogenetic protein signalling in human pulmonary vascular development. *J Pathol* **214**(1), 85–95 (2008).
56. Alejandre-Alcazar, M.A., Shalamanov, P.D., Amarie, O.V., *et al.* Temporal and spatial regulation of bone morphogenetic protein signaling in late lung development. *Dev Dyn* **236**(10), 2825–2835 (2007).
57. Stabile, H., Mitola, S., Moroni, E., *et al.* Bone morphogenic protein antagonist Drm/gremlin is a novel proangiogenic factor. *Blood* **109**(5), 1834–1840 (2007).
58. Seki, T., Hong, K.H., and Oh, S.P. Nonoverlapping expression patterns of ALK1 and ALK5 reveal distinct roles of each receptor in vascular development. *Lab Invest* **86**(2), 116–129 (2006).
59. Guo, X. and Wang, X.F. Signaling cross-talk between TGF-beta/BMP and other pathways. *Cell Res* **19**(1), 71–88 (2009).
60. Hall, S.M., Gorenflo, M., Reader, J., *et al.* Neonatal pulmonary hypertension prevents reorganisation of the pulmonary arterial smooth muscle cytoskeleton after birth. *J Anat* **196**(Pt 3), 391–403 (2000).
61. Favot, L., Hall, S.M., Haworth, S.G., *et al.* Cytoplasmic YY1 is associated with increased smooth muscle-specific gene expression: implications for neonatal pulmonary hypertension. *Am J Pathol* **167**, 1497–1509 (2005).
62. Hislop, A.A., Mak, J.C., Reader, J.A., *et al.* Muscarinic receptor subtypes in the porcine lung during postnatal development. *Eur J Pharmacol* **359**(2–3), 211–221 (1998).
63. Liu, S.F., Hislop, A.A., Haworth, S.G., *et al.* Developmental changes in endothelium-dependent pulmonary vasodilatation in pigs. *Br J Pharmacol* **1106**(2), 324–330 (1992).
64. Tulloh, R.M., Hislop, A.A., Boels, P.J., *et al.* Chronic hypoxia inhibits postnatal maturation of porcine intrapulmonary artery relaxation. *Am J Physiol* **1272**(5 Pt 2), H2436–H2445 (1997).
65. Hislop, A.A., Springall, D.R., Buttery, L.D., *et al.* Abundance of endothelial nitric oxide synthase in newborn intrapulmonary arteries. *Arch Dis Child Fetal Neonatal Ed* **73**(1), F17–F21 (1995).

66. Nandi, M., Leiper, J., Arrigoni, F., *et al.* Developmental regulation of GTP-CH1 in the porcine lung and its relationship to pulmonary vascular relaxation. *Pediatr Res* **59**(6), 767–772 (2006).
67. Vallance, P. and Leiper, J. Cardiovascular biology of the asymmetric dimethylarginine: dimethylarginine dimethylaminohydrolase pathway. *Arterioscler Thromb Vasc Biol* **24**(6), 1023–1030 (2004).
68. Arrigoni, F.I., Vallance, P., Haworth, S.G., *et al.* Metabolism of asymmetric dimethylarginines is regulated in the lung developmentally and with pulmonary hypertension induced by hypobaric hypoxia. *Circulation* **107**(8), 1195–1201 (2003).
69. Sanchez, L.S., de la Monte, S.M., Filippov, G., *et al.* Cyclic-GMP-binding, cyclic-GMP-specific phosphodiesterase (PDE5) gene expression is regulated during rat pulmonary development. *Pediatr Res* **43**(2), 163–168 (1998).
70. Evans, A.M., Osipenko, O.N., Haworth, S.G., *et al.* Resting potentials and potassium currents during development of pulmonary artery smooth muscle cells. *Am J Physiol* **275**(3 Pt 2), H887–H899 (1998).
71. Porter, L.P., McNamee, J.E., and Wolf, M.B. Endothelin-1 induces endothelial barrier failure in the cat hindlimb. *Shock* **11**(2), 111–114 (1999).
72. Eibl, G., Forgacs, B., Hotz, H.G., *et al.* Endothelin A but not endothelin B receptor blockade reduces capillary permeability in severe experimental pancreatitis. *Pancreas* **25**(2), e15–e20 (2002).
73. Schutte, H., Mayer, K., Burger, H., *et al.* Endogenous nitric oxide synthesis and vascular leakage in ischemic-reperfused rabbit lungs. *Am J Respir Crit Care Med* **164**(3), 412–418 (2001).
74. Wojciak-Stothard, B., Tsang, L.Y., and Haworth, S.G. Rac and Rho play opposing roles in the regulation of hypoxia/reoxygenation-induced permeability changes in pulmonary artery endothelial cells. *Am J Physiol Lung Cell Mol Physiol* **288**(4), L749–L760 (2005).
75. Chiu, Y.J., Kusano, K., Thomas, T.N., *et al.* Endothelial cell-cell adhesion and mechanosignal transduction. *Endothelium* **11**(1), 59–73 (2004).
76. Minshall, R.D., Sessa, W.C., Stan, R.V., *et al.* Caveolin regulation of endothelial function. *Am J Physiol Lung Cell Mol Physiol* **285**(6), L1179–L1183 (2003).
77. Schubert, W., Frank, P.G., Woodman, S.E., *et al.* Microvascular hyperpermeability in caveolin-1 (–/–) knock-out mice. Treatment with a specific

nitric-oxide synthase inhibitor, L-NAME, restores normal microvascular permeability in Cav-1 null mice. *J Biol Chem* **277**(42), 40091–40098 (2002).

78. Carmeliet, P., Ferreira V., Breier G., *et al.* Abnormal blood vessel development and lethality in embryos lacking a single VEGF allele. *Nature* **380**(6573), 435–439 (1996).
79. Gebb, S.A. and Shannon, J.M. Tissue interactions mediate early events in pulmonary vasculogenesis. *Dev Dyn* **217**(2), 159–169 (2000).
80. Yao, Y., Nowak, S., Yochelis, A., *et al.* Matrix GLA protein, an inhibitory morphogen in pulmonary vascular development. *J Biol Chem* **282**(41), 30131–30142 (2007).
81. van Tuyl, M., Liu, J., Wang, J., *et al.* Role of oxygen and vascular development in epithelial branching morphogenesis of the developing mouse lung. *Am J Physiol Lung Cell Mol Physiol* **288**(1), L167–L178 (2005).
82. Akeson, A.L., Wetzel, B., Thompson, F.Y., *et al.* Embryonic vasculogenesis by endothelial precursor cells derived from lung mesenchyme. *Dev Dyn* **217**(1), 11–23 (2000).
83. Alvarez, D.F., Huang, L., King, J.A., *et al.* Lung microvascular endothelium is enriched with progenitor cells that exhibit vasculogenic capacity. *Am J Physiol Lung Cell Mol Physiol* **294**(3), L419–L430 (2007).
84. Ng, Y.S., Rohan, R., Sunday, M.E., *et al.* Differential expression of VEGF isoforms in mouse during development and in the adult. *Dev Dyn* **220**, 112–121 (2001).
85. Maniscalco, W.M., Watkins, R.H., D'Angio, C.T., *et al.* Hyperoxic injury decreases alveolar epithelial cell expression of vascular endothelial growth factor (VEGF) in neonatal rabbit lung. *Am J Respir Cell Mol Biol* **16**(5), 557–567 (1997).
86. Bhatt, A.J., Amin, S.B., Chess, P.R., *et al.* Expression of vascular endothelial growth factor and Flk-1 in developing and glucocorticoid-treated mouse lung. *Pediatr Res* **47**(5), 606–613 (2000).
87. McGrath-Morrow, S.A., Cho, C., Cho, C., *et al.* Vascular endothelial growth factor receptor 2 blockade disrupts postnatal lung development. *Am J Respir Cell Mol Biol* **32**(5), 420–427 (2005).
88. Balasubramaniam, V., Mervis, C.F., Maxey, A.M., *et al.* Hyperoxia reduces bone marrow, circulating, and lung endothelial progenitor cells in the

developing lung: implications for the pathogenesis of bronchopulmonary dysplasia. *Am J Physiol Lung Cell Mol Physiol* **292**(5), L1073–L1084 (2007).

89. Kunig, A.M., Balasubramaniam, V., Markham, N.E., *et al.* Recombinant human VEGF treatment enhances alveolarization after hyperoxic lung injury in neonatal rats. *Am J Physiol Lung Cell Mol Physiol* **289**(4), L529–L535 (2005).
90. Asikainen, T.M., Chang, L.Y., Coalson, J.J., *et al.* Improved lung growth and function through hypoxia-inducible factor in primate chronic lung disease of prematurity. *FASEB J* **20**(10), 1698–1700 (2006).
91. Asikainen, T.M., Waleh, N.S., Schneider, B.K., *et al.* Enhancement of angiogenic effectors through hypoxia-inducible factor in preterm primate lung *in vivo*. *Am J Physiol Lung Cell Mol Physiol* **291**(4), L588–L595 (2006).
92. Hislop, A.A., Lee, R.J., McGregor, C.G., *et al.* Lung growth after transplantation of an adult lobe of lung into a juvenile rat. *J Thorac Cardiovasc Surg* **115**(3), 644–651 (1998).
93. Hsia, C.C. and Johnson, R.L., Jr. Further examination of alveolar septal adaptation to left pneumonectomy in the adult lung. *Respir Physiol Neurobiol* **151**(2–3), 167–177 (2006).
94. Fehrenbach, H., Voswinckel, R., Michl, V., *et al.* Neoalveolarisation contributes to compensatory lung growth following pneumonectomy in mice. *Eur Respir J* **31**(3), 515–522 (2008).
95. Zhang, Q., Bellotto, D.J., Ravikumar, P., *et al.* Postpneumonectomy lung expansion elicits hypoxia-inducible factor-1alpha signaling. *Am J Physiol Lung Cell Mol Physiol* **293**(2), L497–L504 (2007).
96. Zhang, Q., Moe, O.W., Garcia, J.A., *et al.* Regulated expression of hypoxia-inducible factors during postnatal and postpneumonectomy lung growth. *Am J Physiol Lung Cell Mol Physiol* **290**(5), L880–L889 (2006).
97. Nolen-Walston, R.D., Kim, C.F., Mazan, M.R., *et al.* Cellular kinetics and modeling of bronchioalveolar stem cell response during lung regeneration. *Am J Physiol Lung Cell Mol Physiol* **294**(6), L1158–L1165 (2008).
98. Fernandez, L.G., Mehta, C.K., Kron, I.L., *et al.* Reinitiation of compensatory lung growth after subsequent lung resection. *J Thorac Cardiovasc Surg* **134**(5), 1300–1305 (2007).
99. Sakurai, M.K., Lee, S., Arsenault, D.A., *et al.* Vascular endothelial growth factor accelerates compensatory lung growth after unilateral pneumonectomy. *Am J Physiol Lung Cell Mol Physiol* **292**(3), L742–L747 (2007).

100. Burlingham, W.J., Love, R.B., Jankowska-Gan, E., *et al.* IL-17-dependent cellular immunity to collagen type V predisposes to obliterative bronchiolitis in human lung transplants. *J Clin Invest* **117**(11), 3498–3506 (2007).
101. Du, L., Sullivan, C.C., Chu, D., *et al.* Signaling molecules in nonfamilial pulmonary hypertension. *N Engl J Med* **348**(6), 500–509 (2003).
102. Morrell, N.W., Yang, X., Upton, P.D., *et al.* Altered growth responses of pulmonary artery smooth muscle cells from patients with primary pulmonary hypertension to transforming growth factor-beta(1) and bone morphogenetic proteins. *Circulation* **104**(7), 790–795 (2001).
103. Teichert-Kuliszewska, K., Kutryk, M.J., Kuliszewski, M.A., *et al.* Bone morphogenetic protein receptor-2 signaling promotes pulmonary arterial endothelial cell survival: implications for loss-of-function mutations in the pathogenesis of pulmonary hypertension. *Circ Res* **98**(2), 209–217 (2006).
104. Jeffery, T.K., Upton, P.D., Trembath, R.C., *et al.* BMP4 inhibits proliferation and promotes myocyte differentiation of lung fibroblasts via Smad1 and JNK pathways. *Am J Physiol Lung Cell Mol Physiol* **288**(2), L370–L378 (2005).
105. Nishimura, T., Vaszar, L.T., Faul, J.L., *et al.* Simvastatin rescues rats from fatal pulmonary hypertension by inducing apoptosis of neointimal smooth muscle cells. *Circulation* **108**(13), 1640–1645 (2003).
106. Hansmann, G., Wagner, R.A., Schellong, S., *et al.* Pulmonary arterial hypertension is linked to insulin resistance and reversed by peroxisome proliferator-activated receptor-gamma activation. *Circulation* **115**(10), 1275–1284 (2007).
107. Takayama, Y., May, P., Anderson, R.G., *et al.* Low density lipoprotein receptor-related protein 1 (LRP1) controls endocytosis and c-CBL-mediated ubiquitination of the platelet-derived growth factor receptor beta (PDGFR beta). *J Biol Chem* **280**(18), 18504–18510 (2005).
108. Schermuly, R.T., Dony, E., Ghofrani, H.A., *et al.* Reversal of experimental pulmonary hypertension by PDGF inhibition. *J Clin Invest* **115**(10), 2811–2821 (2005).
109. Ghofrani, H.A., Seeger, W., and Grimminger, F. Imatinib for the treatment of pulmonary arterial hypertension. *N Engl J Med* **353**(13), 1412–1413 (2005).
110. Cowan, K.N., Heilbut, A., Humpl, T., *et al.* Complete reversal of fatal pulmonary hypertension in rats by a serine elastase inhibitor. *Nat Med* **6**(6), 698–702 (2000).

111. McMurtry, M.S., Archer, S.L., Altieri, D.C., *et al.* Gene therapy targeting survivin selectively induces pulmonary vascular apoptosis and reverses pulmonary arterial hypertension. *J Clin Invest* **115**(6), 1479–1491 (2005).
112. Fantozzi, I., Platoshyn, O., Wong, A.H., *et al.* Bone morphogenetic protein-2 upregulates expression and function of voltage-gated K+ channels in human pulmonary artery smooth muscle cells. *Am J Physiol Lung Cell Mol Physiol* **291**(5), L993–L1004 (2006).
113. Bonnet, S., Michelakis, E.D., Porter, C.J., *et al.* An abnormal mitochondrial-hypoxia inducible factor-1alpha-Kv channel pathway disrupts oxygen sensing and triggers pulmonary arterial hypertension in fawn hooded rats: similarities to human pulmonary arterial hypertension. *Circulation* **113**(22), 2630–2641 (2006).
114. Nebigil, C.G., Launay, J.M., Hickel, P., *et al.* 5-hydroxytryptamine 2B receptor regulates cell-cycle progression: cross-talk with tyrosine kinase pathways. *Proc Natl Acad Sci USA* **97**(6), 2591–2596 (2000).
115. Zhao, Y.D., Courtman, D.W., Deng, Y., *et al.* Rescue of monocrotaline-induced pulmonary arterial hypertension using bone marrow-derived endothelial-like progenitor cells: efficacy of combined cell and eNOS gene therapy in established disease. *Circ Res* **96**(4), 442–450 (2005).
116. Wang, X.X., Zhang, F.R., Shang, Y.P., *et al.* Transplantation of autologous endothelial progenitor cells may be beneficial in patients with idiopathic pulmonary arterial hypertension: a pilot randomized controlled trial. *J Am Coll Cardiol* **49**(14), 1566–1571 (2007).
117. Raoul, W., Wagner-Ballon, O., Saber, G., *et al.* Effects of bone marrow-derived cells on monocrotaline- and hypoxia-induced pulmonary hypertension in mice. *Respir Res* **8**, 8 (2007).
118. Massaro, D., Alexander, E., Reiland, K., *et al.* Rapid onset of gene expression in lung, supportive of formation of alveolar septa, induced by refeeding mice after calorie restriction. *Am J Physiol Lung Cell Mol Physiol* **292**(5), L1313–L1326 (2007).
119. Otto, W.R. Lung epithelial stem cells. *J Pathol* **197**(4), 527–535 (2002).
120. Giangreco, A., Shen, H., Reynolds, S.D., *et al.* Molecular phenotype of airway side population cells. *Am J Physiol Lung Cell Mol Physiol* **286**(4), L624–L630 (2004).
121. Summer, R., Kotton, D.N., Sun, X., *et al.* Translational physiology: origin and phenotype of lung side population cells. *Am J Physiol Lung Cell Mol Physiol* **287**(3), L477–L483 (2004).

122. Irwin, D., Helm, K., Campbell, N., *et al.* Neonatal lung side population cells demonstrate endothelial potential and are altered in response to hyperoxia-induced lung simplification. *Am J Physiol Lung Cell Mol Physiol* **293**(4), L941–L951 (2007).
123. Thebaud, B., Ladha, F., Michelakis, E.D., *et al.* Vascular endothelial growth factor gene therapy increases survival, promotes lung angiogenesis, and prevents alveolar damage in hyperoxia-induced lung injury: evidence that angiogenesis participates in alveolarization. *Circulation* **112**(16), 2477–2486 (2005).
124. Mondrinos, M.J., Koutzaki, S., Jiwanmall, E., *et al.* Engineering three-dimensional pulmonary tissue constructs. *Tissue Eng* **12**(4), 717–728 (2006).

Chapter 3
Epithelial–Mesenchymal Interactions During Lung Development and Their Potential Relevance to Lung Repair

Alexandra Firsova, Tomoko Hyakumura, Timothy J. Cole and Richard Mollard

Together with their precursors and descendants, epithelial and mesenchymal cells interact throughout development to ultimately define the physiology and architecture of the mature and functional lung. It is hypothesised that during development, certain classes of both epithelial and mesenchymal stem and progenitor cells are retained in an undifferentiated state so they may be reactivated later in adult life to repair the damaged lung. Furthermore, it is hypothesised that several classes of molecules intimately associated with the developmental process are reutilised later during times of repair. These molecules include members of the fibroblast growth factor and transforming growth factor families, their receptors, agonist and antagonist signalling systems, transcriptional superfamily members and components of the extracellular matrix and basement membrane. How these various molecules orchestrate a morphogenetic signal coordinating both appropriate proliferation and differentiation of the epithelium and restructuring or maintenance of the supportive mesenchymal mass in order to effect repair of damaged tissue, however, remains unknown. This chapter will detail some of the cellular and molecular mechanisms interpreting epithelial–mesenchymal interactions from the moment of gastrulation to the establishment of the early branching lung anlage, currently described in the literature. As continual research is applied to this field, it is predicted these signalling pathways instructive to the formation and allocation of epithelial and mesenchymal stem and progenitor cells during development will be exploited to provide advanced interventional therapeutics for treating congenital respiratory diseases and respiratory damage following environmental insult (see Chapter 2).

Keywords: Early lung development; molecular signalling; epithelial; mesenchymal.

Outline

1. Formation of the Lung Epithelium and Mesenchyme

In the mouse, the first sign of gastrulation, or formation of the three primitive germ layers, occurs between E6.75 and E6.85 when a circumferential groove appears in the presumptive primitive streak region of the embryonic epiblast/ectoderm.[1] The primitive streak extends anteriorly from the posterior end of the epiblast forming regions of organising activity in temporal order known as the early- and mid-gastrula organisers (EGO and MGO, respectively) and the node.[2–5] Fate mapping studies have demonstrated that cells of the posterior epiblast migrating through the EGO and MGO give rise to axial mesendoderm, a condensed stripe of cells expressing *Foxa2*, *Brachyury* (*T*), *Goosecoid* (*Gsc*) and *Mixl1*.[6–8] Axial mesendoderm contains progenitors of the anterior definitive endoderm (ADE) and lateral plate mesoderm (LPM) and as such progenitors of both foregut endoderm (e.g. lung, liver and pancreas) and their associated mesoderms.[3,4,6,9,10] In the absence of subsequent signalling from the node, appropriate ADE fate specification and positional information is not maintained and hence morphogenesis of the lung and other organs does not commence as embryos generally die prior to E9.5.[5]

Examples of molecules providing signalling information from the node to the ADE are Nodal and Fibroblast growth factor 8 (Fgf 8). *Nodal* is first expressed throughout the epiblast and visceral endoderm at E6.5 and is then restricted to the posterior proximal epiblast around the node at E8.0.[11,12] *Fgf8*, one member of a large gene family of growth factors (see below), is similarly expressed in the epiblast at the presumptive proximo-posterior aspect of the embryo, prior to appearance of the primitive streak, and regulates *Nodal* expression around the node.[11] *Nodal* null mutants display pre-streak developmental arrest while $Fgf8^{-/-}$ embryos form neither endoderm nor mesoderm and do not survive past E9.5.[11,13,14] *Fgf8* and *Nodal* serial hypomorphs and compound mutants have further defined the importance of these molecules during the subsequent inductive events of the node upon the LPM and then subsequently in left–right symmetry specification of foregut endoderm and as such lung epithelial precursors (see below).

2. Respiratory Endodermal Specification and Primary Epithelial Bud Outgrowth

During their anterior migration along the ventral surface of the embryo, ADE derivatives of the mesendoderm maintain *Foxa2* expression and downregulate *T* expression, while mesoderm contributing to the lateral plate retains *T* expression.[6,15,16] As respiratory endodermal precursors contained within the ADE migrate to the ventro-lateral wall of the developing foregut, their specification is reliant upon signals derived from co-migrating pro-cardiogenic mesodermal cells found within the region of the anterior LPM.[17–22] At approximately the seven-somite stage (E8.25), the liver and pancreatic markers *Albumin* and *Pdx1* are expressed in the ventral *Foxa2* expressing foregut endoderm.[23–25] At the eight-somite stage, Fgfs 1 and/or 2 liberated from the pro-cardiogenic mesoderm, interact(s) with their receptors Fgfr1 and Fgfr4 within the anterior ventral foregut endoderm to activate *Nkx2.1*, thus marking future thyroid and lung endoderm within this same domain and in an overlapping fashion.[25,26] Lung, liver and pancreatic marker activation is temporally controlled and regulated by the anterio-posterior proximity of the cardiac Fgf source. Local concentrations of Fgf change with time according to morphogenetic movements of the developing heart, with higher concentrations of Fgf being required for *Nkx2.1* expression, lower Fgf concentrations being required for *Albumin* expression and the absence of Fgf signalling being required for *Pdx1* expression[25,27] (Fig. 1). Separation of *Nkx2.1* and *Albumin* expression domains then occurs with advancing somite age. *Nkx2.1* expression is maintained throughout the developing lung endoderm at E10.5, being expressed at high levels in the left and right bronchi, and then becoming progressively restricted to alveolar type II and Clara cells with further development.[25,26] In these more mature respiratory derivatives, Nkx2.1 signalling regulates *surfactant protein d (Sftpd)* expression and with Foxa2 and Gata6 regulates transcriptional activation of further lung differentiation markers and proteins critical to normal lung function such as *Sftpb* and uteroglobulin (*Scgb1a1* or *CC10*), and *Stfpa* and *Sftpc*, respectively.[23,28–33]

The importance of *Nkx2.1* expression for appropriate lung function or alternative progression to genetic disease has been demonstrated at several levels. At approximately 20 somites, or E9.0-9.5 in the mouse, respiratory development is marked morphologically for the first time by the appearance of the laryngotracheal groove within the ventral wall of the caudal pharyngeal endoderm[34] (Fig. 2). This groove produces a ridge on the external surface of the pharynx to form the laryngotracheal diverticulum. The diverticulum becomes separated from the foregut by tracheoesophogeal folds that fuse to form a partition, dividing the foregut into the ventral portion, the laryngotracheal tube, and a dorsal portion, the oesophagus.[35] *Nkx2.1* null mutants display tracheoesophogeal fistula and hypoplastic lungs that fail to undergo correct branching following lobar outgrowth.[36] Furthermore, due to the lack of Nkx2.1 signalling, null mutant respiratory endoderm does not express *Sftpc*. Such studies reveal the importance not only of Fgf signalling from the respiratory associated

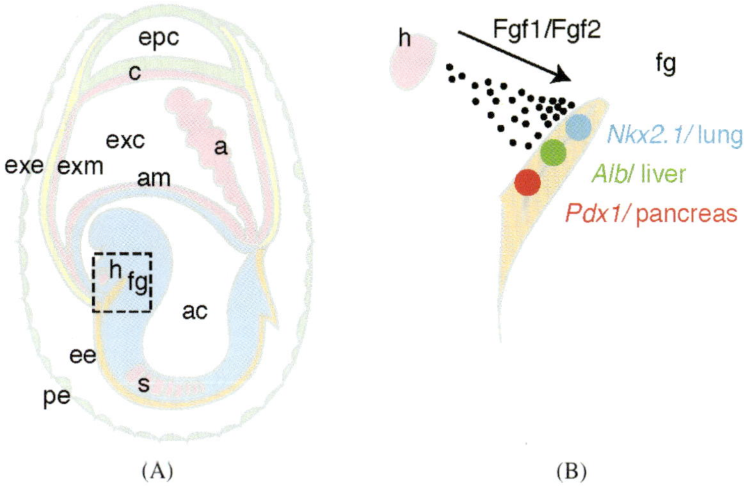

Figure 1: Specification of lung endoderm within the foregut. (A) An eight somite stage embryo about to commence turning. (B) Enlargement of broken box in (A) and schematic simplification of cardiac to foregut Fgf signalling. A concentration gradient of Fgf1 and/or Fgf2 released from the cardiac mesoderm in coordination with morphogenetic movements of the foregut specifies regional identity of the lung, liver and pancreatic endoderm (at this time, it is expected that individual lung, liver and pancreatic precursors will exist in overlapping and mixed domains, not as the simplified discrete domains depicted here). Higher Fgf concentrations specify lung, lower concentrations specify liver while the absence of Fgf1 and Fgf2 signalling is permissive to pancreas formation. a, amion; ac, amniotic cavity; am, amniotic mesoderm; c, chorion; ee, embryonic endoderm; epc, ectoplacental cone; exc, exocoelomic cavity; exe, extraembryonic endoderm; exm, extraembryonic mesoderm; fg, foregut; h, heart; pe, parietal endoderm; s, somite.

cardiac mesenchyme to elicit appropriate specification and morphogenesis of respiratory epithelium though Nkx2.1 signalling, but also underline the requirement of the precise dosage requirement of this signal for appropriate proliferation, structural organisation, differentiation and function of the lung epithelium.

3. Left and Right Bud Outgrowth and Left–Right Patterning

(See also Chapter 2)

Following specification within the foregut endoderm and formation of the laryngotracheal tube, distal and ventral respiratory precursor endoderm

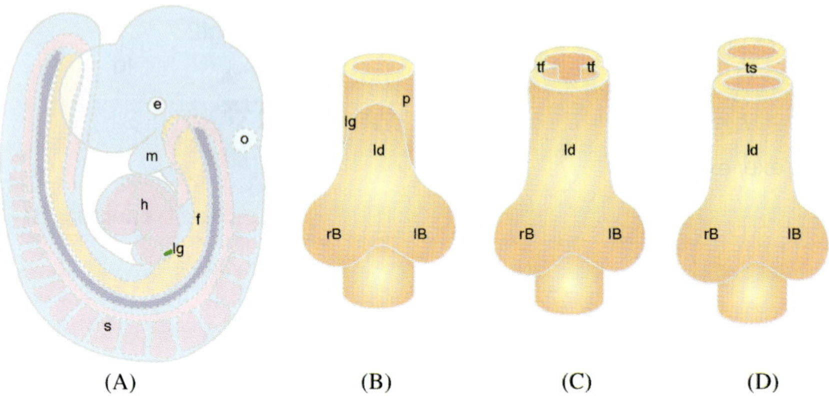

Figure 2: Formation of the laryngotracheal groove and tracheoesophageal septum. (A) At approximately the 20 somite stage, the laryngotracheal groove appears in the ventral wall of the caudal foregut. (B) The groove forms a ridge in the pharynx called the laryngotracheal diverticulum. (C) Tracheoesophageal folds form between the pharynx and the laryngotracheal diverticulum. (D) At approximately the 25 somite stage, the tracheoesophageal folds fuse to form the tracheoesophageal septum, and this septum partitions the oesophagus from the trachea as the lung buds invade the surrounding mesenchyme. e, eye; f, foregut; h, heart; lB, left lung bud; ld, laryngotracehal diverticulum; lg, laryngotracehal groove; m, mandible of the first branchial arch; o, otic vesicle; p, pharynx; rB, right lung bud; s, somite; tf, tracheoesphogeal fold; ts, tracheoesophageal septum.

invades the surrounding splanchnic derivatives of the lateral plate mesoderm to form two primitive lung outgrowths (left and right) at approximately E9.5.[37] In mice, these left and right outgrowths form, respectively, one left lobe and four right lobes (see Fig. 3; apical, azygos, cardiac and diaphragmatic). Several molecular signalling systems have been identified as necessary for bud outgrowth, with one of the major regulators being the retinoic acid (RA) signalling system (see also Chapter 7). Retinoic acid receptor alpha and retinoid X receptor alpha (Rara/Rxra) heterodimers present in the foregut endoderm and mesoderm, together with some redundancy involving Rarb and Rxrb, interpret signalling of all-*trans*-retinoic-acid (a*t*RA) originating from within the foregut endoderm prior to E9.5 to regulate specification of respiratory precursors contained within the ADE.[38–41] In the absence of such signalling, foregut endoderm fails to produce left and/or right lung outgrowths resulting in either left lung agenesis or lung hypoplasia. How RA signalling intercalates with specification of other systems

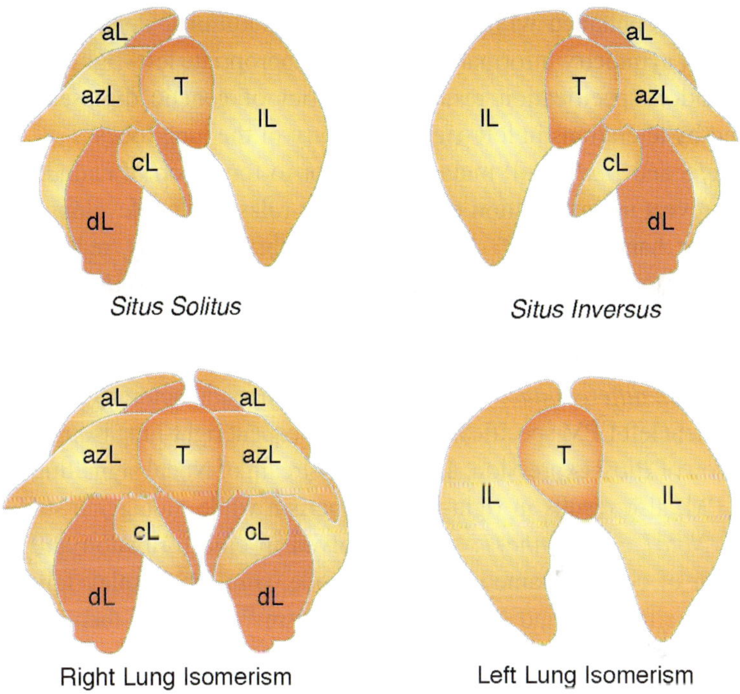

Figure 3: Schematical representation of right and left lung lobes. *Situs solitus* represents the normal condition, *situs inversus* represents inversion of the normal condition, right lung isomerism represents the replacement of the left lobe with a mirror imaged right lobe and left lung isomerism represents the replacement of the right lobe with a mirror imaged left lobe. aL, apical (cranial) lobe; azL, azygos (accessory/intermediate) lobe; dl, diaphramatic (caudal) lobe; cl, cardiac (medial) lobe; T, trachea.

affecting ADE fate and the precise timing of its signal is at present unknown, but appears to commence at approximately E8.5 in the mouse.

4. Endodermal Specification and Left–Right Symmetry (See also Chapter 2)

As discussed above, anterior–posterior positional specification of the ADE is effected by signals elicited from the early-, mid- and late-streak stages and continually through the eight-somite stage. In addition to anterior–posterior positional specification at the late-streak stage, signalling from the node throughout the LPM also provides lung endodermal precursors contained within the ADE, with additional dorso-ventral and left–right symmetry information.[5] Lack of appropriate left–right symmetry information specification results in a series of genetic diseases characterised by alterations in left–right symmetry of the lung lobes.

Fgf8 and *Nodal* null mutants die prior to formation of endoderm and hence reveal no information regarding their roles in lung development. On the other hand, *Fgf8*$^{neo/-}$ hypomorphic embryos display reduced Fgf8 signalling to the left LPM, as evidenced by reduced *Nodal* expression within the LPM, and a subsequent right-lung isomerism.[42,43] (Fig. 3). Similarly, reduced Nodal signalling in *Nodal*$^{\Delta 600/-}$ hypomorphic embryos and *Growth differentiation factor-1* (*Gdf1*) compound heterozygotes results in deficient left LPM specification and a subsequent right-lung isomerism.[14,44] Ablation of *Pitx2* expression, a gene downstream of Nodal signalling within the LPM, also results in right-lung isomerism while mutagenic studies of *Pkd2*, *Noto1* and *Dnahc11*, genes involved in cilia formation and hence function of cells contained within the node, similarly display failed left-LPM specification and right-lung isomerism.[45–49] It has therefore been postulated that ciliated cells resident within the node create a unidirectional flow of Fgf8 signalling across the primitive streak, thus inducing Nodal signalling specifically within the left, and not right, 0 to 2 somite or E7.5–E8.0 LPM, which then acts on the ADE to define left-to-right identity of respiratory and other constituent endodermal precursors.[11,42]

In addition to right-lung isomerism, inappropriate specification of the foregut endoderm through compromised signalling from the LPM (e.g. compromised *Pcsk6*, *Lefty1* and *Gdf1* expression) can also manifest as *situs inversus* or left-lung isomerism[50–52] (Fig. 3). Such studies demonstrate that specification of the lung epithelium and its subsequent lobar patterning originates from signals within the node that affect LPM identity and, in turn, lung epithelial left-to-right symmetry.

5. Secondary Branching and Branching Morphogenesis (See also Chapter 2)

After the appearance of the left and right primary buds, serial branch point formation from the primary lobes occurs in an ostensibly stereotypical pattern, encompassing some spatial, temporal and lineage error variants, and being characterised by three overlapping yet distinct morphogenetic modes: domain branching, planar bifurcation and orthogonal bifurcation.[49] Domain branching provides proximo-distal periodicity and circumferential specification to branch points emerging from the primordial outgrowths of each lobe. Planar bifurcation involves the expansion and bifurcation of bud tips from some tertiary buds and subsequent outgrowths in a two-dimensional plane. Orthogonal bifurcation provides some of these bifurcations with an approximate 90° rotational third-dimensional plane, thus forming a rosette in cross-section. Such epithelial patterning encompasses the process known as branching morphogenesis, a process that commences in the pseudoglandular period of lung development at approximately E11.5, continuing through to the canalicular period and thus establishing the highly arborised nature of the adult lung epithelium. During branching morphogenesis, numerous signalling systems between the lung epithelium and mesenchyme are activated to divide the growing buds into areas of proximal morphogenetic stability and distal morphogenetic activity. Areas of morphogenetic activity are believed to contain various stem/progenitor cell populations that are maintained in the adult to be activated in times of repair. The signalling systems specifying and maintaining these systems are thus of interest to better understanding how resident stem cells may be activated following injury to elicit repair.

Early studies of the embryonic mouse lung demonstrated the crucial role of epithelial–mesenchymal interactions for appropriate branching of the lung epithelium and specification of the epithelial and mesenchymal sub-domains.[53,54] Appropriate geometric patterning of the lung, in terms of budding and branching, only occurs if the lung's mesenchyme is present. In the absence of the supporting mesenchyme, epithelial cells lose orientation with each other, fail to bud or branch, transform from a columnar to squamous phenotype and die. When epithelium and mesenchyme from chick lung epithelia are separated from each other and recombinations made between the trachea, terminal buds and vascular airsacs, mesenchyme from non-branching structures such as the airsacs and trachea are observed to suppress branching, whereas mesenchyme from the branching regions such as the terminal buds and parabronchi promote supernumerary bud formation along the trachea.[54,55] Furthermore, recombinations between E13 rat tracheal mesenchyme with E13 lung epithelium result in an anteriorisation of lung structures, with the expression of a tracheal endoderm phenotype, while recombinations of tracheal endoderm with distal lung mesenchyme results in a posteriorisation of tracheal endoderm, with the expression of a lung phenotype.[34,56] These homeotic-like transformations appear to be restricted with a temporal specificity, but demonstrate the inductive power of the mesenchyme upon the epithelium in terms of pattern formation and differentiation.[57]

Evidence has come to support the notion that these heterotypic mesenchymal populations situated at the epithelial–mesenchymal interface release diffusible molecules that regulate basement membrane deposition and alterations in epithelial proliferation rates.[58–65] As such, within areas of developing clefts, the mesenchyme induces areas of morphogenetic stability through relatively low epithelial proliferation and high basement membrane content, while in areas of distal bud outgrowth, distinct mesenchymal subsets induces morphogenetic activity through higher epithelial proliferation and lower basement membrane content. Basement membrane constituents recognised to play a major role establishing separate branching and non-branching domains include fibronectin, collagen IV, laminin and nidogen. Soluble signalling pathways instigating and maintaining the distal respiratory phenotype during mouse lung organogenesis have also been described and involve Fgf, Sprouty (Spry), Bone morphogenetic protein (Bmp), Sonic hedgehog (Shh) and Wnt family members. Because of their likely therapeutic value to future applications aiming to induce repair following respiratory injury, these soluble signalling systems and their roles in instructing proliferation, differentiation and regulation of the lung architecture during branching morphogenesis will be the subject of the remaining sections.

6. Fgf Signalling Within the Epithelium and Mesenchyme During Lung Branching Morphogenesis

(See also Chapter 2)

Some Fgfs have been described above to participate in crucial interactions governing early respiratory endodermal specification. Those described to play instructive roles during later stage lung branching morphogenesis include Fgf1, 2, 7, 9 and 10. There are 23 Fgf family members and, with the exception of Fgf11 to Fgf14, these proteins effect signalling through the binding of a family of tyrosine kinase receptors. The Fgfrs are encoded by four separate genes that transcribe seven major isoforms following common alternative splicing of exon III: *Fgfr1b*, *Fgfr1c*, *Fgfr2b*, *Fgfr2c*, *Fgfr3b*, *Fgfr3c* and *Fgfr4*.[66]

Between E11 and E20, Fgf1 immunoreactivity is found throughout the mesenchyme but not epithelium.[67] Fgf2 immunoreactivity is detected in the epithelia of the presumptive trachea and lung bud, but later in the mesenchyme, and is found associated with the apical cell membranes of budding epithelial cells, being stored in the extracellular matrix and membrane bound proteoglycans.[67,68] *Fgf7* expression is found throughout the lung mesenchyme commencing at E14.5.[69] *Fgf1*$^{-/-}$ and *Fgf2*$^{-/-}$ single and compound homozygote and *Fgf7*$^{-/-}$ mice display no overt lung phenotype.[70,71] The lack of phenotype in these mice has been attributed to a functional redundancy with other family members.[72] *Fgf9* is expressed within the presumptive mesothelium and lung bud epithelium at E10.5. By E12.5 expression is restricted to the mesothelium and then downregulated by E14.5.[73,74] *Fgf9*$^{-/-}$ mice display a normal generation and number of pulmonary lobes and epithelial proliferation, but a reduced mesenchymal proliferation and hypoplastic secondary branching.[73] *Fgf10* is expressed within the mesenchyme through lung development, being localised to discrete cellular subsets within the peripheral mesenchyme, corresponding to the previously described cellular morphogenetic areas.[75] *Fgf10*$^{-/-}$ mutants develop a trachea, but at E11.5 mainstem bronchi, bud outgrowth and correspondingly branching morphogenesis are disrupted.[76,77] *Mlcv-nLac-24* enhancer trap *Fgf10* allelic hypomorphic lungs similarly display hypoplasia, and altered epithelial (decreased *Nkx2.1* and *Sftpb* expression) and smooth muscle differentiation.[78] *Fgf18* expression is not detected within the foregut epithelium or mesenchyme between E8.5 and E10.5, but is detected within the lung mesenchyme between E12.5 to E18.5.[79,80] *Fgf18*$^{-/-}$ mice display no overt foetal phenotype, but fail to appropriately form alveoli.[81]

Fgfr1 is expressed within the endoderm and mesoderm of the presumptive trachea and lung bud at E8.5 and subsequently diffusely through the mesenchyme.[82,83] *Fgfr1b* and *Fgfr1c* isoform expression in the developing lung has not yet been fully described. Most *Fgfr1*$^{-/-}$ mice die prior to E9.5, while an *Fgfr1b* in-frame stop codon or deletion of the *Fgfr1-IIIb* exon results in no apparent phenotype.[84–87] *Fgfr2b* is expressed in the developing lung epithelium throughout organogenesis while *Fgfr2c* is expressed within the mesenchyme.[82,88,89] A dominant negative *Fgfr2b* transgenic isoform inhibits lung branching morphogenesis (but not bud outgrowth), while *Fgfr2b* null mutagenesis results in lung agenesis.[82,89,90] *Dermo1-Cre* mediated ablation of *Fgfr1* and *Fgfr2* in lung mesenchyme commencing at E10.5 demonstrates a functional redundancy with compound heterozygotes and *Fgfr1*$^{-/-}$*Fgfr2*$^{+/-}$ mice displaying no phenotype, while *Fgfr1*$^{+/-}$ *Fgfr2*$^{-/-}$ and compound homozygotes mice display an allelic dose response hypoplasia.[91] *Fgfr3* transcripts are not detectable in the developing lung until E16.5, with the protein localising postnatally to the mesenchyme.[92,93] *Fgfr4* transcripts are associated with formation of the presumptive lung buds from the foregut endoderm at the 7–9 somite stage, just prior to the appearance of *Nkx2.1* in the same foregut endoderm domain.[25] Neither *Fgfr3*$^{-/-}$ nor *Fgfr4*$^{-/-}$ mice show an obvious lung phenotype, yet compound homozygotes, although showing no overt branching or early developmental lung phenotype, fail to form alveoli postnatally.[93]

Due to their redundancy during development, it is unclear how Fgf1, 2 and 7 may be relevant to repair in times of injury. However, *in vitro* binding and expression studies suggest that mesenchymal Ffg1, Fgf3 and Fgf7 may act through epithelial Fgfr2b to regulate lung epithelial proliferation and bud outgrowth, hence controlling stem/progenitor cell activity.[66,94] Due to the clarity of knockout phenotypes and associated studies, it is now accepted that Fgf9 secreted by the presumptive mesothelium and acting through mesenchymal Fgfr1 and Fgfr2, regulates Fgf10 activity within discrete mesenchymal domains distal to growing buds[91] (Fig. 4). Fgf10 acting through epithelial Fgfr2b directs epithelial proliferation and hence bud outgrowth and simultaneously differentiation.[74,78,95,96] Lineage tracing analysis has demonstrated that the *Fgf10* expressing nascent mesenchyme is reallocated from the distal tips as the bud extends outwards to be located subjacent to the tubules as they extend.[97] Here these cells express *smooth muscle actin* (*Acta2*). As such, multiple Fgf signalling mechanisms

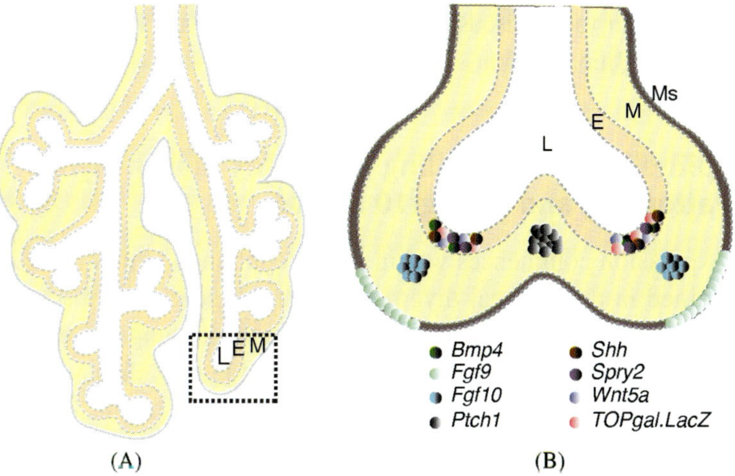

Figure 4: Simplified schematical representation of several signalling systems contributing to epithelial and mesenchymal interactions. (A) An approximate E12.5 lung bud. (B) The epithelium within the box indicated in (A) has branched. The location of several signalling systems involved in this process is indicated. E, epithelium, L, lumen; M, mesenchyme; Ms presumptive mesothelium.

from within the presumptive mesothelium and through the nascent mesenchyme coordinately regulate activity of both epithelial and smooth muscle progenitors. If stem/progenitor cells specified during development are retained in a similar functional context through to adult life (which is a widely held yet still to be proven belief), appropriately applying an exogenous Fgf signal or stimulating an endogenous Fgf signal may prove of benefit to repair of the damaged lung.

A number of mediators have been identified to maintain and confine the Fgf10 signal within the distal and discrete mesenchymal populations identified as precursors to smooth muscle progenitors and relay this signal to effect epithelial behavioural outcome. As such, what effect these molecules will have upon broader cellular outcomes following attempts to amplify progenitors of therapeutic value during the repair of lung injury ought to be considered. Signalling pathways that modify and interpret the

Fgf signal and to be discussed in this context are those involving Shh, Dicer, Spry, Bmp and Wnt.

7. Epithelial Shh Signalling Regulates Mesenchymal Patterning and the Fgf10 Signal

Shh expression commences within the foregut endoderm at E8.0 and is maintained in the tracheal diverticulum as it emerges from the ventral wall of the foregut, and then also later in the distal lung bud epithelium during the pseudoglandular stage.[98,99] *Shh*$^{-/-}$ foetuses display inhibited oesophageal and tracheal septation from the laryngotracheal tube, with the oesophageal epithelium appearing as stomach epithelium and showing continuity with the lung epithelium.[100,101] Although fistula formation resembles that observed in *Nkx2.1*$^{-/-}$ foetuses, *Shh* expression is unaffected in *Nkx2.1*$^{-/-}$ lungs and *Nkx2.1* expression is detectable in *Shh*$^{-/-}$ foetuses, suggesting parallel yet independent effector pathways.[100] In *Shh*$^{-/-}$ lungs, the normally focal mesenchymal *Fgf10* expression domains distal to the lung epithelium are lost, instead extending throughout the mesenchyme, and normal *Acta2* expression is lacking. As such, the control of lobation and branch point formation are lost and the lungs appear either as one large sac or a simple bifurcation.[100,101] Specification of lung endodermal progenitors thus requires Shh, and continued *Shh* expression in these progenitors is prerequisite to the appropriate interpretation and regulation of Fgf10 signalling cues.

In the mouse, Shh binding to Patched1 (Ptc1) and Ptc2 receptors removes unliganded Ptc-mediated suppression of the intracellular signal transducer Smoothened.[102] Smoothened then relieves suppression of the transcriptional activators Gli1, Gli2 and Gli3 to positively effect regulation of downstream signalling target genes. From E11.5 to E13.5, *Ptc1* is expressed at highest levels within the mesenchyme adjacent to distal buds and at lower levels in the proximal mesenchyme, while *Ptc2* is expressed at higher levels in the epithelium.[75,99,100,103] Most studies described to date concerning the Ptc receptors during early lung development have focussed on *Ptc1* expression. During branching morphogenesis and in a feedback regulatory fashion that ultimately sequesters the Shh signal, epithelial Shh upregulates *Ptc1* expression in the mesenchyme and as such, increases mesenchymal proliferation, while *Ptc2* expression in the epithelium is unaffected by Shh signalling.[100,104] Ptc1 acts through Hedgehog interacting protein (Hip1) also in a feedback loop to further dampen the Shh signal by downregulating *Shh* expression.[105] In the absence of Hip1, *Ptc1* expression is upregulated in the mesenchyme, *Fgf10* expression is downregulated and branch point formation is lost. Hence, either loss or expansion of the *Fgf10* expression domain due to alterations in the Shh signaling pathway results in altered lung branching because focal localisation of the Fgf10 epithelial morphogenic source from within the mesenchyme is lost.

All three *Gli* genes are expressed in the foregut mesoderm and later in lung mesenchyme in specific overlapping patterns.[106] *Gli2*$^{-/-}$ foetuses display one right and one left lobe, lung hypoplasia and oesophageal and tracheal stenosis, demonstrating in a similar fashion to *Shh* and *Hip1*, a role in lung progenitor specification/activity for appropriate lobar patterning and domain branching.[107] Reducing *Gli3* dosage in *Gli2*$^{-/-}$ mice increases the severity of the phenotype, resulting in tracheoesophageal fistula and a more severe lung phenotype, while *Gli2*$^{-/-}$ *Gli3*$^{-/-}$ compound mutants develop no trachea or lungs and die by E10.5.[107] Epithelial *Shh* expression thus appears to play a role upregulating mesenchymal *Ptc1*, focally inhibiting mesenchymal *Fgf10*, and hence providing the appropriate feedback required for directed epithelial bud outgrowth and the maintenance of discrete mesenchymal Fgf10 sources[108] (Fig. 4).

Regulation of *Shh* expression within the epithelium is effected at least in part through a redundancy in Foxa1 and Foxa2 signalling. *Sftpc-Cre* mediated deletion of *Foxa2* (*Foxa2*$^{\Delta/\Delta}$) in *Foxa1*$^{-/-}$ mice results in disrupted branching, tubule dilatation and inhibited epithelial differentiation.[109] The *Sftpc* promoter appears to be active from E10 onwards and *Foxa1*$^{-/-}$ and *Foxa2*$^{\Delta/\Delta}$ simple null lungs appear similar to controls.[109,110] *Shh* and *Acta2* expression are reduced in compound mutants, while *Nkx2.1* expression is unaffected.[103,109] Because *Foxa1* and *Foxa2* expression are unaffected in *Shh* null mice, epithelial *Foxa1* and *Foxa2* transcripts play overlapping roles regulating activity of *Shh* within the epithelium, and hence an upstream regulatory role specifying epithelial stem cell/progenitor behaviour and the identity of stem cells/progenitors contained within the mesenchyme.[109]

8. Epithelial Dicer Signalling Regulates Mesenchymal *Fgf10* Expression

Dicer is an RNA II endonuclease located within the foregut endoderm at E9.0 and the morphogenetically active lung distal bud epithelium and mesenchyme at E11.5.[111] It plays an essential role cleaving small interfering and micro-RNA (siRNA and miRNA, respectively) species for liberation of their mature and functional forms. Targeted excision of the *Dicer*^{flox} allele within the lung epithelium mediated through *Shh*^{cre} conditional expression between E10.5 and E11.5 results in normal lobation and continued epithelial growth, but a failure of branching commencing at approximately E12.5.[111] *Fgf10* expression is not restricted focally within the distal mesenchyme in *Dicer*^{flox} epithelial conditional mutants, being expanded throughout the distal mesenchyme. Epithelial *Dicer* expression is thus postulated to normally restrict mesenchymal *Fgf10* expression, which, in turn, appropriately directs epithelial progenitor outgrowth.

9. Epithelial Sprouty Modifies the Mesenchymal Fgf10 Signal

Sprouty genes encode tyrosine receptor kinase inhibitors and three of the four mammalian gene family members are expressed in the mouse (*Spry1*, *Spry2* and *Spry4*[112–115]). Between E11.5 and E14.5, all three *Spry* genes are expressed in the lung distal tip epithelium, while some *Spry2* and *Spry4* expression is found in the adjacent mesenchyme. Blocking *Spry2* function with antisense oligonucleotides increases epithelial proliferation, while *Spry2*$^{-/-}$ lungs display a type of posterior transformation of more anterior domains, producing ectopic proximal-like branch points.[49,116] Spry2 is believed to regulate the periodicity of branch formation by initially restricting the branch number of the two ventral domains in the left lobe and diaphragmatic lobes at approximately E12.5.[49] Mechanistically, mesenchymally derived Fgf10 upregulates epithelial *Spry2*, which in turn inhibits epithelial Fgfr2b signalling, thus reducing epithelial proliferation locally in the distal tip and hence creating a branch point[113] (Fig. 4). *Spry4*$^{-/-}$ lungs display a temporary growth delay while *Spry2*$^{-/-}$*Spry4*$^{-/-}$ compound homozygotes die by E12.5 and display a more significant reduction in airway branching.[117] Although functional redundancy is suggested to occur between Spry2 and Spry4, this remains to be formally tested.[117] As such, appropriate regulation of distal epithelial stem cell/progenitor amplification and allocation from mesenchymal signalling may require consideration of the Spry signalling system.

10. Interaction of the Epithelial Bmp and Mesenchymal Fgf Signals

Bmps comprise a multigene family of approximately 22 members belonging to the transforming growth factor beta (Tgf β) superfamily.[118] Bmps signal through serine/threonine heterodimers comprised of a type I receptor (Bmpr1a, Bmpr1b, Acvr1 or Acvr1b) and a type II receptor (Bmpr2, Acvr2a or Acvr2b). Ligand binding of either the type I or type II receptor results in formation of a type I and type II heterotetrameric complex. This complex mediates phosphorylation of intracellular Smad1, Smad5 or Smad8 signalling proteins. The phosphorylated Smad protein is released from the heterotetramer to bind Smad4, and this multi-Smad complex is then transported to the nucleus to regulate transcription of downstream effector pathways. Like Bmps, Tgf βs also signal through Smad proteins to regulate lung development, but through interaction with the Tgf β receptors. The effects of Tg fβ and Smad signalling upon lung development are described elsewhere.[119] Bmp activity is regulated in several fashions. Firstly, Smad6 inhibits Smad1 and Smad5 phosphorylation and hence signal transduction through Bmpr1a and Bmpr1b.[120] Secondly, a number of secreted Bmp antagonists (e.g. Noggin [Nog] and Gremlin) bind Bmp family members (e.g. Bmp2, Bmp4 and Bmp7) and block their interaction with Bmpr1a and Bmpr1b.[121,122]

Bmp4 is expressed within the E9.5 ventral foregut mesoderm and from E10.5 to at least E15.5 within the distal lung epithelial tips and surrounding mesenchyme.[123–125] During this time, *Bmp5* and *Bmp7* are each expressed respectively throughout the lung mesenchyme and epithelium while *Bmp2* transcripts are not detectable.[123,126] *Bmpr1a* is expressed within the E9.5 foregut endoderm and mesoderm, becoming restricted to the distal lung epithelial tips and adjacent mesenchyme between E11.5 and E15.5, while *Bmpr1b* is expressed within the developing foregut commencing at E9.5 and then in the primary bronchial epithelium at E12.5.[127] *Noggin* is expressed in dorsal foregut mesoderm at E9.5, at low levels in distal mesenchyme commencing at E10.5 and in presumptive parabronchial smooth muscle from E11.5 to E15.5, while *Gremlin* is expressed throughout the mesenchyme from at least E11.5, with highest levels present adjacent to the growing epithelial tips, and at lower levels within E11.5 distal tip epithelium and in the E14.5 airway epithelium.[103,124,128]

The addition of Bmp4 to cultured whole E11.5 lung explants increases branching morphogenesis and epithelial proliferation.[128–130] Although misexpression of Bmp4 throughout the epithelium reduces branching morphogenesis (while increasing epithelial proliferation), a stimulatory role for *Bmp4* signalling upon the lung epithelium and branching morphogenesis is supported by the increased branching and proliferation observed following addition of *Gremlin* antisense oligonucleotides to culture, and the decreased branching observed following *Sftpc*-promoter mediated (i) dominant negative *Bmpr1b*; (ii) *XNoggin*; and (iii) *Gremlin* transgene expression within the developing lung epithelium.[123,124,131] With a similar increase in *Sftpc* expression observed following Bmp4 or Gremlin addition to E11.5 whole lung cultures, the latter *Sftpc*-promoter transgenic studies suggest that Bmp4 signalling maintains a distal epithelial bud phenotype that appears prerequisite to the expansion of stem cell/progenitors within the distal bud, and hence their continued budding.[128]

Nog$^{-/-}$ foetuses display oesophageal atresia, tracheoesophageal fistula and truncated and/or fused lobes.[103,125] This *Nog*$^{-/-}$ phenotype is partially rescued by *Bmp4*$^{+/-}$ complementation, suggesting the inhibitory Nog mesenchymal signalling is required to attenuate foregut endodermal Bmp4 signalling to induce appropriate tracheoesophageal septation.[125] Bmp4 thus acts at several levels, specifying events associated with both early foregut and later lung morphogenesis. With respect to morphogenesis of the foregut endoderm, the action of Bmp4 is presumably downstream of *Nkx2.1* expression.[23] With respect to a mechanistic action of *Bmp4* upon later lung epithelial differentiation and morphogenesis, *Fgf10* and *Shh* expression are not affected in *Sftpc*-promoter-*Xnoggin* transgenics and epithelial *Bmp4* expression appears unaltered in $Shh^{-/-}$ mice.[100,124] Furthermore, *Pdx1*-promoter mediated misexpression of *Fgf10* within the lung epithelium induces a distalisation of the proximal lung epithelium involving upregulation of *Sftpc* within the proximal epithelium and a corresponding increase in *Bmp4* expression.[132] Such an increase in *Bmp4* expression accords with that observed after Fgf10 treatment of denuded lung epithelium and the downregulation of *Bmp4* observed following treatment of whole lung explants with *Fgf10* antisense oligonucleotides.[113,133] It has thus been proposed that epithelial Bmp4 and mesenchymal Fgf10 play interacting and complimentary forces regulating epithelial cell outgrowth and differentiation status[132,133] (Fig. 4).

In addition to influencing epithelial differentiation and morphogenesis, epithelial Bmp4 appears to interact with mesenchymal-derived Fgf10 to influence differentiation of parabronchial smooth muscle cells during the pseudoglandular/cannalicular phases of lung development.[97] *Sftpc*-promoter mediated *Bmp4* misexpression throughout the lung epithelium and addition of Bmp4 to E13.5 lung mesenchyme both result in upregulated *Acta2* expression. $Fgf10^{lacz/-}$ hypomorphic E14.5 lungs display reduced expression of *Bmp4* and an associated decrease in the parabronchial smooth muscle cell population. Mesenchymally derived Fgf10 through epithelial Bmp4 is thus hypothesised to provide a signalling system regulating differentiation of the smooth muscle progenitors into the smooth muscle phonotype. Along with its effect upon epithelial proliferation and differentiation, designing therapies to exploit the regulatory effects of Bmp signalling upon both epithelial and mesenchymal stem cells/progenitors following injury may be of merit.

11. Wnt Signalling

Wnt signalling through the Frizzled (Fz) and Arrow/Lrp receptors acts via Dishevelled (Dvl) to stabilise β-catenin (Ctnnb1), thus permitting nuclear translocation, interaction with nuclear T cell factor/lymphoid enhancer factor (Tcf/Lef) and transcriptional activation of downstream target genes.[134] The complexity of the Wnt signalling pathway is illustrated by the presence in mice of 19 *Wnt* genes, ten *Fz* genes and three *Dvl* genes, and the ability of Wnt proteins to activate both canonical receptor and non-canonical protein kinase C and c-Jun signalling pathways.[135] Within the lung, both Wnt2a and Wnt2b are expressed at high levels in the E10.5 to E13.5 distal mesenchyme.[123,136,137] *Wnt5a* is expressed primarily in the distal pulmonary epithelium at E12.5 and at lower levels in the surrounding mesenchyme.[138,139] *Wnt7b* is expressed throughout the lung epithelium at E11.5, being subsequently restricted to the distal bud epithelium.[100,140]

Wnt2a and Wnt2b activate the canonical signalling pathway and no lung phenotypes have been described in *Wnt2a* or *Wnt2b* mutants.[136,137] Wnt5a activates the non-canonical signalling pathway and *Wnt5a*$^{-/-}$ lungs display an increase in epithelial and mesenchymal proliferation, inhibition of maturation, expansion of distal airways and an increase in *Shh*, *Bmp4, Ptch1* and *Fgf10* expression[138] (Fig. 4). Accordingly, *Sftpc*-promoter mediated *Wnt5a* overexpression induces lobation defects, reduced lung branching and a transient increase in mesenchymal mass.[139] E13.5 epithelial *Shh* and mesenchymal *Ptch1* and *Gli1* expression are decreased, while mesenchymal *Fgf10* expression expands and is increased in areas where *Ptch1* is decreased.[139] Corresponding to the broadened expression pattern of *Fgf10* in the distal mesenchyme, *Bmp4* expression is broadened across the epithelium.[139] Increased *Fgf10* and *Bmp4* expression domains are consistent with a reduced branching and dilation phenotype of the distal airways due to a loss in focal branch point formation.

TOPgal.lacZ and *BATgal.lacZ* reporters demonstrate activity of canonical Wnt signalling pathways that overlap with β-*catenin* expression within the anterior foregut endoderm and E10.5 to E18.5 lung epithelium and parabronchial mesenchyme and smooth muscle.[141–144] *TOPgal.lacZ* expression within the E11.5 and E12.5 lung epithelium is highest in the distal tips, while at E15.5 expression is located predominantly within the proximal lung epithelium (see Fig. 4). *Sftpc*-mediated conditional ablation of epithelial *Ctnnb1* induces reduced secondary branching with ectopic smooth muscle differentiation within distal mesenchyme.[142] Loss of epithelial *Ctnnb1* results in a loss of epithelial *Bmp4* and *Fgfr2* expression and associated Fgf downstream epithelial Erk1/2 activity.[144] *Dermo1*$^{Cre/+}$-mediated conditional inactivation of *Ctnnb1* within lung mesenchyme results in partial right pulmonary isomerism but also a decrease in lung epithelial branching.[145] Decreased branching is accompanied by a reduction in submesothelial Fgf10 and smooth muscle progenitor allocation, while epithelial *Shh*, *Bmp4* and *Fgfr2* expression are not affected.

Wnt signalling can be inhibited by Dickkopf1 (Dkk1), Dkk2 and Dkk3 activity. All three genes are expressed in the lung epithelium with overlapping temporal patterns commencing at E11.5.[146] *Sftpc* promoter mediated upregulation of *Dkk1* and Dkk1 supplementation to E11.5 lungs reduces parabronchial Acta2 immunoreactivity and is associated with reduced cleft formation and cleft fibronectin immunoreactivity.[144,146] The addition of soluble fibronectin to Dkk1-treated explants rescues branching morphogenesis, but also rescues parabronchial Acta2 immunoreactivity, while inducing ectopic distal mesenchyme Acta2 immunoreactivity.[146]

Upstream regulation of canonical Wnt signalling is further mediated by both Fgf9 and Fgf10. *Fgf9*$^{-/}$ lungs and *Dermo1*-mediated mesenchymal *Fgfr1/Fgfr2* or *Ctnnb1* ablation all result in a similar hypoplastic lung phenotype associated with reduced mesenchymal *Wnt2a*.[137] The epithelial hypoplasia is postulated to result as a secondary effect to reduced mesenchymal mass resulting from the mesenchyme specific reduction in Wnt signalling and not an alteration of Fgf signalling to the epithelium. Reduced lung mesenchymal Fgf10 signalling in *Mlcv-nLac-24.Fgf10* hypomorph × *TOPgal.lacZ* foetal lungs results in defective parabronchial smooth muscle formation associated with reduced distal lung epithelial *Ctnnb1* expression and canonical Wnt signalling.[78] Ablation of lung epithelial *Gata6* expression acts to upregulate Ctnnb1 signalling, which in turn leads to decreased *Sftpc* and *Scgb1a1* expression, yet an increase in the number of Sftpc$^+$Scgb1a$^+$ cells proposed to represent multipotent progenitors of the airway epithelium.[147,148] Therefore, canonical Wnt signalling appears to regulate specification or allocation of the epithelial stem cell/progenitor pool available for expansion and, as such, subsequent differentiation. How or whether canonical Wnt signalling specifically regulates smooth muscle specification remains to be formally tested. Deletion of *Wnt7b* exon 3 (*Wnt7b*^{D3}) from within the embryonic but not placental compartment produces *Wnt7b*^{D3} null lungs that display both epithelia and mesenchymal hypoplasia, yet overtly unaffected cell fate and lung architecture.[140,149] Together with the lack of lung phenotype observed following *Wnt2a* and *Wnt2b* null mutagenesis, it has been suggested that a highly redundant signalling system exists among the Wnts, yet it remains most likely that they effect specification of both epithelial and smooth muscle progenitors through feedback loops involving the Fgf10 signalling system. Determining further methods to regulate expansion of both epithelial and mesenchymal progenitor pools, following lung injury through modulation of the canonical Wnt signalling pathway, may thus be of great therapeutic benefit in the future.

12. Conclusion and the Way Forward

In order to develop a means of repairing lung function following injury, a thorough understanding of the mechanisms regulating lung epithelial and mesenchymal stem cell/progenitor specification, proliferation, three-dimensional organisation and functional differentiation is essential. Developmental studies have provided a great deal of information concerning these processes and have continually fortified the notion that, because of their intimate and intertwined relationships, epithelial and mesenchymal molecular signalling systems must be considered together in order to appropriately understand the molecular mechanisms governing behaviour of the individual cell types. Reconciliation of the major signalling pathways affecting epithelial and mesenchymal specification and behaviour, as outlined above, should enable modifications relevant to activation of respiratory specific signalling pathways to generic therapeutics and thus in the future provide a more directed, relevant and efficient approach to correcting functional deficits following respiratory disease and/or injury.

References

1. Hashimoto, K. and Nakatsuji, N. Formation of the primitive streak and mesoderm cells in mouse embryos — detailed scanning electron microscopical study. *Dev Growth Differ* **31**(3), 209–218 (1989).
2. Blum, M., Gaunt, S.J., Cho, K.W., *et al.* Gastrulation in the mouse: the role of the homeobox gene goosecoid. *Cell* **69**(7), 1097–1106 (1992).

3. Beddington, R.S. Induction of a second neural axis by the mouse node. *Development* **120**(3), 613–620 (1994).
4. Sulik, K., Dehart, D.B., Iangaki, T., *et al.* Morphogenesis of the murine node and notochordal plate. *Dev Dyn* **201**(3), 260–278 (1994).
5. Robb, L. and Tam, P.P. Gastrula organiser and embryonic patterning in the mouse. *Semin Cell Dev Biol* **15**(5), 543–554 (2004).
6. Lawson, K.A., Meneses, J.J., and Pedersen, R.A. Cell fate and cell lineage in the endoderm of the presomite mouse embryo, studied with an intracellular tracer. *Dev Biol* **115**(2), 325–339 (1986).
7. Tam, P.P. and Beddington, R.S. Establishment and organization of germ layers in the gastrulating mouse embryo. *Ciba Found Symp* **165**, 27–41; discussion 42–29 (1992).
8. Lewis, S.L. and Tam, P.P. Definitive endoderm of the mouse embryo: formation, cell fates, and morphogenetic function. *Dev Dyn* **235**(9), 2315–2329 (2006).
9. Lawson, K.A., Meneses, J.J., and Pedersen, R.A. Clonal analysis of epiblast fate during germ layer formation in the mouse embryo. *Development* **113**(3), 891–911 (1991).
10. Lawson, K.A. and Pedersen, R.A. Cell fate, morphogenetic movement and population kinetics of embryonic endoderm at the time of germ layer formation in the mouse. *Development* **101**(3), 627–652 (1987).
11. Sun, X., Meyers, E.N., Lewandoski, M., *et al.* Targeted disruption of Fgf8 causes failure of cell migration in the gastrulating mouse embryo. *Genes Dev* **13**(14), 1834–1846 (1999).
12. Schier, A.F. Nodal signaling in vertebrate development. *Annu Rev Cell Dev Biol* **19**, 589–621 (2003).
13. Conlon, F.L., Lyons, K.M., Takaesu, N., *et al.* A primary requirement for nodal in the formation and maintenance of the primitive streak in the mouse. *Development* **120**(7), 1919–1928 (1994).
14. Andersson, O., Reissmann, E., Jornvall, H., *et al.* Synergistic interaction between Gdf1 and Nodal during anterior axis development. *Dev Biol* **293**(2), 370–381 (2006).
15. Ang, S.L., Wierda, A., Wong, D., *et al.* The formation and maintenance of the definitive endoderm lineage in the mouse: involvement of HNF3/forkhead proteins. *Development* **119**(4), 1301–1315 (1993).

16. Kwon, G.S., Viotti, M., and Hadjantonakis, A.K. The endoderm of the mouse embryo arises by dynamic widespread intercalation of embryonic and extraembryonic lineages. *Dev Cell* **15**(4), 509–520 (2008).
17. Theiler, K. *The House Mouse: Development and Normal Stages from Fertilization to Four Weeks of Age* (Springer-Verlag, 1972).
18. Marin, L., Dameron, F., and Relier, J.P. Changes in the cellular environment of differentiating type II pneumocytes. Quantitative study in the perinatal rat lung. *Biol Neonate* **41**(3–4), 172–182 (1982).
19. Narita, N., Bielinska, M., and Wilson, D.B. Wild-type endoderm abrogates the ventral developmental defects associated with GATA-4 deficiency in the mouse. *Dev Biol* **189**(2), 270–274 (1997).
20. Ang, S.L. and Rossant, J. HNF-3 beta is essential for node and notochord formation in mouse development. *Cell* **78**(4), 561–574 (1994).
21. Marvin, M.J., Di Rocco, G., Gardiner, A., *et al.* Inhibition of Wnt activity induces heart formation from posterior mesoderm. *Genes Dev* **15**(3), 316–327 (2001).
22. Wells, J.M. and Melton, D.A. Early mouse endoderm is patterned by soluble factors from adjacent germ layers. *Development* **127**(8), 1563–1572 (2000).
23. Minoo, P., Su, G., Drum, H., *et al.* Defects in tracheoesophageal and lung morphogenesis in Nkx2.1(–/–) mouse embryos. *Dev Biol* **209**(1), 60–71 (1999).
24. Jung, J., Zheng, M., Goldfarb, M., *et al.* Initiation of mammalian liver development from endoderm by fibroblast growth factors. *Science* **284**(5422), 1998–2003 (1999).
25. Serls, A.E., Doherty, S., Parvatiyar, P., *et al.* Different thresholds of fibroblast growth factors pattern the ventral foregut into liver and lung. *Development* **132**(1), 35–47 (2005).
26. Lazzaro, D., Price, M., de Felice, M., *et al.* The transcription factor TTF-1 is expressed at the onset of thyroid and lung morphogenesis and in restricted regions of the foetal brain. *Development* **113**(4), 1093–1104 (1991).
27. Deutsch, G., Jung, J., Zheng, M., *et al.* A bipotential precursor population for pancreas and liver within the embryonic endoderm. *Development* **128**(6), 871–881 (2001).
28. Sawaya, P.L., Stripp, B.R., Whitsett, J.A., *et al.* The lung-specific CC10 gene is regulated by transcription factors from the AP-1, octamer, and hepatocyte nuclear factor 3 families. *Mol Cell Biol* **13**(7), 3860–3871 (1993).

29. Kaestner, K.H., Hiemisch, H., Luckow, B., *et al.* The HNF-3 gene family of transcription factors in mice: gene structure, cDNA sequence, and mRNA distribution. *Genomics* **20**(3), 377–385 (1994).
30. Bohinski, R.J., Di Lauro, R., and Whitsett, J.A. The lung-specific surfactant protein B gene promoter is a target for thyroid transcription factor 1 and hepatocyte nuclear factor 3, indicating common factors for organ-specific gene expression along the foregut axis. *Mol Cell Biol* **14**(9), 5671–5681 (1994).
31. Bruno, M.D., Bohinski, R.J., Huelsman, K.M., *et al.* Lung cell-specific expression of the murine surfactant protein A (SP-A) gene is mediated by interactions between the SP-A promoter and thyroid transcription factor-1. *J Biol Chem* **270**(12), 6531–6536 (1995).
32. Kelly, S.E., Bachurski, C.J., Burhans, M.S., *et al.* Transcription of the lung-specific surfactant protein C gene is mediated by thyroid transcription factor 1. *J Biol Chem* **271**(12), 6881–6888 (1996).
33. Yang, H., Lu, M.M., Zhang, L., *et al.* GATA6 regulates differentiation of distal lung epithelium. *Development* **129**(9), 2233–2246 (2002).
34. Spooner, B.S. and Wessells, N.K. Mammalian lung development: interactions in primordium formation and bronchial morphogenesis. *J Exp Zool* **175**(4), 445–454 (1970).
35. Moore, K.L. The respiratory system. In: Wonsiewicz, M. (ed.), *The Developing Juman* (W.B. Saunders Company, 1988).
36. Yuan, B., Li, C., Kimura, S., *et al.*, Inhibition of distal lung morphogenesis in Nkx2.1(–/–) embryos. *Dev Dyn* **217**(2), 180–190 (2000).
37. Ten Have-Opbroek, A.A. The development of the lung in mammals: an analysis of concepts and findings. *Am J Anat* **162**(3), 201–219 (1981).
38. Mendelsohn, C., Lohnes, D., Decimo, D., *et al.* Function of the retinoic acid receptors (RARs) during development (II). Multiple abnormalities at various stages of organogenesis in RAR double mutants. *Development* **120**(10), 2749–2771 (1994).
39. Mascrez, B., Mark, M., Dierich, A., *et al.* The RXRalpha ligand-dependent activation function 2 (AF-2) is important for mouse development. *Development* **125**(23), 4691–4707 (1998).
40. Mollard, R., Ghyselinck, N.B., Wendling, O., *et al.* Stage-dependent responses of the developing lung to retinoic acid signaling. *Int J Dev Biol* **44**(5), 457–462 (2000a).

41. Mollard, R., Viville, S., Ward, S.J., *et al.* Tissue-specific expression of retinoic acid receptor isoform transcripts in the mouse embryo. *Mech Dev* **94**(1–2), 223–232 (2000b).
42. Meyers, E.N., Lewandoski, M., and Martin, G.R. An Fgf8 mutant allelic series generated by Cre- and Flp-mediated recombination. *Nat Genet* **18**(2), 136–141 (1998).
43. Meyers, E.N. and Martin, G.R. Differences in left-right axis pathways in mouse and chick: functions of FGF8 and SHH. *Science* **285**(5426), 403–406 (1999).
44. Norris, D.P., Brennan, J., Bikoff, E.K., *et al.* The Foxh1-dependent autoregulatory enhancer controls the level of Nodal signals in the mouse embryo. *Development* **129**(14), 3455–3468 (2002).
45. Kitamura, K., Miura, H., Miyagawa-Tomita, S., *et al.* Mouse Pitx2 deficiency leads to anomalies of the ventral body wall, heart, extra- and periocular mesoderm and right pulmonary isomerism. *Development* **126**(24), 5749–5758 (1999).
46. Pennekamp, P., Karcher, C., Fischer, A., *et al.* The ion channel polycystin-2 is required for left-right axis determination in mice. *Curr Biol* **12**(11), 938–943 (2002).
47. Karcher, C., Fischer, A., Schweickert, A., *et al.* Lack of a laterality phenotype in Pkd1 knock-out embryos correlates with absence of polycystin-1 in nodal cilia. *Differentiation* **73**(8), 425–432 (2005).
48. Beckers, A., Alten, L., Viebahn, C., *et al.* The mouse homeobox gene Noto regulates node morphogenesis, notochordal ciliogenesis, and left right patterning. *Proc Natl Acad Sci USA* **104**(40), 15765–15770 (2007).
49. Metzger, R.J., Klein, O.D., Martin, G.R., *et al.* The branching programme of mouse lung development. *Nature* **453**(7196), 745–750 (2008).
50. Meno, C., Shimono, A., Saijoh, Y., *et al.* Lefty-1 is required for left-right determination as a regulator of lefty-2 and nodal. *Cell* **94**(3), 287–297 (1998).
51. Constam, D.B. and Robertson, E.J. SPC4/PACE4 regulates a TGFbeta signaling network during axis formation. *Genes Dev* **14**(9), 1146–1155 (2000).
52. Rankin, C.T., Bunton, T., Lawler, A.M., *et al.* Regulation of left-right patterning in mice by growth/differentiation factor-1. *Nat Genet* **24**(3), 262–265 (2000).
53. Rudnick, D. Developmental capacities of the chick lung in chorioallantoic grafts. *J Exp Zool* **66**, 124–154 (1933).

54. Alescio, T. Observations on organotypic cultures of mouse embryonal lung. *Arch Ital Anat Embriol* **65**, 323–363 (1960).
55. Hilfer, S.R., Rayner, R.M., and Brown, J.W. Mesenchymal control of branching pattern in the fetal mouse lung. *Tissue Cell* **17**(4), 523–538 (1985).
56. Shannon, J.M., Nielsen, L.D., Gebb, S.A., *et al.* Mesenchyme specifies epithelial differentiation in reciprocal recombinants of embryonic lung and trachea. *Dev Dyn* **212**(4), 482–494 (1998).
57. Lawson, K.A. Mesenchyme specificity in rodent salivary gland development: the response of salivary epithelium to lung mesenchyme *in vitro*. *J Embryol Exp Morphol* **32**(2), 469–493 (1974).
58. Alescio, T. and Colombo Piperno, E. A quantitative assessment of mesenchymal contribution to epithelial growth rate in mouse embryonic lung developing *in vitro*. *J Embryol Exp Morphol* **17**(1), 213–227 (1967).
59. Alescio, T. and Di Michele, M. Relationship of epithelial growth to mitotic rate in mouse embryonic lung developing *in vitro*. *J Embryol Exp Morphol* **19**(2), 227–237 (1968).
60. Masters, J.R. Epithelial-mesenchymal interaction during lung development: the effect of mesenchymal mass. *Dev Biol* **51**(1), 98–108 (1976).
61. Goldin, G.V. and Wessells, N.K. Mammalian lung development: the possible role of cell proliferation in the formation of supernumerary tracheal buds and in branching morphogenesis. *J Exp Zool* **208**(3), 337–346 (1979).
62. Goldin, G.V. and Opperman, L.A. Induction of supernumerary tracheal buds and the stimulation of DNA synthesis in the embryonic chick lung and trachea by epidermal growth factor. *J Embryol Exp Morphol* **60**, 235–243 (1980).
63. Goldin, G.V., Hindman, H.M., and Wessells, N.K. The role of cell proliferation and cellular shape change in branching morphogenesis of the embryonic mouse lung: analysis using aphidicolin and cytochalasins. *J Exp Zool* **232**(2), 287–296 (1984).
64. Thomas, T. and Dziadek, M. Expression of collagen alpha 1(IV), laminin and nidogen genes in the embryonic mouse lung: implications for branching morphogenesis. *Mech Dev* **45**(3), 193–201 (1994).
65. Mollard, R. and Dziadek, M. A correlation between epithelial proliferation rates, basement membrane component localization patterns, and morphogenetic potential in the embryonic mouse lung. *Am J Respir Cell Mol Biol* **19**(1), 71–82 (1998).

66. Itoh, N. and Ornitz, D.M. Functional evolutionary history of the mouse Fgf gene family. *Dev Dyn* **237**(1), 18–27 (2008).
67. Han, R.N., Liu, J., Tanswell, A.K., *et al.* Expression of basic fibroblast growth factor and receptor: immunolocalization studies in developing rat fetal lung. *Pediatr Res* **31**(5), 435–440 (1992).
68. Sannes, P.L., Burch, K.K., and Khosla, J. Immunohistochemical localization of epidermal growth factor and acidic and basic fibroblast growth factors in postnatal developing and adult rat lungs. *Am J Respir Cell Mol Biol* **7**(2), 230–237 (1992).
69. Mason, I.J., Fuller-Pace, F., Smith, R., *et al.* FGF-7 (keratinocyte growth factor) expression during mouse development suggests roles in myogenesis, forebrain regionalisation and epithelial-mesenchymal interactions. *Mech Dev* **45**(1), 15–30 (1994).
70. Guo, L., Degenstein, L., and Fuchs, E. Keratinocyte growth factor is required for hair development but not for wound healing. *Genes Dev* **10**(2), 165–175 (1996).
71. Miller, D.L., Ortega, S., Bashayan, O., *et al.* Compensation by fibroblast growth factor 1 (FGF1) does not account for the mild phenotypic defects observed in FGF2 null mice. *Mol Cell Biol* **20**(6), 2260–2268 (2000).
72. Warburton, D., Bellusci, S., De Langhe, S., *et al.* Molecular mechanisms of early lung specification and branching morphogenesis. *Pediatr Res* **57**(5 Pt 2), 26R–37R (2005).
73. Colvin, J.S., White, A.C., Pratt, S.J., *et al.* Lung hypoplasia and neonatal death in Fgf9-null mice identify this gene as an essential regulator of lung mesenchyme. *Development* **128**(11), 2095–2106 (2001).
74. del Moral, P.M., De Langhe, S.P., Sala, F.G., *et al.* Differential role of FGF9 on epithelium and mesenchyme in mouse embryonic lung. *Dev Biol* **293**(1), 77–89 (2006).
75. Bellusci, S., Grindley, J., Emoto, H., *et al.* Fibroblast growth factor 10 (FGF10) and branching morphogenesis in the embryonic mouse lung. *Development* **124**(23), 4867–4878 (1997).
76. Min, H., Danilenko, D.M., Scully, S.A., *et al.* Fgf-10 is required for both limb and lung development and exhibits striking functional similarity to Drosophila branchless. *Genes Dev* **12**(20), 3156–3161 (1998).
77. Sekine, K., Ohuchi, H., Fujiwara, M., *et al.* Fgf10 is essential for limb and lung formation. *Nat Genet* **21**(1), 138–141 (1999).

78. Ramasamy, S.K., Mailleux, A.A., Gupte, V.V., *et al.* Fgf10 dosage is critical for the amplification of epithelial cell progenitors and for the formation of multiple mesenchymal lineages during lung development. *Dev Biol* **307**(2), 237–247 (2007).
79. Maruoka, Y., Ohbayashi, N., Hoshikawa, M., *et al.* Comparison of the expression of three highly related genes, Fgf8, Fgf17 and Fgf18, in the mouse embryo. *Mech Dev* **74**(1–2), 175–177 (1998).
80. Whitsett, J.A., Clark, J.C., Picard, L., *et al.* Fibroblast growth factor 18 influences proximal programming during lung morphogenesis. *J Biol Chem* **277**(25), 22743–22749 (2002).
81. Usui, H., Shibayama, M., Ohbayashi, N., *et al.* Fgf18 is required for embryonic lung alveolar development. *Biochem Biophys Res Commun* **322**(3), 887–892 (2004).
82. Peters, K.G., Werner, S., Chen, G., *et al.* Two FGF receptor genes are differentially expressed in epithelial and mesenchymal tissues during limb formation and organogenesis in the mouse. *Development* **114**(1), 233–243 (1992).
83. Sugi, Y., Sasse, J., Barron, M., *et al.* Developmental expression of fibroblast growth factor receptor-1 (cek-1; flg) during heart development. *Dev Dyn* **202**(2), 115–125 (1995).
84. Deng, C.X., Wynshaw-Boris, A., Shen, M.M., *et al.* Murine FGFR-1 is required for early postimplantation growth and axial organization. *Genes Dev* **8**(24), 3045–3057 (1994).
85. Yamaguchi, T.P., Harpal, K., Henkemeyer, M., *et al.* Fgfr-1 is required for embryonic growth and mesodermal patterning during mouse gastrulation. *Genes Dev* **8**(24), 3032–3044 (1994).
86. Partanen, J., Schwartz, L., and Rossant, J. Opposite phenotypes of hypomorphic and Y766 phosphorylation site mutations reveal a function for Fgfr1 in anteroposterior patterning of mouse embryos. *Genes Dev* **12**(15), 2332–2344 (1998).
87. Zhang, H., Dessimoz, J., Beyer, T.A., *et al.* Fibroblast growth factor receptor 1-IIIb is dispensable for skin morphogenesis and wound healing. *Eur J Cell Biol* **83**(1), 3–11 (2004).
88. Miki, T., Fleming, T.P., Bottaro, D.P., *et al.* Expression cDNA cloning of the KGF receptor by creation of a transforming autocrine loop. *Science* **251**(4989), 72–75 (1991).

89. Arman, E., Haffner-Krausz, R., Gorivodsky, M., *et al.* Fgfr2 is required for limb outgrowth and lung-branching morphogenesis. *Proc Natl Acad Sci USA* **96**(21), 11895–11899 (1999).
90. Peters, K., Werner, S., Liao, X., *et al.* Targeted expression of a dominant negative FGF receptor blocks branching morphogenesis and epithelial differentiation of the mouse lung. *EMBO J* **13**(14), 3296–3301 (1994).
91. White, A.C., Xu, J., Yin, Y., *et al.* FGF9 and SHH signaling coordinate lung growth and development through regulation of distinct mesenchymal domains. *Development* **133**(8), 1507–1517 (2006).
92. Peters, K., Ornitz, D., Werner, S., *et al.* Unique expression pattern of the FGF receptor 3 gene during mouse organogenesis. *Dev Biol* **155**(2), 423–430 (1993).
93. Weinstein, M., Xu, X., Ohyama, K., *et al.* FGFR-3 and FGFR-4 function cooperatively to direct alveogenesis in the murine lung. *Development* **125**(18), 3615–3623 (1998).
94. Igarashi, M., Finch, P.W., and Aaronson, S.A. Characterization of recombinant human fibroblast growth factor (FGF)-10 reveals functional similarities with keratinocyte growth factor (FGF-7). *J Biol Chem* **273**(21), 13230–13235 (1998).
95. Ornitz, D.M., Xu, J., Colvin, J.S., *et al.* Receptor specificity of the fibroblast growth factor family. *J Biol Chem* **271**(25), 15292–15297 (1996).
96. De Moerlooze, L., Spencer-Dene, B., Revest, J.M., *et al.* An important role for the IIIb isoform of fibroblast growth factor receptor 2 (FGFR2) in mesenchymal-epithelial signalling during mouse organogenesis. *Development* **127**(3), 483–492 (2000).
97. Mailleux, A.A., Kelly, R., Veltmaat, J.M., *et al.* Fgf10 expression identifies parabronchial smooth muscle cell progenitors and is required for their entry into the smooth muscle cell lineage. *Development* **132**(9), 2157–2166 (2005).
98. Echelard, Y., Epstein, D.J., St-Jacques, B., *et al.* Sonic hedgehog, a member of a family of putative signaling molecules, is implicated in the regulation of CNS polarity. *Cell* **75**(7), 1417–1430 (1993).
99. Bai, C.B., Auerbach, W., Lee, J.S., *et al.* Gli2, but not Gli1, is required for initial Shh signaling and ectopic activation of the Shh pathway. *Development* **129**(20), 4753–4761 (2002).

100. Pepicelli, C.V., Lewis, P.M., and McMahon, A.P. Sonic hedgehog regulates branching morphogenesis in the mammalian lung. *Curr Biol* **8**(19), 1083–1086 (1998).
101. Litingtung, Y., Lei, L., Westphal, H., *et al.* Sonic hedgehog is essential to foregut development. *Nat Genet* **20**(1), 58–61 (1998).
102. Katoh, Y. and Katoh, M. Integrative genomic analyses on GLI2: mechanism of Hedgehog priming through basal GLI2 expression, and interaction map of stem cell signaling network with P53. *Int J Oncol* **33**(4), 881–886 (2008).
103. Weaver, M., Batts, L., and Hogan, B.L. Tissue interactions pattern the mesenchyme of the embryonic mouse lung. *Dev Biol* **258**(1), 169–184 (2003).
104. van Tuyl, M. and Post, M. From fruitflies to mammals: mechanisms of signalling via the Sonic hedgehog pathway in lung development. *Respir Res* **1**(1), 30–35 (2000).
105. Chuang, P.T. and McMahon, A.P. Branching morphogenesis of the lung: new molecular insights into an old problem. *Trends Cell Biol* **13**(2), 86–91 (2003).
106. Grindley, J.C., Bellusci, S., Perkins, D., *et al.* Evidence for the involvement of the Gli gene family in embryonic mouse lung development. *Dev Biol* **188**(2), 337–348 (1997).
107. Motoyama, J., Liu, J., Mo, R., *et al.* Essential function of Gli2 and Gli3 in the formation of lung, trachea and oesophagus. *Nat Genet* **20**(1), 54–57 (1998).
108. Affolter, M., Bellusci, S., Itoh, N., *et al.* Tube or not tube: remodeling epithelial tissues by branching morphogenesis. *Dev Cell* **4**(1), 11–18 (2003).
109. Wan, H., Dingle, S., Xu, Y., *et al.*, Compensatory roles of Foxa1 and Foxa2 during lung morphogenesis. *J Biol Chem* **280**(14), 13809–13816 (2005).
110. Perl, A.K., Wert, S.E., Nagy, A., *et al.* Early restriction of peripheral and proximal cell lineages during formation of the lung. *Proc Natl Acad Sci USA* **99**(16), 10482–10487 (2002).
111. Harris, K.S., Zhang, Z., McManus, M.T., *et al.* Dicer function is essential for lung epithelium morphogenesis. *Proc Natl Acad Sci USA* **103**(7), 2208–2213 (2006).
112. de Maximy, A.A., Nakatake, Y., Moncada, S., *et al.* Cloning and expression pattern of a mouse homologue of drosophila sprouty in the mouse embryo. *Mech Dev* **81**(1–2), 213–216 (1999).
113. Mailleux, A.A., Tefft, D., Ndiaye, D., *et al.* Evidence that SPROUTY2 functions as an inhibitor of mouse embryonic lung growth and morphogenesis. *Mech Dev* **102**(1–2), 81–94 (2001).

114. Zhang, S., Lin, Y., Itaranta, P., *et al.* Expression of Sprouty genes 1, 2 and 4 during mouse organogenesis. *Mech Dev* **109**(2), 367–370 (2001).
115. Cabrita, M.A. and Christofori, G. Sprouty proteins, masterminds of receptor tyrosine kinase signaling. *Angiogenesis* **11**(1), 53–62 (2008).
116. Tefft, J.D., Lee, M., Smith, S., *et al.*, Conserved function of mSpry-2, a murine homolog of Drosophila sprouty, which negatively modulates respiratory organogenesis. *Curr Biol* **9**(4), 219–222 (1999).
117. Taniguchi, K., Ayada, T., Ichiyama, K., *et al.* Sprouty2 and Sprouty4 are essential for embryonic morphogenesis and regulation of FGF signaling. *Biochem Biophys Res Commun* **352**(4), 896–902 (2007).
118. Chen, D., Zhao, M., and Mundy, G.R. Bone morphogenetic proteins. *Growth Factors* **22**(4), 233–241 (2004).
119. Xing, Y., Li, C., Hu, L., *et al.* Mechanisms of TGFbeta inhibition of LUNG endodermal morphogenesis: the role of TbetaRII, Smads, Nkx2.1 and Pten. *Dev Biol* **320**(2), 340–350 (2008).
120. Imamura, T., Takase, M., Nishihara, A., *et al.*, Smad6 inhibits signalling by the TGF-beta superfamily. *Nature* **389**(6651), 622–626 (1997).
121. Groppe, J., Greenwald, J., Wiater, E., *et al.* Structural basis of BMP signalling inhibition by the cystine knot protein Noggin. *Nature* **420**(6916), 636–642 (2002).
122. Wordinger, R.J., Zode, G., and Clark, A.F. Focus on molecules: gremlin. *Exp Eye Res* **87**(2), 78–79 (2008).
123. Bellusci, S., Henderson, R., Winnier, G., *et al.* Evidence from normal expression and targeted misexpression that bone morphogenetic protein (Bmp-4) plays a role in mouse embryonic lung morphogenesis. *Development* **122**(6), 1693–1702 (1996).
124. Weaver, M., Yingling, J.M., Dunn, N.R., *et al.* Bmp signaling regulates proximal-distal differentiation of endoderm in mouse lung development. *Development* **126**(18), 4005–4015 (1999).
125. Que, J., Choi, M., Ziel, J.W., *et al.* Morphogenesis of the trachea and esophagus: current players and new roles for noggin and Bmps. *Differentiation* **74**(7), 422–437 (2006).
126. King, J.A., Marker, P.C., Seung, K.J., *et al.* BMP5 and the molecular, skeletal, and soft-tissue alterations in short ear mice. *Dev Biol* **166**(1), 112–122 (1994).
127. Dewulf, N., Verschueren, K., Lonnoy, O., *et al.* Distinct spatial and temporal expression patterns of two type I receptors for bone morphogenetic

proteins during mouse embryogenesis. *Endocrinology* **136**(6), 2652–2663 (1995).

128. Shi, W., Zhao, J., Anderson, K.D., *et al.* Gremlin negatively modulates BMP-4 induction of embryonic mouse lung branching morphogenesis. *Am J Physiol Lung Cell Mol Physiol* **280**(5), L1030–1039 (2001).
129. Bragg, A.D., Moses, H.L., and Serra, R., Signaling to the epithelium is not sufficient to mediate all of the effects of transforming growth factor beta and bone morphogenetic protein 4 on murine embryonic lung development, *Mech Dev* **109**(1), 13–26 (2001).
130. Shi, W., Chen, H., Sun, J., *et al.* Overexpression of Smurf1 negatively regulates mouse embryonic lung branching morphogenesis by specifically reducing Smad1 and Smad5 proteins. *Am J Physiol Lung Cell Mol Physiol* **286**(2), L293–300 (2004).
131. Lu, M.M., Yang, H., Zhang, L., *et al.* The bone morphogenic protein antagonist gremlin regulates proximal-distal patterning of the lung. *Dev Dyn* **222**(4), 667–680 (2001).
132. Nyeng, P., Norgaard, G.A., Kobberup, S., *et al.* FGF10 maintains distal lung bud epithelium and excessive signaling leads to progenitor state arrest, distalization, and goblet cell metaplasia. *BMC Dev Biol* **8**, 2 (2008).
133. Weaver, M., Dunn, N.R., and Hogan, B.L. Bmp4 and Fgf10 play opposing roles during lung bud morphogenesis. *Development* **127**(12), 2695–2704 (2000).
134. Fuerer, C., Nusse, R., and Ten Berge, D. Wnt signalling in development and disease. Max Delbruck center for molecular medicine meeting on wnt signaling in development and disease. *EMBO* Rep **9**(2), 134–138 (2008).
135. van Amerongen, R., Mikels, A., and Nusse, R. Alternative wnt signaling is initiated by distinct receptors. *Sci Signal* **1**(35), re9 (2008).
136. Levay-Young, B.K. and Navre, M. Growth and developmental regulation of wnt-2 (irp) gene in mesenchymal cells of fetal lung. *Am J Physiol* **262**(6 Pt 1), L672–683 (1992).
137. Yin, Y., White, A.C., Huh, S.H., *et al.* An FGF-WNT gene regulatory network controls lung mesenchyme development. *Dev Biol* **319**(2), 426–436 (2008).
138. Li, C., Xiao, J., Hormi, K., *et al.* Wnt5a participates in distal lung morphogenesis. *Dev Biol* **248**(1), 68–81 (2002).
139. Li, C., Hu, L., Xiao, J., *et al.* Wnt5a regulates Shh and Fgf10 signaling during lung development. *Dev Biol* **287**(1), 86–97 (2005).

140. Shu, W., Jiang, Y.Q., Lu, M.M., *et al.* Wnt7b regulates mesenchymal proliferation and vascular development in the lung. *Development* **129**(20), 4831–4842 (2002).
141. Maretto, S., Cordenonsi, M., Dupont, S., *et al.* Mapping Wnt/beta-catenin signaling during mouse development and in colorectal tumors. *Proc Natl Acad Sci USA* **100**(6), 3299–3304 (2003).
142. Mucenski, M.L., Wert, S.E., Nation, J.M., *et al.* Beta-Catenin is required for specification of proximal/distal cell fate during lung morphogenesis. *J Biol Chem* **278**(41), 40231–40238 (2003).
143. Okubo, T. and Hogan, B.L. Hyperactive Wnt signaling changes the developmental potential of embryonic lung endoderm. *J Biol* **3**(3), 11 (2004).
144. Shu, W., Guttentag, S., Wang, Z., *et al.* Wnt/beta-catenin signaling acts upstream of N-myc, BMP4, and FGF signaling to regulate proximal-distal patterning in the lung. *Dev Biol* **283**(1), 226–239 (2005).
145. De Langhe, S.P., Carraro, G., Tefft, D., *et al.* Formation and differentiation of multiple mesenchymal lineages during lung development is regulated by beta-catenin signaling. *PLoS ONE* **3**(1), e1516 (2008).
146. De Langhe, S.P., Sala, F.G., Del Moral, P.M., *et al.* Dickkopf-1 (DKK1) reveals that fibronectin is a major target of Wnt signaling in branching morphogenesis of the mouse embryonic lung. *Dev Biol* **277**(2), 316–331 (2005).
147. Kim, C.F., Jackson, E.L., Woolfenden, A.E., Lawrence, S., Babar, I., Vogel, S., Crowley, D., Bronson, R.T., and Jacks, T. Identification of bronchioalveolar stem cells in normal lung and lung cancer. *Cell* **121**(6), 823–835 (2005).
148. Zhang, Y., Goss, A.M., Cohen, E.D., Kadzik, R., Lepore, J.J., Muthukumaraswamy, K., Yang, J., DeMayo, F.J., Whitsett, J.A., Parmacek, M.S., and Morrisey, E.E. A Gata6-Wnt pathway required for epithelial stem cell development and airway regeneration. *Nat Genet* **40**(7), 862–70 (2008).
149. Rajagopal, J., Carroll, T.J., Guseh, J.S., *et al.* Wnt7b stimulates embryonic lung growth by coordinately increasing the replication of epithelium and mesenchyme. *Development* **135**(9), 1625–1634 (2008).

Chapter 4
Epithelial Progenitor Cells of the Mammalian Lung

Roxana M. Teisanu, Joshua C. Snyder and Barry R. Stripp

The adult lung is a quiescent organ whose function is highly dependent upon epithelial integrity. Maintenance and restoration of the airway epithelium relies upon regulated activation of progenitor cells whose identity varies between tracheobronchial, bronchiolar, and alveolar zones. Defects in epithelial repair are associated with a broad spectrum of lung diseases that range from the inability to fight infections to hyper-proliferative diseases such as pulmonary fibrosis and lung cancer. The present chapter summarizes our current understanding of the hierarchical organization, function, and regulation of the stem/progenitor cell hierarchies that maintain the lung epithelium. Much of our current knowledge relies upon the availability of methods that have been widely used to characterize stem cell hierarchies in other highly proliferative systems. These methods and their limitations are described in detail while emphasizing the differences between quiescent and highly proliferative organs. This chapter identifies significant knowledge gaps and identifies future directions that will help advance the field of lung regenerative medicine.

Keywords: Progenitor cell; stem cell; repair; tracheobronchial epithelium; bronchiolar epithelium; alveolar epithelium.

Outline

1. Introduction

1.1. *Heterogeneity of epithelial cells in the adult airway*

The adult mammalian lung is a complex organ that evolved to facilitate gas transfer between inspired air and the blood. To accomplish this process structural components of the lung are organized into a branching system of conducting airways and blood vessels that terminate in the gas exchange unit: the alveolus. The respiratory tract is lined by a structurally heterogeneous epithelium in which cell types are organized into specialized functional zones based upon their progenitor–progeny relationships.[1–3] This epithelium fulfills a range of secondary functions from mucociliary clearance[4] to secretion,[5,6] metabolism, and the detoxification of either systemic or inhaled xenobiotic chemicals.[7–9] Epithelial cell types can be distinguished based upon their unique histological and ultrastructural characteristics and gene expression profiles. The lining of tracheal and bronchial airways is a pseudo-stratified epithelium composed predominantly of basal, ciliated, and secretory cells. In addition, less abundant nonciliated cell types also exist, such as neuroendocrine, brush, and intermediate cells.[10–12] The most distal conducting airways, termed bronchioles, are lined by a simple cuboidal epithelium consisting of Clara cells, ciliated cells and rare neuroendocrine cells.[13] Neuroendocrine cells are commonly organized into clusters which are associated with nonciliated cells that lack the typical ultrastructural features of Clara cells.[14] The epithelium lining alveoli is composed of alveolar type II cells (ATII), which secrete pulmonary surfactant, and alveolar type I cells (ATI), which are highly specialized epithelial cells with a large surface area to facilitate gas transfer[3,15] (see also Chapter 6).

1.2. *Epithelial function*

With every inspiration the lung epithelium is exposed to micro-organisms and their pyrogenic by-products, viral particles, oxidative pollutants, and particulate matter. Despite these persistent environmental challenges the epithelial lining turns over at a very low rate due to innate hosts' defense mechanisms that preserve tissue integrity.[16–19] To prevent undue injury and an unwarranted immunological response, the airway epithelium provides for physical clearance of inhaled stimulants through mucociliary clearance.[20] In the event that mucociliary clearance is overwhelmed, the epithelium also plays a pivotal role in regulation of the innate inflammatory response. Recent studies highlight the fact that tight regulation of the NF-kappaB signalling pathway in epithelial cells is required for appropriate control of the inflammatory response and communication with resident lung macrophages required for an effective response.[21–25]

In contrast, these processes are likely compromised in the setting of chronic lung diseases, such as cystic fibrosis and chronic obstructive pulmonary disease, resulting in chronic epithelial injury, defective repair, and subsequent remodelling of the epithelial–mesenchymal trophic unit.[26] More recently, we have demonstrated that defective epithelial repair can phenocopy aspects of chronic lung disease, such as bronchiolar and alveolar extracellular matrix (ECM) deposition and an augmented inflammatory response suggesting a role for altered progenitor cell behavior in disease etiology.[27,28] In other organ systems, it has also been shown that deregulation of the stem cell compartment leads to acute and chronic diseases.[29,30] In the haematopoietic system, for example, increased proliferation of either stem or progenitor cells can lead to disorders such as leukaemia and lymphoma,[30] while defects in cellular differentiation can lead to severe immunological abnormalities.[31] Although evidence of a direct cause–effect relationship between different disease states and the status of the resident stem cell or the niche is currently lacking, it has been suggested that diseases such as lung cancer can arise from resident progenitor cells[29,30] (see also Chapters 4 and 17). Understanding the hierarchical organization of progenitor cells within the lung and the relationship between these cells and changes in their behaviour with lung disease becomes critically important. This chapter summarizes what is known of progenitor cells within each of the three major epithelial compartments: the tracheal–bronchial epithelium, the intra-pulmonary airways, and the gas-exchange area.

2. Progenitor Cells and Lineage Specification in the Developing Lung

(See also Chapters 2 and 3)

Lineage relationship amongst different cell types and progenitor–progeny relationship in the extra- and intrapulmonary epithelium have been thoroughly studied in mouse models of lung development. One of the earliest events in lung development is formation of the two lung buds from the foregut endoderm. Growth of the lung primordia and subsequent branching morphogenesis, sacculation, and alveolarization are dependent upon appropriate interactions with the surrounding foregut mesoderm.[32] Signalling between the two compartments is mediated by a variety of growth factors, amongst which members of the FGF,[33,34] Tgf-β/BMP,[35] and Wnt[36,37] families of paracrine regulatory factors play central roles. The tracheal epithelium arises from different foregut endoderm progenitors located anterior to the initial lung buds forming within the foregut endoderm. Initially the trachea and the oesophagus form a common tubular structure that is subsequently partitioned into two separate tubular organs.[32] Despite the distinct origins of extrapulmonary versus intrapulmonary airways, the finding that distal lung mesenchyme can redirect primitive tracheal epithelium to a distal fate suggests that early endodermal progenitor cells are multipotent and that their fate is tightly regulated by paracrine interactions with surrounding mesenchyme.[38,39] Thus, epithelial cells of the conducting and alveolar airways, including basal, ciliated, secretory, neuroendocrine, and alveolar type I and II, arise from common endodermal progenitors. However, a growing body of evidence suggests that once lung endoderm has been specified in the embryonic foregut, the ensuing process of lung development yields distinct lineages that are maintained independently of one another in the late embryonic and postnatal periods. This developmental process has been reviewed extensively elsewhere.[32]

3. Adult Stem Cells and Tissue Maintenance by a Stem Cell Hierarchy

The term "adult tissue stem cell" was coined to describe cells with the ability to proliferate and generate differentiated progeny within a tissue, yet retain the capacity for long-term self-renewal. Unlike embryonic stem cells, a pluripotent cell type that can be maintained in culture by providing medium additives but lack a stable *in vivo* counterpart, adult tissue stem cells generally have more restricted differentiation potential and can be maintained for prolonged periods *in vivo*.[40,41] Cells with these characteristics were first described in the haematopoietic system. The capacity of haematopoietic stem cells to generate all specialised cell types of the haematopoietic lineage and undergo long-term self-renewal was validated through the use of *in vivo* transplantation assays.[42,43] Similar cells have since been described in other organs including the intestine,[44–46] nervous system,[47] epidermis,[48,49] and epithelial of the cornea,[50] mammary gland,[51–53] lung,[17,18,54,55] liver,[56,57] pancreas,[58] and prostate.[59] However, tissue-specific differences in the kinetics of cell replacement coupled with anatomic constraints and differences in cellular complexity have made identification and classification of stem cells difficult in many tissues. Accordingly, a range of other properties have been used to define the stem cell compartment within a tissue and the mechanisms contributing to their regulation. Such mechanisms may be less uniformly applicable and in some cases can only be applied to stem cells of a single tissue type. Relevant properties of stem cells and their more specialised derivatives are discussed below in the context of understanding stem cell behaviour within the lung and other tissues, and assays commonly used for their identification and characterization.

The classical view of a stem cell hierarchy is that rare stem cells are believed to proliferate infrequently and divide asymmetrically. Asymmetric division yields one daughter that is identical to the parent cell, and another that has more limited capacity for self-renewal. The latter has the ability to generate large numbers of differentiating progeny and was therefore termed the "transit-amplifying" (TA) cell.[41,60,61] Despite this, not all putative tissue stem cells have been shown to undergo asymmetric division. Moreover, among those tissue stem cells that have been shown to undergo asymmetric cell divisions, regulatory mechanisms have been defined for just a few.[62,63] The best described mechanism defines self-renewal of drosophila germline stem cells, for which asymmetric divisions are tightly regulated through adhesion complexes formed between stem cells and somatic cells that form the stem cell niche.[64] Even though there is support for the concept that asymmetric cell division may in some cases control long-term maintenance of tissue stem cells, far less is known of mechanisms that govern the lifespan and behaviour of TA cells. Tissue-specific differences are observed in the lifespan of TA cells, which has potential to impact the activity of tissue stem cells. The lifespan of TA cells is dictated by two variables: the probability of self-renewal versus differentiation and their proliferative frequency.[65,66] The impact that the probability of TA cell differentiation has on tissue homeostasis is evident in a comparison of TA cells of the intestinal epithelium and the interfolicular epidermis. Both of these tissues harbour TA pools that proliferate frequently for normal tissue maintenance but differ in their longevity. TA cells of the intestinal epithelium exhibit a relatively high probability of generating post-mitotic progeny leading to their rapid depletion and continuous requirement for stem cell proliferation to maintain regenerative capacity. In contrast, TA cells of the interfolicular epidermis have a relatively low probability of generating post-mitotic progeny leading to their long-term maintenance and infrequent activation of stem cells.[65,66] These examples serve to highlight how changes in the probability of TA self-renewal versus differentiation have potential to dramatically impact the requirement for stem cell activation in tissue maintenance and help to explain tissue-specific differences in stem cell behaviour.

4. Approaches to Characterise and Hierarchically Organise Progenitor Cells

A number of approaches have been developed to identify progenitor cells, understand their behavior *in vivo*, and to distinguish tissue stem cells from TA cells within a stem cell hierarchy. The more commonly used are discussed briefly below:

Lineage analysis

Progenitor–progeny relationships can be defined *in vivo* through the use of labelling methods to tag cells and determine their fate. Three principal approaches have been used. The most straightforward approach is to label the nascent DNA of S-phase cells through incorporation of nucleotide analogues. Commonly used nucleotide analogues include tritiated thymidine (TdR) or bromodeoxyuridine (BrdU), although other halogenated thymidine analogues have also been used, such as chlorodeoxyuridine (CldU) and iododeoxyuridine (IdU).[1,3,67–69] Delivery of these labelled DNA precursors into the systemic circulation results in their delivery to proliferating cells and incorporation into DNA during S-phase. Incorporation into nuclear DNA is typically assessed by either autoradiographic or immunohistochemical detection. When coupled with either ultrastructural analysis by transmission electron microscopy or immunophenotypic analysis at the light microscopic level, it is possible to determine the phenotype of proliferating cells and with an appropriate chase period in the absence of label, the fate of their progeny. Caveats with this approach are that experiments involving pulse-chase strategies do not allow for the retrospective identification of the parental cell that initially incorporated labelled DNA precursors, and that continuing proliferation of cells leads to label dilution which renders this method useless for lineage analysis.

An alternative method that allows analysis of lineage relationships is that of lineage labelling using recombinant retroviral vectors.[70,71] Mixed populations of retroviral vectors carrying different reporter genes allow for the analysis of clonality and lineage relationships. A caveat with this approach is the inability to retrospectively define the phenotype of the founding cell for a population of tagged cells that are identified after a defined "chase" period. Many of these technical hurdles have been overcome through the use of lineage tracing methods involving Cre/LoxP technologies.[41,72] Cre recombinase is a bacteriophage enzyme that catalyzes homologous recombination between adjacent LoxP sequences. Any DNA sequence that is engineered to include two adjacent copies of the LoxP sequence can serve as a substrate for Cre. Recombination catalyzed by Cre can lead to either deletion or reversal of the intervening sequence depending upon whether the LoxP elements are in the same or opposing relative orientation, respectively. A prerequisite for lineage analysis using the Cre/LoxP system is that progenitor cell-specific genes have been identified and their promoters characterised allowing the generation of transgenic mice capable of directing cell type-specific expression of Cre. Mice carrying the Cre-expressing transgene are crossed with a reporter line in which a LoxP-flanked transcriptional terminator sequence is present between an upstream ubiquitous promoter and a downstream reporter gene. In this case, Cre-mediated recombination activates the expression of the downstream reporter, thus introducing a lineage tag within the cell that can be traced among daughter cells. When this strategy is coupled with use of ligand-regulated forms of Cre such as CreER, lineage tags can be introduced into defined cell types in a temporally controlled fashion.[72,73]

In vivo label retention

The concept that tissue stem cells proliferate infrequently compared to their transit-amplifying progeny formed the basis for assays to define these populations based upon this distinguishing characteristic. Infrequently proliferating cells have been identified in organs with high cellular turnover by labelling nuclear DNA with nucleotide analogues as discussed in detail above, followed by a chase period that dilutes the label within frequently cycling cells.[74,75] This assay has been used to define "label-retaining" cells (LRC). One prerequisite of such experiments is that the stem cell actively proliferates at the time of the pulse labelling. However, recent evidence in the intestinal stem cell compartment challenges the label-retaining dogma and shows that there is a population of non-label retaining stem cells at the bottom of the crypt, distinct from the classical LRC, found four cells upstream from the bottom of the crypt. Lineage-tracing analysis demonstrates that these stem cells have the ability to generate the entire epithelium of the crypt and the villus, including the +4 epithelial cells.[45] Another limitation of this approach is its reliance on the difference in proliferation rate between the highly proliferative TA compartment and the relative quiescence of the stem cell. Thus, in organs with very low cellular turnover such as the lung, pancreas, and liver, where both compartments proliferate infrequently in steady state, various injury models have been used to reveal the difference in proliferative potential of stem versus TA cell populations.[17,18,55,58,76,77] In these organs, chemical injury and/or tissue resection was used to reveal a population of label retaining cells with stem cell properties.

In vitro and transplantation assays

In vitro assays are dependent upon the existence of a set of positive and negative selection markers that allow prospective isolation of stem cell-enriched populations. Once such enrichment is possible, the self-renewal and differentiating properties of these cells can be studied in clonogenic, serial-passage, and limited dilutions analysis assays.[78,79] However, *in vitro* experiments have the caveat of being dependent upon subjective medium components and are very susceptible to false interpretation due to inadequate culturing conditions that can bias progenitor cell renewal or differentiation independently of whether these cells function as tissue stem cells *in vivo*. Transplantation assays are rooted in the haematopoietic stem cell field where functional validation of stem cell behaviour can be verified through analysis of the capacity for long term reconstitution of the entire haematopoietic system from one single cell.[42,43,78] This type of approach has been effectively employed in both the mammary gland and prostate to demonstrate that single cells isolated based on specific cell surface markers can reconstitute the entire gland when transplanted in immuno-compromised hosts.[51,52,59]

5. Progenitor Cells of Tracheobronchial Airways

Analysis of clonogenic, differentiation, and proliferative potential of progenitor cells in the tracheobronchial epithelium have benefited from the use of *in vitro*[80] and transplantation[81] assays as well as *in vivo* models of epithelial injury and repair.[82] Transplantation studies in the trachea have been quite successful due to the relative ease of cell isolation, identification of cell surface markers for fractionation, and surgical transplantation compared to their more distal counterparts. Early work demonstrated that progenitor cells could be isolated from the tracheal epithelium, fractionated according to their ability to interact with the lectin *Griffonia simplicifolia* I (GSI)-B_4, transplanted into denuded rat tracheas, and differentiated appropriately to restore the cellular diversity of the tracheal epithelium.[81,83] From this study more sophisticated approaches have been utilised to more thoroughly define the progenitor capacity of tracheal epithelial cells. In a classical study, Engelhardt and colleagues isolated human bronchial epithelial cells, transduced these cells with retroviral constructs expressing a reporter gene for use in lineage tracing, and transplanted the cells into a denuded rat trachea. By analyzing the cell types represented in each clone, this study demonstrated that a single progenitor cell can give rise to all the cells of the airway surface epithelium in addition to the cells of the sub-mucosal glands.[84] These results suggest that a progenitor cell with more stem-like properties exists within the tracheobronchial compartment. In addition to these observations, this study is one of only a few to investigate human airway progenitor cell behaviour. Unfortunately, the stochastic nature of retroviral transduction precluded a determination of the molecular or cellular identity of the initiating clonogenic cell. Each of these studies relied upon isolating cells from intact tissues, which could bias the experimental outcome based upon cellular viability during the isolation procedure. Therefore, a series of *in vivo* studies was initiated to better characterise the stem cell hierarchy in the tracheal epithelium.

Borthwick and colleagues (see Ref. 85) repeatedly injured the tracheal epithelium through exposure to SO_2 or polydocanol and simultaneously treated with BrdU to label proliferating cells. To determine if a stem cell niche exists in the tracheal epithelium, label-retaining cells in the repairing epithelium were quantified after various chase periods. These data demonstrated that label-retaining cells existed in the upper and lower trachea in the sub-mucosal glands and at the junction of the inter-cartilaginous rings, respectively. This study also demonstrated that keratin 5 (*Krt5)* positive subsets of basal cells could be identified in tissue based upon activity of the β-galactosidase (β-GAL) reporter under control of the human *Krt5* promoter.[16] *Krt5* expressing cells were never immunolocalised with the label-retaining cells. However, their localization to sub-mucosal gland ducts and junctions of inter-cartilaginous rings was suggestive evidence that the label-retaining cell may be *Krt5* positive. This study provided *in vivo* evidence that a stem cell niche may exist in the tracheal epithelium. The precise molecular and cellular identity of these label-retaining cells remains unknown and recent studies have questioned label retention as a relevant assay for defining stem cell potential.[45] Collectively, these studies demonstrated that epithelial progenitors exist in the tracheal epithelium, with a subpopulation of these cells sharing characteristics of resident adult tissue stem cells.

To determine the cellular identity and molecular profile of the tracheal epithelial stem cell, a combination of *in vivo* lineage tracing and *in vitro* assays were utilised. These studies have demonstrated that subsets of tracheal and bronchial basal epithelial cells can be characterised by their multipotent differentiation potential. In 2004, Schoch and colleagues as well as Hong and colleagues reported that the expression of keratins 5 and 14, respectively, demarcate cells with multipotent differentiation potential.[86–88] Using a ubiquitously expressing β-GAL transgenic mouse, Shoch and colleagues demonstrated that the mouse tracheal epithelium contains a population of cells capable of clonal expansion and multipotent differentiation when cultured *in vitro*, confirming the previous observations made by Engelhardt and colleagues in human xenograft transplantation.[84] Based on the previous observation that *Krt5* expression defines a subset of basal epithelial cells in the tracheal epithelium, a transgenic reporter mouse expressing EGFP under the control of the human *Krt5* promoter was used to fractionate tracheal basal cells. Fluorescent activated cell sorting and *in vitro* culturing indicated that EGFP positive cells display multipotent differentiation capacity and increased clonogenic potential relative to EGFP negative epithelial cells.[86] In contrast, Hong and colleagues used an *in vivo* lineage tracing strategy. A bitransgenic mouse line expressing the fused CRE-ERt under control of the *Krt14* promoter crossed with the ROSA26-flox-stop recombination substrate was utilised to establish differentiation and clonogenic potential. In this study, it was demonstrated that following naphthalene ablation of secretory cells, an abundant facultative transit amplifying cell of proximal airways, *Krt14* expressing basal cells display either multipotent or unipotent differentiation capacity in the tracheal epithelium[88] and multipotent differentiation capacity in the bronchial epithelium.[87] The relationship between the *Krt5* and *Krt14* expressing basal cells is currently unknown. Further fractionation of basal cells in conjunction with *in vitro* and transplantation assays will be required before this relationship can be effectively studied. However, data to date suggest that basal cells are a heterogeneous population that includes a subset with greater clonogenic and differentiation potential suggesting that this subpopulation of basal cells may be the equivalent of a local tissue stem cell.

If cell-based therapeutic approaches are to be implemented, faithful cell surface markers that allow for fractionation of progenitor cells must be identified. Several methods have been utilised to fractionate tracheal and bronchial epithelial cells. Previous work in the haematopoietic system fractionated stem cells on the basis of the phenotypic ability to rapidly efflux Hoechst dye, termed "side population" (SP) cells.[89] As such, several studies in the tracheal and bronchial epithelium have adopted this approach. Unfortunately, in mouse this method has been highly irreproducible indicating that the SP cell may be either epithelial or mesenchymally derived.[90,91] Despite this, a recent study used this approach to fractionate human SP tracheal and bronchial epithelial cells and analyze their differential potential *in vitro*. The results of this study indicate that in human SP cells are capable of differentiating into epithelial cells. However, whether the SP enriches for cells with clonogenic or differentiation potential remains to be determined.[92] Regardless, it is evident that adopting cell surface markers or phenotypic profiles used to fractionate stem cells in other compartments does not guarantee enrichment of airway progenitors. Recent work has also demonstrated that human airway epithelial progenitors can be isolated, grown in three-dimensional culture (see also Chapter 9), efficiently transduced, and are capable of restoration of the airway epithelium in denuded tracheas.[93] In addition, these progenitor cells can be fractionated based on the expression of aquaporin-3 to enrich for progenitor cell capacity.[94] Despite the current inability to reproducibly purify subpopulations of progenitor cells, a seminal study recently demonstrated that cell-based therapeutic intervention in lung disease is now a reality. In a proof of principle study, Macchiarini and colleagues conducted the world's first human main bronchus transplant using a donor denuded trachea seeded with recipient airway epithelial cells. The patient had presented with a stenosis of the left main bronchus and severely attenuated FVC and FEV_1 levels due to turberculosis infection. Four months following transplantation, the epithelium was completely restored, FVC and FEV_1 levels returned to normal, and no signs of allograft rejection were apparent.[95] Though the study cannot rule out invasion of the recipient's own epithelial cells into the transplanted trachea, it yields great promise for the future of lung regenerative medicine. Collectively, these data indicate that despite the low rate of epithelial turnover in the tracheal and bronchial epithelium, this compartment maintains a subpopulation of epithelial cells with incredible differentiation and clonogenic potential. Future studies are needed to adequately harness the potential of these cells in the treatment of lung disease.

6. Progenitor Cells of Bronchiolar Airways

The identity of bronchiolar progenitor cells and their differentiation potential was initially studied using TdR labelling methods to mark proliferative cells in the normal and repairing epithelium. Evans and colleagues, using Ozone and NO_2 injury models to selectively injure ciliated cells in rat airways, demonstrated that epithelial renewal is accomplished through a wave of proliferation. Pulse-chase experiments involving systemically delivered [^{3}H]-thymidine (TdR) were used to demonstrate that nonciliated bronchiolar (Clara) cells represented the only proliferative cell type that responded to ciliated cell injury and that daughter cells gave rise to either mature Clara or ciliated cells.[2] Furthermore, the authors very thoroughly describe the morphological changes that secretory cells go through in order to enter the cell cycle and replenish the pool of terminally differentiated ciliated cells. Thus, the name "Type A Clara cell" was used in reference to Clara cells that have lost both secretory granules and smooth endoplasmic reticulum, ultrastructural changes that were accompanied by cell cycle progression through S-phase.[1] These findings indicated that secretory cells are multifunctional, fulfilling roles in secretion and metabolism in the resting state, yet proliferating in response to injury to effect renewal of the epithelium. This duality of function for airway progenitor cells is in contrast to the obligate progenitor function observed among transit-amplifying cells in classical stem cell hierarchies, thus the term "facultative transit-amplifying" (TA) cells.

The existence of bronchiolar stem cells and their contribution to repair has been suggested from analysis of airway injury models involving selective ablation of mature Clara cells. Clara cells of mammalian airways constitutively express phase I metabolizing enzymes such as cytochrome P450 that render them susceptible to injury by their enzymatic substrates. Clara cells of the mouse airway express cytochrome P450 isoenzymes 2B2 and 2F2 that metabolise naphthalene to toxic metabolites leading to selective cell death.[96] This model of naphthalene-induced airway injury has been used to investigate mechanisms of repair following depletion of this abundant pool of facultative TA cells.[18,55,97–100] Clara cell ablation is followed by a proliferative response, the magnitude, spatial context, and kinetics of which are dictated by the extent of injury and airway location, leading to renewal of the epithelium.[98,101] Repair of the Clara cell-depleted bronchiolar epithelium is accomplished through activation of putative tissue stem cells that localise to two discrete microenvironments including bronchoalveolar duct junctions (BADJ) and neuroepithelial bodies (NEB).[17,98] Using this injury repair model, Hong *et al.*, and later Giangreco *et al.*, identified naphthalene-resistant CCSP-expressing cells that incorporate labelled DNA precursors into nuclear DNA during the early repair response and show a long-term label-retaining phenotype indicative of an infrequently cycling tissue stem cell.[18,55] The term "variant CCSP-expressing" (vCE) cell was used to describe this population of naphthalene-resistant cells from naphthalene-sensitive Clara cells. A direct role for vCE cells in the repair of naphthalene-injured airways was further supported through the use of a transgenic mouse model allowing ablation of the entire CCSP expressing population.[18,97] In this model, conditional ablation of all CCSP-expressing progenitor cells, both naphthalene-sensitive Clara cells and the putative bronchiolar stem cell, was accompanied by complete abrogation of bronchiolar repair.[18,97] More recently naphthalene-resistant vCE cells residing at the BADJ were found to exhibit the unique molecular property of expressing the alveolar type II epithelial cell marker gene pro-surfactant protein-C.[100] Cells with this CCSP/Pro-SPC expressing phenotype have been proposed as a multipotent bronchioalveolar stem cell, based on *in vitro* data suggesting that they can generate cells expressing either airway or alveolar marker genes when cultured *in vitro.*[100] It has not been determined whether resident bronchiolar stem cells, either those defined based on their resistance to naphthalene or other putative stem cells, have the capacity to generate and renew alveolar epithelium in addition to that of the bronchiole.

Localization of naphthalene-resistant vCE cells to defined microenvironments within bronchioles suggests that these regions are analogous to stem cell niches described in other organs.[48,102] However, what actually constitutes the stem cell niche and what local factors regulate maintenance and activation of the stem cell population remain unknown. Neuroepithelial body-associated CCSP-expressing cells have the distinguishing properties of both naphthalene-resistance and unique electrophysiological properties, suggesting that local microenvironmental factors regulate cellular behaviour.[103] Properties of NEB-associated CCSP-expressing cells that may confer resistance to naphthalene-induced airway injury include reduced expression of phase I xenobiotic metabolizing enzymes such as CYP2f2.[17] Collectively, these data argue that extrinsic factors controlled by the local microenvironment impact the behaviour of bronchiolar progenitor cells for maintenance of the stem cell phenotype, and that this phenotype confers intrinsic resistance to environmental agents such as naphthalene.

Insight into molecular pathways that regulate the behaviour of bronchiolar progenitor cells comes from studies involving the use of mouse models carrying genetic perturbations of specific signalling pathways. Experiments using a Cre recombinase-activated form of the K-ras gene were the first to suggest that activation of this transducer of pro-mitotic signalling pathways had the potential to impact the behaviour of bronchiolar progenitor cells.[104] Epithelial hyperplasia coupled with an increase in the abundance of CCSP/SP-C dual positive cells suggested that active K-ras signalling led to expansion of cells with some characteristics of the bronchiolar stem cell prior to formation of adenomas and peripheral lung adenocarcinomas. In addition to providing insight into possible mechanisms regulating stem cell pool size, this study also provided support for the idea that tissue stem cells may be the cell of origin for lung cancer (see also Chapter 17). In contrast, activation of Wnt/β-catenin signalling, a pathway directly linked to epithelial carcinogenesis and stem cell maintenance in the gut, led to expansion of bronchiolar stem cells without evidence of hyperplastic/neoplastic transformation.[105] In this study, potentiation of β-catenin signalling at the time of bronchiolar epithelial cell specification that occurs during the late embryonic period of lung development resulted in the expansion of a cell population that is naphthalene-resistant, highly proliferative in response to airway injury, and co-expresses CCSP and SP-C. The ability of Wnt/β-catenin pathway to expand the pool of airway stem cells through altering the maturation of developmental progenitor cells suggested that downregulation of this pathway in the developing airway is necessary for establishment of the normal balance between stem cells and specialised cell types of the bronchiolar epithelium.[106,107] Changes in the abundance of the putative bronchiolar stem cell population defined by co-expression of CCSP and SP-C were also seen in mice homozygous for a null allele of the Gata-6 gene.[107] Gata-6 deficient mice displayed enhanced Wnt/β-catenin signalling in the developing lung suggesting that β-catenin can function as a common transducer for multiple signalling pathways that have potential to regulate airway progenitor cell behaviour. However, the finding that specific deletion of β-catenin within the airway epithelium late in development did not impact epithelial maturation or the capacity to undergo repair following injury suggests that β-catenin is not necessary for the bronchiolar stem cell activation or maintenance.[108] These findings suggest that multiple signalling pathways have the potential to impact stem cell pool size and differentiation potential. Other potential molecular regulators of the stem cell compartment studied to date include p38α MAP kinase, PI-3 kinase, Bmi-1, and PTEN.[109–112] However, even though deregulation of these pathways impacts the behaviour of stem/progenitor cells *in vivo*, the pathway(s) involved in physiological regulation of the stem cell within the niche have not yet been defined.

Further efforts to characterise cellular components of the bronchiolar stem cell hierarchy and their regulation have involved isolation and fractionation of airway progenitor cells. Caveats with this approach are that dissociation of lung tissue eliminates spatial cues that aid in the definitive identification of the bronchiolar stem cell defined by naphthalene resistance, and the common problem that gene expression analysis suffers from potential artifacts introduced during the isolation procedure. However, some of these difficulties have been overcome through the use of genetically modified mouse models to introduce lineage tags and regulate progenitor cell pool size. Methods used for preparation of lung cells have either enriched for certain epithelial cell types through the use of defined protease cocktails coupled with application to lung tissue in a manner that restricts enzymatic activity to a localised compartment,[8,113] or have used methods for isolation of all lung cells.[114]

Enrichment of different members of the bronchiolar stem cell hierarchy is highly dependent on the existence of specific positive and negative selection strategies that exploit either biochemical or molecular characteristics to effect fractionation of a mixed population of cells.[90,100,114–116] Known markers for tissue stem cells, such as stem cell antigen-1 (Sca-1), have been used most frequently.[90,100] A strategy based on positive selection for Sca-1 and CD34, and negative selection for CD45 and CD31, was used to enrich cells that were immunoreactive for both CCSP and SP-C from crude preparations of cells isolated using a method for enrichment of alveolar epithelial cells.[97] When placed in culture, these cells have the ability to express aquaporin 5, CCSP and SP-C, markers for alveolar and airway epithelium, leading to the proposal that these cells have potential for either alveolar or airway differentiation, and hence may represent a bronchioalveolar stem cell (BASC). However, the use of antibodies directed to these same cell surface antigens on mixed cell preparations enriched for either total epithelial cells[116] or total lung cells[114] results in a different outcome. Bronchiolar progenitor cells were isolated by a fractionation procedure in which mixed cell preparations enriched for epithelial cells were depleted of cells with the cell surface markers CD45, CD34, and CD31, and positive selection imposed for cells with a Sca-1^{low} phenotype.[116] In this study, Sca-1^{low} cells could be further fractionated based upon their autofluorescence (AF) characteristics yielding AF^{high} and AF^{low} subfractions that were enriched in Clara cells and a candidate bronchiolar stem cell fraction, respectively. Support for the presence of bronchiolar stem cells within the AF^{low} subfraction was provided by the demonstration that transgenic mice with an expanded bronchiolar stem cell pool have a dramatic loss of AF^{high} cells with a corresponding increase in the abundance of the AF^{low} fraction. Lineage tracing experiments were also used to demonstrate that genetic tags introduced into CCSP-expressing cells are represented within cells contained within both AF^{high} and AF^{low} fractions of the Sca-1^{low}, $CD34^{negative}$ fraction of airway cells. In this study, the majority of CCSP-immunoreactive cells were found to show SPC-immunoreactivity and these dual immunoreactive cells were present in both AF^{high} and AF^{low} fractions of the Sca-1^{low} population. It was concluded that gene expression is altered within isolated Clara cells leading to promiscuous expression of SP-C.

A significant hurdle to overcome is the lack of novel marker genes that define the stem cell and more abundant facultative TA cell compartments. Recent studies in the gut have provided novel molecular markers that distinguish subsets of intestinal progenitor cells with stem cell activity.[45] However, validation of these marker genes is complex and requires rigorous *in vivo* lineage tracing. A number of strategies have been used to broaden the repertoire of molecular markers for analysis of epithelial cell differentiation in the developing and adult lung. Microarray analysis of lung tissue during naphthalene injury and repair led to the identification of novel genes that define the naphthalene-sensitive Clara cell population. Many of these genes, including Claudin10, FMO3, PON1, AO3, are expressed in a developmental sequence within maturing airway secretory cells.[117] However, these strategies have not revealed genes that specifically define the bronchiolar stem cell due to the rarity of this cell population in the lung.

7. Progenitor Cells of the Alveolar Compartment

The gas exchange area of the lung is lined by alveolar epithelial cells. There are two types of alveolar epithelial cells with different functional and morphological features. Alveolar type I (ATI) cells are thin, flat cells that line the alveolus and juxtapose the endothelial cells, together forming the gas exchange area. They are the predominant epithelial cell type in the alveolar compartment and can be identified based on T1α (RTI40), caveolin 1 and aquaporin 5 expression.[15] Alveolar type II (ATII) cells are larger, cuboidal cells whose main function is surfactant production and they can differentiate into ATI cells.[3] Alveolar type II cells can be distinguished at the ultrastructural level by the presence of lamellar bodies and microvilli.[118] Although in the adult mouse ATII produce surfactant proteins A, B and C (SP-A, SP-B and SP-C), SP-C expression is considered to be a unique marker for mature ATII cells.[119,120] Despite this, SP-C is one of the earliest genes expressed in the lung epithelium.[15] In the developing mouse lung, cells expressing SP-C appear at the tips of epithelial branches as early as E 10.5 indicating that expression of this gene product has potential to mark immature lung progenitor cells in addition to mature ATII cells.[32] These cells are highly proliferative in the developing lung and give rise to both mature epithelial cell types of the alveolar compartment. However, the alveolar epithelium turns over very slowly in uninjured mature alveolus.[121]

A number of injury models have been used as a tool to investigate the mechanisms of alveolar repair. Commonly used agents to elicit alveolar injury include hyperoxia[5,122] and NO_2[3] resulting in predominantly ATI and endothelial cell injury. Repair mechanisms have been investigated largely through the use of strategies to label proliferative cells such as the use of labelled DNA precursors like TdR or BrdU.[3] In such models, ATII cells represent the only labelled epithelial cell population immediately after *in vivo* delivery of labelled nucleotide analogues.[3] However, studies using a labelling period followed by a chase in the absence of label have shown that proliferating ATII cells can either self-renew or give rise to ATI cells, thus demonstrating the progenitor–progeny relationship between these two cell types.[3] Very similar findings have been observed following isolation and culture of ATII cells, in which their capacity to proliferate and generate ATI cells has been clearly demonstrated.[119,123] Further *in vitro* studies have also raised the possibility that ATI cells may under some conditions exhibit the capacity to proliferate and contribute to epithelial renewal.[124] However, the question from these experiments is whether cells cultured *in vitro* that exhibit some features of ATI cells, either morphological and/or molecular, are truly representative of their *in vivo* counterpart.

Studies similar to those performed in conducting airways identifying putative local tissue stem cells have not been performed in the alveolar compartment. Despite this, it is clear that not all ATII cells proliferate in response to alveolar damage and there is evidence of heterogeneity among ATII cell population.[122,125,126] It has been proposed that a bronchioalveolar stem cell exists which, at least *in vitro*, has the capacity for long-term maintenance and the generation of progeny that express some genes indicative of both bronchiolar and alveolar differentiation.[100] However, a cell type capable of contributing to both bronchiolar and alveolar epithelial lineages has not been demonstrated *in vivo*. Furthermore, evidence arguing that bronchiolar and alveolar lineages are maintained as distinct regenerative compartments in the adult lung has been generated using *in vivo* lineage-tracing studies.[116] These *in vivo* observations in animal models support the finding in human patients suffering from distal lung diseases with either alveolar wall destruction or fibrosis, that the alveolar compartment has a limited ability to repair and restore normal function.

8. Conclusions and Future Directions

Much of what we currently know of epithelial repair mechanisms results from controlled *in vivo* studies involving the use of animal models. Among the advantages of using animal models is the ability to study complex processes that are regulated through the action of multiple factors in genetically defined inbred strains that can be manipulated further to reveal cellular and molecular mechanisms of lung repair. From these studies it is clear that normal maintenance and repair of the epithelium lining airways is accomplished through the action of region-specific stem/progenitor cells. Further developments in understanding the biology of these cell types would benefit from the establishment of improved *in vitro* and transplantation models to define intrinsic and extrinsic regulators of cellular behaviour. Another hurdle to be overcome is extrapolation of findings in model organisms to humans. Even though it is likely that general principles of epithelial maintenance revealed in animal studies are relevant to understanding this process in humans, it is also likely that species differences exist that are key to understanding the pathophysiology of human disease and strategies that might lead to therapeutic intervention.

Acknowledgements

Barry Stripp was supported by National Institute of Health grants ES08964, HL064888, HL090146, and HL089141.

References

1. Evans, M.J., Cabral-Anderson, L.J., and Freeman, G. Role of the Clara cell in renewal of the bronchiolar epithelium. *Lab Invest* **38**, 648–653 (1978).

2. Evans, M.J., Johnson, L.V., Stephens, R.J., and Freeman, G. Renewal of the terminal bronchiolar epithelium in the rat following exposure to NO_2 or O_3. *Lab Invest* **35**, 246–257 (1976).
3. Evans, M.J., Cabral, L.J., Stephens, R.J., and Freeman, G. Transformation of alveolar type 2 cells to type 1 cells following exposure to NO_2. *Exp Mol Pathol* **22**, 142–150 (1975).
4. Lourenco, R.V., Klimek, M.F., and Borowski, C.J. Deposition and clearance of 2 micron particles in the tracheobronchial tree of normal subjects — smokers and nonsmokers. *J Clin Invest* **50**, 1411–1420 (1971).
5. Massaro, D. and Massaro, G.D. Biochemical and anatomical adaptation of the lung to oxygen-induced injury. *Fed Proc* **37**, 2485–2488 (1978).
6. Massaro, G.D., Amado, C., Clerch, L., and Massaro, D. Studies on the regulation of secretion in Clara cells with evidence for chemical nonautonomic mediation of the secretory response to increased ventilation in rat lungs. *J Clin Invest* **70**, 608–613 (1982).
7. Buonarati, M., Morin, D., Plopper, C., and Buckpitt, A. Glutathione depletion and cytotoxicity by naphthalene 1,2-oxide in isolated hepatocytes. *Chem Biol Interact* **71**, 147–165 (1989).
8. Chichester, C.H., Philpot, R.M., Weir, A.J., Buckpitt, A.R., and Plopper, C.G. Characterization of the cytochrome P-450 monooxygenase system in nonciliated bronchiolar epithelial (Clara) cells isolated from mouse lung. *Am J Respir Cell Mol Biol* **4**, 179–186 (1991).
9. Devereux, T.R. Alveolar type II and Clara cells: isolation and xenobiotic metabolism. *Environ Health Perspect* **56**, 95–101 (1984).
10. Reid, L.M. and Jones, R. Mucous membrane of respiratory epithelium. *Environ Health Perspect* **35**, 113–120 (1980).
11. Pack, R.J., Al-Ugaily, L.H., and Morris, G. The cells of the tracheobronchial epithelium of the mouse: a quantitative light and electron microscope study. *J Anat* **132**, 71–84 (1981).
12. Boers, J.E., Ambergen, A.W., and Thunnissen, F.B. Number and proliferation of basal and parabasal cells in normal human airway epithelium. *Am J Respir Crit Care Med* **157**, 2000–2006 (1998).
13. Boers, J.E., Ambergen, A.W., and Thunnissen, F.B. Number and proliferation of Clara cells in normal human airway epithelium. *Am J Respir Crit Care Med* **159**, 1585–1591 (1999).

14. Boers, J.E., den Brok, J.L., Koudstaal, J., Arends, J.W., and Thunnissen, F.B. Number and proliferation of neuroendocrine cells in normal human airway epithelium. *Am J Respir Crit Care Med* **154**, 758–763 (1996).
15. Herzog, E.L., Brody, A.R., Colby, T.V., Mason, R., and Williams, M.C. Knowns and unknowns of the alveolus. *Proc Am Thorac Soc* **5**, 778–782 (2008).
16. Borthwick, D.W., Shahbazian, M., Krantz, Q.T., Dorin, J.R., and Randell, S.H. Evidence for stem-cell niches in the tracheal epithelium. *Am J Respir Cell Mol Biol* **24**, 662–670 (2001).
17. Reynolds, S.D., Giangreco, A., Power, J.H., and Stripp, B.R. Neuroepithelial bodies of pulmonary airways serve as a reservoir of progenitor cells capable of epithelial regeneration. *Am J Pathol* **156**, 269–278 (2000).
18. Hong, K.U., Reynolds, S.D., Giangreco, A., Hurley, C.M., and Stripp, B.R. Clara cell secretory protein-expressing cells of the airway neuroepithelial body microenvironment include a label-retaining subset and are critical for epithelial renewal after progenitor cell depletion. *Am J Respir Cell Mol Biol* **24**, 671–681 (2001).
19. Rawlins, E.L. and Hogan, B.L. Ciliated epithelial cell lifespan in the mouse trachea and lung. *Am J Physiol Lung Cell Mol Physiol* **295**, L231–L234 (2008).
20. Knowles, M.R. and Boucher, R.C. Mucus clearance as a primary innate defense mechanism for mammalian airways. *J Clin Invest* **109**, 571–577 (2002).
21. Cheng, D.S., *et al.* Airway epithelium controls lung inflammation and injury through the NF-^{2}B pathway. *J Immunol* **178**, 6504–6513 (2007).
22. Sadikot, R.T., *et al.* Targeted immunomodulation of the NF-kappaB pathway in airway epithelium impacts host defense against Pseudomonas aeruginosa. *J Immunol* **176**, 4923–4930 (2006).
23. Elizur, A., *et al.* Clara cells impact the pulmonary innate immune response to LPS. *Am J Physiol Lung Cell Mol Physiol* **293**, L383–L392 (2007).
24. Elizur, A., *et al.* Tumor necrosis factor-alpha from macrophages enhances LPS-induced clara cell expression of keratinocyte-derived chemokine. *Am J Respir Cell Mol Biol* **38**, 8–15 (2008).
25. Reynolds, S.D., *et al.* CCSP regulates cross talk between secretory cells and both ciliated cells and macrophages of the conducting airway. *Am J Physiol Lung Cell Mol Physiol* **293**, L114–L123 (2007).

26. Holgate, S.T., *et al.* Epithelial-mesenchymal communication in the pathogenesis of chronic asthma. *Proc Am Thorac Soc* **1**, 93–98 (2004).
27. Reynolds, S.D., *et al.* Airway injury in lung disease pathophysiology: selective depletion of airway stem and progenitor cell pools potentiates lung inflammation and alveolar dysfunction. *Am J Physiol Lung Cell Mol Physiol* **287**, L1256–L1265 (2004).
28. Snyder, J.C., Zemke, A.C., and Stripp, B.R. Reparative capacity of airway epithelium impacts deposition and remodeling of extracellular matrix. *Am J Respir Cell Mol Biol* **40**, 633–642 (2008).
29. Zhu, L., *et al.* Prominin 1 marks intestinal stem cells that are susceptible to neoplastic transformation. *Nature* **457**, 603–607 (2008).
30. Dick, J.E. Stem cell concepts renew cancer research. *Blood* **112**, 4793–4807 (2008).
31. Bagby, G.C., Lipton, J.M., Sloand, E.M., and Schiffer, C.A. Marrow failure. *Hematology Am Soc Hematol Educ Program* 318–336 (2004).
32. Cardoso, W.V. and Lu, J. Regulation of early lung morphogenesis: questions, facts and controversies. *Development* **133**, 1611–1624 (2006).
33. Sekine, K., *et al.* Fgf10 is essential for limb and lung formation. *Nat Genet* **21**, 138–141 (1999).
34. Bellusci, S., Grindley, J., Emoto, H., Itoh, N., and Hogan, B.L. Fibroblast growth factor 10 (FGF10) and branching morphogenesis in the embryonic mouse lung. *Development* **124**, 4867–4878 (1997).
35. Weaver, M., Dunn, N.R., and Hogan, B.L. Bmp4 and Fgf10 play opposing roles during lung bud morphogenesis. *Development* **127**, 2695–2704 (2000).
36. Loscertales, M., Mikels, A.J., Hu, J.K., Donahoe, P.K., and Roberts, D.J. Chick pulmonary Wnt5a directs airway and vascular tubulogenesis. *Development* **135**, 1365–1376 (2008).
37. Yin, Y., *et al.* An FGF-WNT gene regulatory network controls lung mesenchyme development. *Dev Biol* **319**, 426–436 (2008).
38. Shannon, J.M. Induction of alveolar type II cell differentiation in fetal tracheal epithelium by grafted distal lung mesenchyme. *Dev Biol* **166**, 600–614 (1994).
39. Shannon, J.M., Nielsen, L.D., Gebb, S.A., and Randell, S.H. Mesenchyme specifies epithelial differentiation in reciprocal recombinants of embryonic lung and trachea. *Dev Dyn* **212**, 482–494 (1998).

98. Stripp, B.R., Maxson, K., Mera, R., and Singh, G. Plasticity of airway cell proliferation and gene expression after acute naphthalene injury. *Am J Physiol* **269**, L791–L799 (1995).
99. Stevens, T.P., McBride, J.T., Peake, J.L., Pinkerton, K.E., and Stripp, B.R. Cell proliferation contributes to PNEC hyperplasia after acute airway injury. *Am J Physiol* **272**, L486–L493 (1997).
100. Kim, C.F., *et al.* Identification of bronchioalveolar stem cells in normal lung and lung cancer. *Cell* **121**, 823–835 (2005).
101. Van Winkle, L.S., Buckpitt, A.R., Nishio, S.J., Isaac, J.M., and Plopper, C.G. Cellular response in naphthalene-induced Clara cell injury and bronchiolar epithelial repair in mice. *Am J Physiol* **269**, L800–L818 (1995).
102. Fuchs, E., Tumbar, T., and Guasch, G. Socializing with the neighbors: stem cells and their niche. *Cell* **116**, 769–778 (2004).
103. De Proost, I., *et al.* Functional live cell imaging of the pulmonary neuroepithelial body microenvironment. *Am J Respir Cell Mol Biol* **39**, 180–189 (2008).
104. Jackson, E.L., *et al.* Analysis of lung tumor initiation and progression using conditional expression of oncogenic K-ras. *Genes Dev* **15**, 3243–3248 (2001).
105. Reynolds, S.D., *et al.* Conditional stabilization of beta-catenin expands the pool of lung stem cells. *Stem Cells* **26**, 1337–1346 (2008).
106. Mucenski, M.L., *et al.* Beta-catenin regulates differentiation of respiratory epithelial cells *in vivo*. *Am J Physiol Lung Cell Mol Physiol* **289**, L971–L979 (2005).
107. Zhang, Y., *et al.* A Gata6-Wnt pathway required for epithelial stem cell development and airway regeneration. *Nat Genet* **40**, 862–870 (2008).
108. Zemke, A.C. *et al.* β-Catenin is not necessary for maintenance of repair of the bronchiolar epithelium. *Am J Respir Cell Mol Biol* **41**(5), 535–543 (2009).
109. Yang, Y., *et al.* Phosphatidylinositol 3-kinase mediates bronchioalveolar stem cell expansion in mouse models of oncogenic K-ras-induced lung cancer. *PLoS ONE* **3**, e2220 (2008).
110. Ventura, J.J., *et al.* p38alpha MAP kinase is essential in lung stem and progenitor cell proliferation and differentiation. *Nat Genet* **39**, 750–758 (2007).
111. Dovey, J.S., Zacharek, S.J., Kim, C.F., and Lees, J.A. Bmi1 is critical for lung tumorigenesis and bronchioalveolar stem cell expansion. *Proc Natl Acad Sci USA* **105**, 11857–11862 (2008).

112. Yanagi, S., *et al.* Pten controls lung morphogenesis, bronchioalveolar stem cells, and onset of lung adenocarcinomas in mice. *J Clin Invest* **117**, 2929–2940 (2007).
113. Corti, M., Brody, A.R., and Harrison, J.H. Isolation and primary culture of murine alveolar type II cells. *Am J Respir Cell Mol Biol* **14**, 309–315 (1996).
114. McQualter, J.L., *et al.* Endogenous fibroblastic progenitor cells in the adult mouse lung are highly enriched in the Sca-1 positive cell fraction. *Stem Cells* online (December 2008).
115. Summer, R., *et al.* SP (Side Population) Cells and Bcrp1 Expression in Lung. *Am J Physiol Lung Cell Mol Physiol* **285**, 97–104 (2003).
116. Teisanu, R.M., Lagasse, E., Whitesides, J.F., and Stripp, B.R. Prospective isolation of bronchiolar stem cells based upon immunophenotypic and autofluorescence characteristics. *Stem Cells* **27**, 612–622 (2009).
117. Zemke, A.C., *et al.* Molecular staging of epithelial maturation using secretory cell-specific genes as markers. *Am J Respir Cell Mol Biol* **40**, 340–348 (2009).
118. Hatasa, K. and Nakamura, T. Electron microscopic observations of lung alveolar epithelial cells of normal young mice, with special reference to formation and secretion of osmiophilic lamellar bodies. *Z Zellforsch Mikrosk Anat* **68**, 266–277 (1965).
119. Shannon, J.M., Jennings, S.D., and Nielsen, L.D. Modulation of alveolar type II cell differentiated function *in vitro*. *Am J Physiol* **262**, L427–L436 (1992).
120. Fehrenbach, H., *et al.* Keratinocyte growth factor-induced hyperplasia of rat alveolar type II cells *in vivo* is resolved by differentiation into type I cells and by apoptosis. *Eur Respir J* **14**, 534–544 (1999).
121. Evans, M.J., Stephens, R.J., Cabral, L.J., and Freeman, G. Cell renewal in the lungs of rats exposed to low levels of NO_2. *Arch Environ Health* **24**, 180–188 (1972).
122. Yee, M., *et al.* Type II epithelial cells are critical target for hyperoxia-mediated impairment of postnatal lung development. *Am J Physiol Lung Cell Mol Physiol* **291**, L1101–L1111 (2006).
123. Dobbs, L.G., Pian, M.S., Maglio, M., Dumars, S., and Allen, L. Maintenance of the differentiated type II cell phenotype by culture with an apical air surface. *Am J Physiol* **273**, L347–L354 (1997).

124. Danto, S.I., Shannon, J.M., Borok, Z., Zabski, S.M., and Crandall, E.D. Reversible transdifferentiation of alveolar epithelial cells. *Am J Respir Cell Mol Biol* **12**, 497–502 (1995).
125. Roper, J.M., *et al. In vivo* exposure to hyperoxia induces DNA damage in a population of alveolar type II epithelial cells. *Am J Physiol Lung Cell Mol Physiol* **286**, L1045–L1054 (2004).
126. Roper, J.M., Staversky, R.J., Finkelstein, J.N., Keng, P.C., and O'Reilly, M.A. Identification and isolation of mouse type II cells on the basis of intrinsic expression of enhanced green fluorescent protein. *Am J Physiol Lung Cell Mol Physiol* **285**, L691–L700 (2003).

Chapter 5
Safety of Exogenous Stem Cells

Elizabeth K. Sage, Michael R. Loebinger
and Sam M. Janes

Adult stem cells posses the ability to undergo both self-renewal and differentiation into multiple lineages. In chronic lung injury these cells are activated in response to tissue damage, migrate to the site of injury and contribute to both structural and functional repair. In acute injury they reduce the pulmonary inflammatory response via a number of mechanisms to cause downregulation of pro-inflammatory cytokines and a reduction in pathological lung damage. However, there are significant concerns that they are at risk of undergoing malignant transformation. Murine bone marrow-derived stem cells have been shown to acquire chromosomal abnormalities *in vitro* that corresponds to fibrosarcoma formation *in vivo*, a finding that was not reproduced using human adult stem cells. Despite these concerns a number of clinical studies have already been undertaken and initial results, although varied, seem promising for their use in a clinical setting with a good short-term safety profile. There have been no reports to date of an increase in the number of malignancies, but further long-term safety data is needed.

Keywords: Mesenchymal stem cells; malignant transformation; safety; chromosomal instability; clinical trials.

Outline

1. Introduction
2. Exogenous Stem Cells
3. Potential Malignant Transformation
4. Clinical Trials

1. Introduction

Lung disease is a common cause of morbidity and mortality worldwide and for a significant number of respiratory diseases, therapeutic options are limited. Acute lung injury (ALI) and acute respiratory distress syndrome (ARDS) are major causes of acute respiratory failure and the only treatment strategy shown to be of prognostic benefit is low tidal volume ventilation.[1] Despite this, it is a leading cause of morbidity and mortality within the hospital setting.[2] Interstitial lung disease is characterised by progressive lung fibrosis and debilitating breathlessness and is poorly responsive to current medical therapies, with the only real treatment option being lung transplantation. Idiopathic pulmonary fibrosis (IPF) is usually fatal with only a 30% five-year survival.[3]

Recent advances in stem cell biology have implicated adult stem cells in tissue regeneration and homeostasis, and the traditional view of them being lineage-restricted has been superseded by evidence of their multipotency.[4,5] With increasing excitement about the regenerative potential of stem cells both as therapeutic agents within their own right and as vectors for the delivery of other therapeutic agents, clinical trials have already been undertaken. However, with the defining characteristics of unlimited self-renewal there is concern over the potential for malignant transformation. This chapter focuses largely on the safety of bone marrow-derived stem cells (BMSCs) and their current clinical applications.

2. Exogenous Stem Cells

Stem cells are undifferentiated cells that have the capacity for both unlimited self-renewal and differentiation into more specialised daughter cells. Classically there are two types of stem cells: embryonic and adult. Adult stem cells are found in discrete niches within adult tissue and have the potential to repair damaged tissue by replacing specialised cells. Initially they were thought to be lineage-restricted and produce only progeny of a specific cell type appropriate to its location. The most widely studied adult stem cell is the bone marrow-derived stem cell (BMSC), which consists of haematopoietic stem cells (HSCs) and mesenchymal stem cells (MSCs). HSCs develop into all mature blood cells whilst MSCs differentiate into tissue stromal cells including bone, fat and cartilage.

Studies performed earlier this century suggested that, far from being lineage-restricted, adult stem cells showed significant plasticity and were, under certain conditions, able to produce functional phenotypes of other tissues[4] including the lung. A number of studies using a diverse range of experimental conditions[6–12] and cell types[4,6,7,10,13,14] resulted in a wide variety of results, with some showing significant engraftment of stem cells within the lung[4] and others showing none (see Table 1)[15] (see also Chapter 14).

As well as the uncertainty over the extent of stem cell engraftment in distant sites, the mechanism by which this occurs is also unclear. *In vitro* embryonic and adult stem cells have been shown to fuse, producing phenotypic change in the resultant cell[16] and *in vivo* studies have shown that repopulation of hepatocytes in murine models is as a result of cell fusion.[17] However, this mechanism has not been elucidated in lung engraftment.[18,19] An alternative explanation for engraftment of donor-derived cells within

Table 1: Studies looking at stem cell engraftment within epithelial tissues.

Source	Cell Type	Injury Model	Result
Krause *et al.* (2001)[4]	HSC	Radiation	Type II pneumocyte engraftment
Kotton *et al.* (2001)[6]	MSC	Bleomycin	Type I pneumocyte engraftment
Anjos-Afonso *et al.* (2004)[8]	MSC	No injury	Lung epithelial cell engraftment
Ortiz *et al.* (2003)[12]	MSC	Bleomycin	Type II pneumocyte engraftment
Rojas *et al.* (2005)[13]	MSC	Bleomycin	Lung epithelial cell engraftment
Jiang *et al.* (2002)[14]	MAPC	No injury	Lung, liver and gut engraftment
Theise *et al.* (2002)[7]	BMSC	Radiation	Type II pneumocyte engraftment
Yamada *et al.* (2004)[9]	BMSC	Lipopolysaccharide	Lung epithelial cell engraftment
Hashimoto *et al.* (2004)[10]	BMSC	Bleomycin	Lung epithelial cell engraftment
Ishizawa *et al.* (2004)[11]	BMSC	Elastase	Lung engraftment
Wagers *et al.* (2002)[15]	HSC	Radiation	No lung, liver, gut or renal engraftment

HSC, Haematopoeitic stem cells; MSC, mesenchymal stem cells; MAPC, multipotent adult progenitor cells; BMSC, bone marrow-derived stem cells.

the lung is transdifferentiation — a process that is mediated by both soluble factors (chemokines) and direct cell–cell contact.[20]

Despite these controversies the current consensus is that whilst there is some engraftment of donor-derived stem cells as lung epithelial tissue, it is likely to be at much lower rates than previously suggested and is unlikely to be useful for directed stem cell therapy.[21,22] However, there is evidence that the transplantation of adult stem cells following lung injury has a beneficial effect on the disease process.[11–13,23] In fact these studies have been so promising that intra-arterial infusion of adult stem cells are currently undergoing clinical trials in cardiology,[24] and the use of endothelial-like progenitor cells (EPCs) transduced with human endothelial nitric oxide-synthase (eNOS) for the treatment of pulmonary arterial hypertension (PAH) are in phase I trials[25] (see Chapters 10–12).

3. Potential Malignant Transformation

With the growing excitement about the potential therapeutic uses of adult stem cells there are also concerns regarding their safety. By the very nature of their ability to undergo unlimited self-renewal and differentiation they are at high risk of undergoing malignant transformation. Recently there has been discussion as to whether tumour cells in fact originate from cancer stem cells and that these stem cells within the tumour are responsible for the uncontrolled cell division and growth that characterise the malignant process (see Chapter 17).

A previous study[26] has suggested that BMSCs could be responsible for the development of gastric carcinoma. In a murine model BMSCs were recruited to areas of chronic infection such as that induced by the presence of *Helicobacter felis*, and after initial repopulation of this area they progressed through stages of dysplasia to the development of intraepithelial malignancy. Because of the potential uses of adult stem cells for regenerative medicine and their ongoing use in clinical trials, it is essential to ascertain the characteristics of these cells with regard to their malignant potential.

To study this possibility, murine BMSCs were cultured *in vitro* and then transplanted into immuno-compromised mice and assessed for tumorigenesis. Initially *in vitro* the BMSCs showed both cell proliferation and differentiation into chondrocytes, adipocytes and osteocytes under appropriate conditions, an ability that persisted *in vivo*. However, after prolonged passage they lost their differentiation potential and developed tumours that were histologically consistent with fibrosarcomas. These fibrosarcomas were found in multiple organs and when these were removed and transplanted into secondary recipient mice, further colonies developed. An increase in the number of chromosomal abnormalities was identified in those cells undergoing a higher number of passages suggesting that malignant change correlates with accumulated chromosomal abnormalities during increased passage time.[27] Similar findings were noted by Tolar *et al.*, who found that infusion of donor-derived MSCs into immunodeficient mice resulted in sarcoma formation in both the lung and extremities, and was associated with increased mortality.[28] In both of these studies the presence of cytogenetically abnormal clonal cell expansion was noted prior to *in vivo* infusion, which suggests that the critical transformation event takes place *in vitro*.

In a study by Aguilar *et al.*, systemic delivery of murine MSCs from an immuno-compromised donor resulted in tissue engraftment within the lungs that stained both for collagen and bone markers. This tissue developed further obliterating normal lung tissue and was histologically consistent with a well-differentiated osteosarcoma. In these mice tumours were found only within the lung parenchyma and not at distant sites. MSCs from immuno-competent donors did not result in the same level of tumour formation, suggesting that the strain difference in donor cells may be important in determining the potential for malignant transformation.[29] What is interesting to note is that infusion of human donor MSCs did not produce the same result and the donor cells were cleared from the lung after initial engraftment.

With increasing evidence of the tumourigenic potential of murine adult stem cells it is important to elucidate whether this also occurs with human MSCs. Human cells have two main control points in the cell cycle: senescence and crisis phases. Senescence is characterised by moderate telomere shortening and arrest of the cell cycle. If this stage is bypassed then cells continue to grow and telomeres become critically short, causing the cell to enter the crisis phase in which chromosomal instability provokes significant apoptosis and cell death.

Rubio *et al.* isolated ten human MSC samples from subjects with non-malignant disease and injected them into immunodeficient mice. At four months none of the mice showed evidence of tumour formation. Karyotyping performed both pre- and post-senescence showed chromosomal abnormalities in up to 30% of the post-senescence MSCs, but there was still no detectable tumour growth three months after injection into the murine hosts. Cells were allowed to go on and enter crisis phase during which stage they showed evidence of chromosomal abnormalities consistent with tumour cells, most notably in a pattern consistent with that seen in sarcomas. These tumour-like cells showed high levels of c-myc expression, suggesting that c-myc overexpression may have a role in the bypass of senescence and contribute to malignant transformation. In contrast to the results seen with pre- and post-senescence MSCs, injection of the transformed tumour-like cells (transduced with green fluorescent protein (GFP)) into immunodeficient mice resulted in subjects developing signs and symptoms consistent with a malignant process. Tumours were detected at disparate sites via both fluorescent staining and immunohistochemical staining, suggesting they were of donor origin. As a result of these studies it was postulated that during prolonged *in vitro* expansion, human MSCs bypass senescence and spontaneously acquire tumourigenic potential via uncontrolled telomerase activity.[30]

In contrast to these findings, Bernardo *et al.*[31] failed to show malignant transformation in human stem cells following prolonged *in vitro* culture. Samples taken from ten healthy haematopoietic stem cell donors were propagated *in vitro* for up to 12 weeks and then assessed for telomerase activity, telomere length and the presence of the p53 mutation. In all cases the donor cells were cultured without expressing increased telomerase activity and showed a reduction in telomere length and the absence of any chromosomal mutation, suggesting carcinogenic potential. In fact they exhibited a gradual decline in their proliferative capacity, resulting in a persistent senescence phase rather than bypassing this phase and entering the crisis phase that has been previously described as resulting in malignant transformation. This is in support of previous studies suggesting a lack of telomerase activity in human MSCs.[32] It is possible that susceptibility to malignant transformation previously described in murine MSCs may be species-specific, as those authors were unable to obtain similar results in human cells.[27,29]

4. Clinical Trials

Despite some concerns over the safety of stem cell therapy their potential clinical application is currently being assessed in a wide range of diseases. The use of bone marrow transplant in haematological malignancies has been carried out for decades and there is an increasing body of evidence supporting their use in non-haematological diseases including acute and chronic myocardial dysfunction; acute cerebrovascular accidents; peripheral vascular disease; pulmonary arterial hypertension and a wide range of autoimmune disorders. The most clinically promising evidence to date is in the treatment of cardiovascular disease.

4.1. *Acute myocardial infarction*

The current gold standard treatment for patients with ST-segment elevation myocardial infarction (STEMI) is percutaneous coronary intervention with or without stent insertion or thrombolytic therapy. Despite optimal treatment, survival of the acute event often leads to chronic problems with ischaemic cardiac failure that results in significant morbidity and mortality. There have been numerous multicentre randomised controlled clinical trials using a variety of BMSCs that have yielded sometimes conflicting results, which could be at least in part due to their heterogeneity and small patient numbers. However, overall, the evidence would seem to point to a moderate but nonetheless statistically significant improvement in cardiac function.[33,34]

The BOOST (Bone Marrow Transfer to Enhance ST Elevation Infarct Regeneration Trial), TOPCARE-AMI (Transplantation of Progenitor Cells and Regeneration Enhancement in Acute Myocardial Infarction) and REPAIR-AMI (Reinfusion of Enriched Progenitor Cells and Infarct Remodelling in Acute Myocardial Infarction) trials compared current gold standard treatment with intracoronary injection of undifferentiated BMSCs 3–7 days post STEMI. There was a statistically significant improvement in left ventricular ejection fraction (LVEF) after four[35] and six months[36] in those patients receiving stem cell therapy. This was sustained at 12 months[37,38] but not at 18 months.[39] In contrast, the ASTAMI (Autologous Stem Cell Transplantation in Acute Myocardial Infarction) trial, which was a similar study design, showed no significant difference in either LVEF or infarction size at six months.[40] Although not specifically designed for this purpose, none of the studies reported an increase in the number of deaths from cancers. In the one reported case of death from lung cancer, reassessment of the imaging and notes revealed that the disease was present prior to inclusion in the study.

4.2. *Chronic coronary artery disease*

Cardiac disease resulting from previous ischaemic events is characterised by areas of non-contracting myocardium as a response to hypoxic stress, often seen on echocardiography as regional wall motion abnormalities. There have been a number of randomised controlled clinical trials, again using a variety of BMSCs, looking at the effect of intracoronary injection on both LVEF and infarct size.

The IACT (Intracoronary Autologous Bone Marrow Cell Transplantation in Chronic Coronary Artery Disease) and TOPCARE-CHD (Transplantation of Progenitor Cells and Recovery of Left Ventricular Function in Patients with Chronic Ischaemic Heart Disease) trials both showed that intracoronary infusion of CPCs and undifferentiated BMSCs resulted in significant improvement of LVEF, reduction in infarct size and lower levels of serum brain natriuretic peptide (BNP) compared to placebo.[41,42] These results were replicated following the intracoronary infusion of mesenchymal stem cells along with an improvement in symptoms according to the New York Heart Association (NYHA) grading system, and reduction in wall motion defects.[43] Again a meta-analysis concluded that there was a modest improvement in chronic ischaemic heart disease following intracoronary stem cell injection with no reports of tumour formation.[33]

4.3. *Autoimmune disorders*

HSC transplantation for severe autoimmune diseases started in the late 1990s based on the rationale that stem cell infusion induces resetting of the immune system. Initial studies looked at the role of autologous HSC infusion for the treatment of a variety of autoimmune conditions including systemic sclerosis and multiple sclerosis. Treatment was performed with both myeloablative and non-myeloablative regimens with a better safety profile being found with the non-myeloablative regimen,[44–48] possibly due to a shorter duration of chemotherapy-induced bone marrow suppression. There have already been reports of higher rates of myelodysplasia and leukaemia with autologous HSC treatment,[44,48] making it difficult to support its use in the treatment of non-malignant disease. However, there is still considerable interest in the use of non-myeloablative regimens and there are a number of ongoing randomised controlled trials looking at HSC transplantation for autoimmune diseases.

4.4. *Pulmonary arterial hypertension*

(See also Chapters 10–12)

A rat monocrotaline-induced pulmonary arterial hypertension (PAH) model was used to look at the effect of endothelial-like progenitor cells (EPCs) on vascular remodelling and pulmonary haemodynamics. EPCs were shown to prevent the development of PAH whilst those transduced with human endothelial NO-synthase (eNOS) resulted in reversal of established disease and significantly improved survival.[49,50] An initial prospective randomised controlled trial in humans compared current standard therapy with infusion of non-transduced EPCs in addition to standard therapy, and showed a statistically significant improvement in the six-minute walk distance, mean pulmonary artery pressures and cardiac output. There were no adverse events associated with stem cell therapy although long-term safety is not currently known.[51] The uses of EPCs in the treatment of PAH are currently undergoing phase 1 clinical trials.

5. Conclusion

Although the initial enthusiasm about adult stem cells engrafting in damaged epithelial tissues and forming mature, functioning cells capable of repair has been dampened somewhat by further studies, there is evidence to suggest that exogenous stem cells can be beneficial in some disease processes. Although the mechanisms are far from clear there are already a number of clinical trials using exogenous adult stem cells in cardiovascular, skeletal, immunological and pulmonary disease. The results so far have been varied and this may be in part due to the heterogeneous nature of the cells being infused.

In addition to varying clinical results there is significant concern over the long-term safety of stem cells due to their potential for malignant change. A number of studies in murine models have shown telomere shortening and the development of chromosomal instability resulting in sarcoma-like tumour formation after *in vitro* culture. However, this has not been reliably shown in human stem cells and it is postulated that this process is species-specific. Because of this uncertainty, collection of safety data from ongoing clinical trials should do much to allay these theoretical concerns.

Acknowledgements

Michael Loebginer is a MRC Clinical Training Fellow. Sam Janes is a MRC Clinician Scientist.

References

1. Ventilation with lower tidal volumes as compared with traditional tidal volumes for acute lung injury and the acute respiratory distress syndrome. The Acute Respiratory Distress Syndrome Network. *N Engl J Med* **342**, 1301–1308 (2000).
2. Rubenfeld, G.D., *et al.* Incidence and outcomes of acute lung injury. *N Engl J Med* **353**, 1685–1693 (2005).
3. Perez, A., Rogers, R.M., and Dauber, J.H. The prognosis of idiopathic pulmonary fibrosis. *Am J Respir Cell Mol Biol* **29**, S19–S26 (2003).
4. Krause, D.S., *et al.* Multi-organ, multi-lineage engraftment by a single bone marrow-derived stem cell. *Cell* **105**, 369–377 (2001).
5. Tarnowski, M. and Sieron, A.L. Adult stem cells and their ability to differentiate. *Med Sci Monit* **12**, RA154–RA163 (2006).
6. Kotton, D.N., *et al.* Bone marrow-derived cells as progenitors of lung alveolar epithelium. *Development* **128**, 5181–5188 (2001).
7. Theise, N.D., *et al.* Radiation pneumonitis in mice: a severe injury model for pneumocyte engraftment from bone marrow. *Exp Hematol* **30**, 1333–1338 (2002).

8. Anjos-Afonso, F., Siapati, E.K., and Bonnet, D. *In vivo* contribution of murine mesenchymal stem cells into multiple cell-types under minimal damage conditions. *J Cell Sci* **117**, 5655–5664 (2004).
9. Yamada, M., *et al.* Bone marrow-derived progenitor cells are important for lung repair after lipopolysaccharide-induced lung injury. *J Immunol* **172**, 1266–1272 (2004).
10. Hashimoto, N., Jin, H., Liu, T., Chensue, S.W., and Phan, S.H. Bone marrow-derived progenitor cells in pulmonary fibrosis. *J Clin Invest* **113**, 243–252 (2004).
11. Ishizawa, K., *et al.* Bone marrow-derived cells contribute to lung regeneration after elastase-induced pulmonary emphysema. *FEBS Lett* **556**, 249–252 (2004).
12. Ortiz, L.A., *et al.* Mesenchymal stem cell engraftment in lung is enhanced in response to bleomycin exposure and ameliorates its fibrotic effects. *Proc Natl Acad Sci USA* **100**, 8407–8411 (2003).
13. Rojas, M., *et al.* Bone marrow-derived mesenchymal stem cells in repair of the injured lung. *Am J Respir Cell Mol Biol* **33**, 145–152 (2005).
14. Jiang, Y., *et al.* Pluripotency of mesenchymal stem cells derived from adult marrow. *Nature* **418**, 41–49 (2002).
15. Wagers, A.J., Sherwood, R.I., Christensen, J.L., and Weissman, I.L. Little evidence for developmental plasticity of adult hematopoietic stem cells. *Science* **297**, 2256–2259 (2002).
16. Ying, Q.L., Nichols, J., Evans, E.P., and Smith, A.G. Changing potency by spontaneous fusion. *Nature* **416**, 545–548 (2002).
17. Wang, X., *et al.* Cell fusion is the principal source of bone-marrow-derived hepatocytes. *Nature* **422**, 897–901 (2003).
18. Herzog, E.L., Chai, L., and Krause, D.S. Plasticity of marrow-derived stem cells. *Blood* **102**, 3483–3493 (2003).
19. Harris, R.G., *et al.* Lack of a fusion requirement for development of bone marrow-derived epithelia. *Science* **305**, 90–93 (2004).
20. Loebinger, M.R., Aguilar, S., and Janes, S.M. Therapeutic potential of stem cells in lung disease: progress and pitfalls. *Clin Sci (Lond)* **114**, 99–108 (2008).
21. Bruscia, E.M., *et al.* Engraftment of donor-derived epithelial cells in multiple organs following bone marrow transplantation into newborn mice. *Stem Cells* **24**, 2299–2308 (2006).

22. Bruscia, E.M., *et al.* Assessment of cystic fibrosis transmembrane conductance regulator (CFTR) activity in CFTR-null mice after bone marrow transplantation. *Proc Natl Acad Sci USA* **103**, 2965–2970 (2006).
23. Aliotta, J.M., *et al.* Bone marrow production of lung cells: the impact of G-CSF, cardiotoxin, graded doses of irradiation, and subpopulation phenotype. *Exp Hematol* **34**, 230–241 (2006).
24. Burt, R.K., *et al.* Clinical applications of blood-derived and marrow-derived stem cells for nonmalignant diseases. *JAMA* **299**, 925–936 (2008).
25. Ward, M. and Langleben, D. *Phase I Trial to Establish Safety of Autologous Progenitor Cell-Based Gene Therapy Delivery of heNOS in Patients With Severe Pulmonary Arterial Hypertension (PAH) Refractory to Conventional Treatment.* http://clinicaltrials.gov/ct2/show/NCT00469027 (2009).
26. Houghton, J., *et al.* Gastric cancer originating from bone marrow-derived cells. *Science* **306**, 1568–1571 (2004).
27. Miura, M., *et al.* Accumulated chromosomal instability in murine bone marrow mesenchymal stem cells leads to malignant transformation. *Stem Cells* **24**, 1095–1103 (2006).
28. Tolar, J., *et al.* Sarcoma derived from cultured mesenchymal stem cells. *Stem Cells* **25**, 371–379 (2007).
29. Aguilar, S., *et al.* Murine but not human mesenchymal stem cells generate osteosarcoma-like lesions in the lung. *Stem Cells* **25**, 1586–1594 (2007).
30. Rubio, D., *et al.* Spontaneous human adult stem cell transformation. *Cancer Res* **65**, 3035–3039 (2005).
31. Bernardo, M.E., *et al.* Human bone marrow derived mesenchymal stem cells do not undergo transformation after long-term *in vitro* culture and do not exhibit telomere maintenance mechanisms. *Cancer Res* **67**, 9142–9149 (2007).
32. Zimmermann, S., *et al.* Lack of telomerase activity in human mesenchymal stem cells. *Leukemia* **17**, 1146–1149 (2003).
33. Abdel-Latif, A., *et al.* Adult bone marrow-derived cells for cardiac repair: a systematic review and meta-analysis. *Arch Intern Med* **167**, 989–997 (2007).
34. Lipinski, M.J., *et al.* Impact of intracoronary cell therapy on left ventricular function in the setting of acute myocardial infarction: a collaborative systematic review and meta-analysis of controlled clinical trials. *J Am Coll Cardiol* **50**, 1761–1767 (2007).
35. Schachinger, V., *et al.* Intracoronary bone marrow-derived progenitor cells in acute myocardial infarction. *N Engl J Med* **355**, 1210–1221 (2006).

36. Wollert, K.C., *et al.* Intracoronary autologous bone-marrow cell transfer after myocardial infarction: the BOOST randomised controlled clinical trial. *Lancet* **364**, 141–148 (2004).
37. Schachinger, V., *et al.* Transplantation of progenitor cells and regeneration enhancement in acute myocardial infarction: final one-year results of the TOPCARE-AMI Trial. *J Am Coll Cardiol* **44**, 1690–1699 (2004).
38. Schachinger, V., *et al.* Improved clinical outcome after intracoronary administration of bone-marrow-derived progenitor cells in acute myocardial infarction: Final 1-year results of the REPAIR-AMI trial. *Eur Heart J* **27**, 2775–2783 (2006).
39. Schaefer, A., *et al.* Impact of intracoronary bone marrow cell transfer on diastolic function in patients after acute myocardial infarction: results from the BOOST trial. *Eur Heart J* **27**, 929–935 (2006).
40. Lunde, K., *et al.* Intracoronary injection of mononuclear bone marrow cells in acute myocardial infarction. *N Engl J Med* **355**, 1199–1209 (2006).
41. Strauer, B.E., *et al.* Regeneration of human infarcted heart muscle by intracoronary autologous bone marrow cell transplantation in chronic coronary artery disease: the IACT study. *J Am Coll Cardiol* **46**, 1651–1658 (2005).
42. Assmus, B., *et al.* Transcoronary transplantation of progenitor cells after myocardial infarction. *N Engl J Med* **355**, 1222–1232 (2006).
43. Chen, S., *et al.* Intracoronary transplantation of autologous bone marrow mesenchymal stem cells for ischemic cardiomyopathy due to isolated chronic occluded left anterior descending artery. *J Invasive Cardiol* **18**, 552–556 (2006).
44. Nash, R.A., *et al.* High-dose immunosuppressive therapy and autologous hematopoietic cell transplantation for severe systemic sclerosis: long-term follow-up of the US multicenter pilot study. *Blood* **110**, 1388–1396 (2007).
45. Saccardi, R., *et al.* Autologous stem cell transplantation for progressive multiple sclerosis: update of the European Group for Blood and Marrow Transplantation autoimmune diseases working party database. *Mult Scler* **12**, 814–823 (2006).
46. Binks, M., *et al.* Phase I/II trial of autologous stem cell transplantation in systemic sclerosis: procedure related mortality and impact on skin disease. *Ann Rheum Dis* **60**, 577–584 (2001).
47. Farge, D., *et al.* Autologous stem cell transplantation in the treatment of systemic sclerosis: Report from the EBMT/EULAR Registry. *Ann Rheum Dis* **63**, 974–981 (2004).

48. Samijn, J.P., *et al.* Intense T cell depletion followed by autologous bone marrow transplantation for severe multiple sclerosis. *J Neurol Neurosurg Psychiatry* **77**, 46–50 (2006).
49. Zhao, Y.D., *et al.* Rescue of monocrotaline-induced pulmonary arterial hypertension using bone marrow-derived endothelial-like progenitor cells: efficacy of combined cell and eNOS gene therapy in established disease. *Circ Res* **96**, 442–450 (2005).
50. Kanki-Horimoto, S., *et al.* Implantation of mesenchymal stem cells overexpressing endothelial nitric oxide synthase improves right ventricular impairments caused by pulmonary hypertension. *Circulation* **114**, I181–I185 (2006).
51. Wang, X.X., *et al.* Transplantation of autologous endothelial progenitor cells may be beneficial in patients with idiopathic pulmonary arterial hypertension: a pilot randomized controlled trial. *J Am Coll Cardiol* **49**, 1566–1571 (2007).

Chapter 6

Lung Stem Cell Bioprocessing: Fundamentals of Producing Lung Cells

Siti Norhayati Ismail, Julia M. Polak
and Anastasios Mantalaris

Progress on lung stem cell bioprocessing and tissue engineering has been slow due to the lung's complex architecture and difficulties in differentiating stem cells toward the various pneumocytic lineages, in particular type II pneumocytes. Recent advances on the use of stem cells appear promising in providing a reliable supply of clinically relevant cell numbers of lung cells for potential therapeutic applications. The development of bioprocess technologies for the successful transfer of the current laboratory-based practice of stem cell and tissue culture to the clinic as therapeutics necessitates the application of engineering principles and practices to achieve control, reproducibility, automation, validation, safety of the process and the product to be suitable for clinical applications. Furthermore, the utilisation of bioreactors addresses issues of mass transport and provides a three-dimensional growth environment. A variety of cell sources are being investigated for their suitability in generating lung epithelial cells. This review will discuss the complexity of the lung structure, evaluate the various cell sources, and present an up-to-date assessment of the bioreactor technology available that can contribute to the successful differentiation of stem cells into airway epithelial cells.

Keywords: Animal models; bioreactors; 3D cell cultures; human early trials; lung cell expansion; lung disorders; metabolites; stem cell bioprocessing; translational research.

Outline

1. The Clinical Need

Respiratory diseases account for more than 845,000 emergency hospital admissions each year, behind only injury and poisoning. Of the 580,000 deaths each year in the UK, one in five is due to respiratory diseases, with 35,000 deaths being attributed to lung cancer, followed by pneumonia and chronic obstructive pulmonary disease (COPD).[1] Many chronic lung diseases, such as COPD, remain without cure and are only treatable with lung transplantation. The demand for organ transplants is high, yet the shortage of donor organs severely limits this clinical approach. Consequently, research in respiratory medicine involves the interdisciplinary fields of tissue engineering and stem cell therapy, which have emerged as promising and potential new alternatives for the treatment of lung diseases.

2. Lung Architecture

The lung, which develops from the laryngo-tracheal groove, comprises of at least two types of progenitor cells (see also Chapter 4).[2] The architecture of the lung is complex (see Chapters 2 and 3), with a spatial cellular arrangement from the proximal cartilaginous airway (trachea and bronchi) to the terminal gas exchange unit (alveoli), containing unique types of cell lineages with different structure and function, making them a challenge in regenerative medicine. Lung development capable of gas exchange requires that the epithelium undergoes cell proliferation, branching morphogenesis (differentiating into at least 40 distinct types of lung cell lineages), and alveolar saccule formation, thus generating sufficient surface area.[3] The proximal parts of the lung, the cartilaginous airways, are lined by pseudostratified epithelium and the upper airways with ciliated columnar cells and mucus-secreting cells (goblet cells). The lower airways are lined with cuboidal non-ciliated cell type, known as Clara cells, which are important for detoxifying inhaled pollutants and secrete Clara cell secretory protein.[4] Lastly, the alveoli are lined with type I and II epithelial cells. This chapter will focus on the robust production of alveolar epithelial cells, specifically type I and type II pneumocytes, which are involved in gas exchange and surfactant secretion.

The alveoli are lined by a simple squamous epithelium comprising two morphologically and functionally different types of cells: type I and type II pneumocytes, which play an essential role in pulmonary physiology (Fig. 1). Being in contact with the external environment, they are

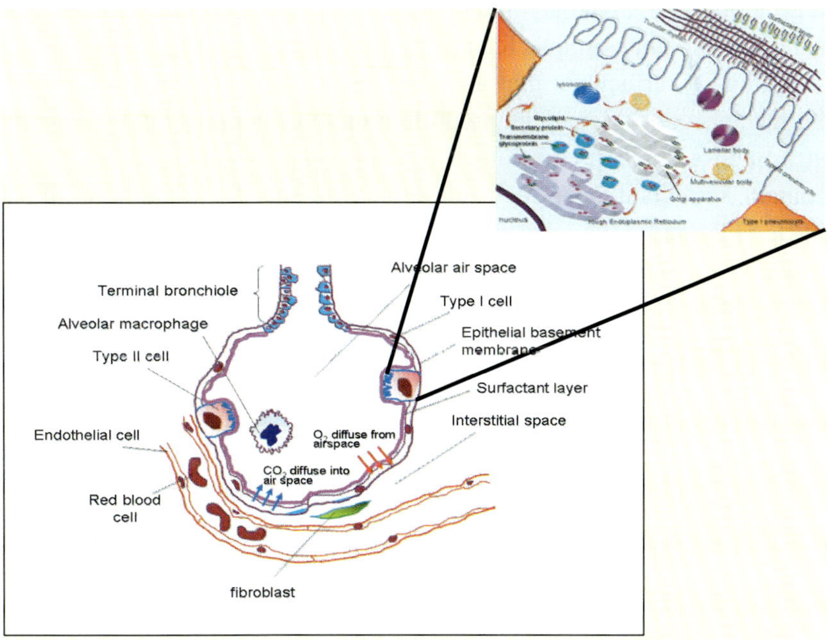

Figure 1: Schematic of the alveolar structure and type II pneumocytes (inset).

frequently injured and a continuous renewal of the alveolar epithelium is necessary to maintain cellular homeostasis.[5] Unlike other epithelial surfaces (*i.e.* skin and gastrointestinal tract), airway and alveolar epithelia have a very slow cell turnover and minimal regenerative capacity.[6] Although the lung has typically been thought of as a complex organ with limited regenerative capacity, there is substantial evidence that reparative and compensatory growth can nonetheless occur.[7] The alveoli architecture has been optimised to maximise gas exchange of oxygen and carbon dioxide. This is achieved through the alveolar structures that require the spatial arrangement of the different pneumocyte cells, in particular type I and type II cells.

Type I pneumocytes are large elongated cells covering approximately 95% of the blood–gas exchange area. In contrast, type II pneumocytes are cuboidal cells, rich in organelles with prominent nucleolus and vacuolated cytoplasm that play an important function in homeostasis of the alveolar unit.[5] They are considered to be the progenitor cells of type I pneumocytes and are, therefore, responsible for the regeneration of alveolar epithelium following alveolar epithelial cell damage.[8] Another unique feature of type II cells is the synthesis and secretion of surfactants that are essential to prevent alveolar collapse. Surfactants consist of a high proportion of phospholipids (approximately 90%) and a small proportion of proteins (approximately 10%), which contain surfactant apoprotein, described as specific protein A (SPA), SPB, SPC, and SPD.[9] Surfactant phospholipids are synthesised in the endoplasmic reticulum, stored in lamellar inclusion bodies, the secretary organelle characteristic of the type II cell.[10–12] Type II cells are not only responsible for the production and secretion of surfactant proteins, but also for the clearance of surfactant from the alveolar surface and its reutilisation for re-secretion.[13] Demonstrating that the differentiated cells synthesise and secrete surfactant proteins provides a way forward for the generation of normal lung epithelial cells *in vitro* that mimic the *in vivo* microenvironment. Type II cells also act as progenitor cells for the generation and proliferation of new specialised adult cells and may form part of the local response to inflammation.

Although much is known about the function of type II cells in terms of synthesis of surfactant lipids and proteins, proliferative capacity, and other biologic properties, the general functions of type I cells are relatively unexplored because there exist no suitable marker molecules that could be used for their definitive identification. Useful markers must discriminate between type I and type II cells; this has been particularly problematic because type II cells are known to be progenitors of type I cells after cell injury.[14]

3. Cell Expansion and Sources

Current stem cell research is aimed at repairing damaged or diseased tissue through the manipulation of endogenous or exogenous stem cells.[15] However, different cell types, from stem cells to adult cells, exhibit different proliferation and differentiation capabilities, as well as having different bioprocessing requirements.[16] This chapter deals primarily with exogenous stem cells (for other cell types, see Chapters 4 and 5).

Several studies demonstrate that different regions within the respiratory system contain different stem cells and have different repair mechanisms.[17] To date, the cells reported to be enriched for stem/progenitor cell activity include basal cells, secretory (Clara) cells, and cells residing in submucosal glands.[18–21] Specifically, a lung-specific side population (SP) has been discovered within the pulmonary tissue. SP cells are identified by their ability to efflux Hoechst dye via ATP binding cassette half-transporter (Bcrp-1/ABCG2), and they have been shown to exhibit properties similar to those of haematopoietic stem cells.[22] Furthermore, putative endogenous epithelial progenitor cells have been located within the adult lung in the basal layer of the upper airways, within or near pulmonary neuroendocrine cell, as well as at the bronchoalveolar junction (see also Chapters 4 and 5).[15,23–28] However, information regarding their self-renewal capacity, clonality and lineage formation is still limited.

The utilisation of exogenous progenitor/stem cells has recently provided hope that they could be used to enhance lung repair or regeneration,[29,30] particularly in patients suffering from chronic lung diseases. Specifically, stem cells have been demonstrated to repair lung injury;[31–34] however, regeneration of lung tissue remains more challenging. Exogenous stem/progenitor cells, as opposed to primary cells, retain their

1. Lung Regeneration

Two animal models, alveolar regeneration in lung emphysema and compensatory lung growth after pneumonectomy, are well established and characterised models for lung regeneration research.

1.1. *Lung emphysema model*

Chronic obstructive pulmonary disease (COPD) is becoming a major cause of death worldwide.[1] Pulmonary emphysema, one of the main features in COPD, is characterised by the progressive and irreversible loss of pulmonary alveoli. Despite recent advances in new drugs and the understanding of this disease, its treatment remains palliative and no therapy can reverse the destroyed alveoli. Therefore, COPD is the most important target in lung regeneration research.

Several animal models have been developed to study COPD;[2–4] the elastase model; cigarette smoke exposure model;[5,6] calorie restriction model;[7,8] vascular apoptosis model;[9] genetic model;[10] and others (Table 1). The elastase-induced emphysema model is the most frequently used to study lung regeneration (Fig. 1). Intratracheal administration of elastase induces acute inflammation, neutrophil accumulation, and an increase in permeability. Three weeks after elastase instillation, the acute inflammation is over, and the remains are the destroyed lung parenchyma. Regenerative studies are performed after the development of the emphysematous change.

1.1.1. *Reagents for lung regeneration*

Retinoic acid (RA)

(See also Chapter 13)

RA is known to be involved in lung development, especially alveologenesis.[11] RA regulates embryonic branching morphogenesis[12] and genes involved in lung development, and promotes alveolar septation. Deletion of RA receptors in mice demonstrates the failure of alveologenesis, the formation of normal alveoli and alveolar elastic fibres.[13] RA synthesizing enzymes and endogenous RA exist in post-natal lungs.[14] Elastin synthesis in lung fibroblasts is increased by RA treatment.[15] RA-receptor and

Table 1: Lung regeneration in pulmonary emphysema models.

Treatment	Route	Animal (Strain)	Model	Summary
All-trans retinoic acid		Rat (Sprague-Dawley)	Elastase	14 days of RA treatment restored Lm.[18,26,27]
	i.p.	Mouse (tight-skin/ C57BL/6)	Genetic	12 days of RA treatment increased the number and reduced the size of alveoli.[19]
	i.p.	Guinea pig	Cigarette smoke	RA treatment failed to reverse cigarette smoking-induced emphysematous changes.[29]
	i.p.	Rabbit (NZW)	Elastase	14 days of RA treatment had no effect on lung compliance and surface area.[30]
	i.p.	Mouse (FVB)	Elastase	14 days of RA treatment did not restore Lm, or elastin mRNA expression.[31]
	i.n.	Rat (Fischer 344)	Elastase	3 weeks of RA inhalation did not improve Lm and alveolar surface area.[23]
	i.p.	Mouse (C57BL/6)	Elastase	14 days of RA treatment restored Lm. Bone marrow-derived cells contributed in the newly regenerated alveoli.[21]
	i.p.	Mouse (N/A)	Dexamethasone	RA treatment restored Lm.[167]
	i.p.	Mouse (TO)	Dexamethasone	10 days of RA treatment restored Lm and alveolar surface area.[22]

(Continued)

Table 1: *(Continued)*

Treatment	Route	Animal (Strain)	Model	Summary
	i.n. or i.p.	Mice (B6C3F1 or A/J)	Cigarette smoke	3 weeks of RA treatment did not improve Lm and alveolar surface area.[33]
	i.p.	Mouse (TO, NIHS, or ICR)	Dexamethasone	10 days of RA injection restored Lm and surface area of the lungs.[22,25]
G-CSF	i.p.	Mouse (C57BL/6)	Elastase	14 days of murine G-CSF treatment restored Lm. Circulating endothelial progenitor cells increased.[21]
M-CSF	i.p.	Mouse (C57BL/6)	Elastase	3 weeks of human M-CSF injection aggravated Lm.[59]
HGF	i.p.	Mouse (C57BL/6)	Elastase	14 days of RA treatment increased circulating endothelial progenitor cells, and restored Lm.[48]
	i.n.	Mouse (C57BL/6)	Elastase	Intranasal inhalation of HGF twice a week for 2 weeks improved Lm and lung static compliance. Lung endogenous stem cells were involved.[51]
	i.v. (gene transfer using HVJ)	Rat (Sprague-Dawley)	Elastase	Intravenous injection of HGF expression vector with HVJ-envelope increased pulmonary vasculature, and improved lung function.[49]

(Continued)

Table 1: *(Continued)*

Treatment	Route	Animal (Strain)	Model	Summary
	i.v. (HGF-secreting adipose tissue-derived stromal cells)	Rat (Sprague-Dawley)	Elastase	Intravenous injection of cultured ASCs improved Lm.[50]
FGF-2 microsphere	i.t.	Dog (Beagle)	Elastase	4 weeks after the intrabronchial administration of FGF-2 microsphere, enhanced Lm.[168]
Adrenomedullin	s.c.	Mouse (C57BL/6)	Elastase	14 days of continuous infusion improved Lm and lung static compliance.[68]
Simvastatin	i.p.	Mouse (C57BL/6)	Elastase	12 days of simvastatin treatment restored Lm.[72]

RA: Retinoic acid; Lm: mean linear intercept; G-CSF: granulocyte colony-stimulating factor; M-CSF: macrophage colony-stimulating factor; HGF: hepatocyte growth factor; HVJ: haemagglutinating virus of Japan; ASC: adipose tissue-derived stromal cell; FGF: fibroblast growth factor.

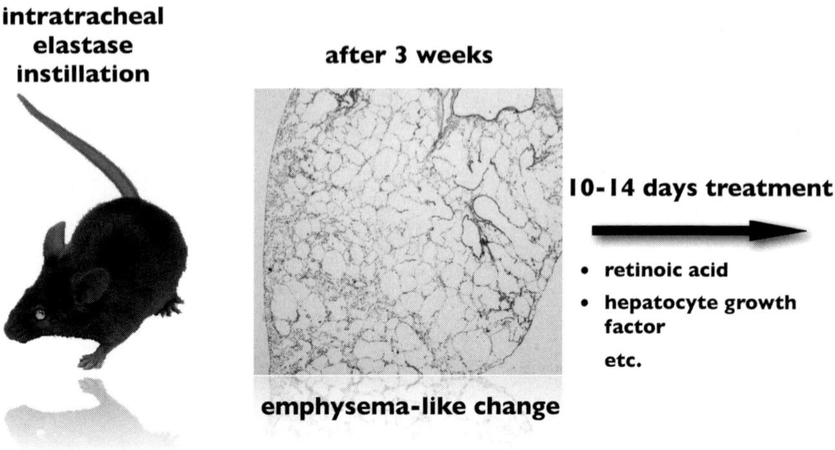

Figure 1: Alveolar regeneration in lung emphysema model.

RA-binding protein are also present in the developing lungs.[16,17] These suggest the importance of RA in developing lung morphology.

In 1997, Massaro *et al.* showed that all-trans-retinoic-acid (ATRA) reversed the anatomic and functional signs in a rat pulmonary emphysema model.[18] Since then, many studies were performed in this area.[19–25] Fourteen studies using RA in emphysema models have been reported; interestingly, eight of these successfully observed lung regeneration after RA treatment,[18,19,21,22,26–28] and the remaining six reported the failure of RA-induced lung regenerative capacity.[23,29–33] The reasons for this discrepancy might be 1) the difference in animal species, and 2) the difference in RA dose threshold. As described in the compensatory lung growth section, small animals, such as rodents, have a better capacity for lung regeneration because its somatic growth continues throughout the lifespan. This may affect the results of RA treatment. Another factor is the required RA dose for lung regeneration. Stinchcombe *et al.* evaluated the effect of RA on three different strains of mice, TO, ICR, and NIHS, and found that the RA dose threshold for inducing alveolar regeneration differed for each strain.[25]

Based on these animal studies, two clinical studies using RA in patients with pulmonary emphysema were performed.[34,35] Although oral administration of RA modulated protease/antiprotease balance, matrix metalloproteinase-9/tissue inhibitor of metalloproteinase-1, in COPD patients,[36] no statistical change was observed in lung function or CT density.[34,35] Since most COPD patients are elderly and their lungs are mature, they may have a low number of stem cells, and only low capacity for regeneration.

Hepatocyte growth factor (HGF)

HGF was originally isolated as a potent mitogen for mature hepatocytes in primary cultures.[37,38] HGF is a pleiotropic growth factor that has been shown to have mitogenic, morphogenic, and protective effects via tyrosine kinase phosphorylation of its receptor, c-Met, after pulmonary injuries[39–41] or during lung development.[42] In particular, it is a potent mitogen for alveolar type II epithelial cells *in vivo*[43] and *in vitro*.[44] In addition, HGF also activates migration and proliferation of endothelial cells and induces angiogenesis. The angiogenic activity of HGF is mediated through its direct actions on endothelial cells,[45,46] and its indirect actions that occur through an increase of endothelial cell mitogens.[46,47]

Because of the above potential, the effect of HGF in lung regeneration has been extensively studied. Intraperitoneal administration of HGF significantly increases the Sca-1^+/Flk-1^+ fraction in peripheral mononuclear cells in mice, and induces proliferation of both bone marrow-derived and resident endothelial cells in the alveolar wall, resulting in a reversal of elastase-induced pulmonary emphysema in mice.[48] Transfection of cDNA encoding human HGF demonstrated an efficient expression of HGF in alveolar endothelial and epithelial cells, and resulted in a more extensive pulmonary vasculature and inhibition of alveolar wall cell apoptosis in a rat emphysema model.[49] Intravenous injection of adipose tissue-derived stromal cells, which secrete HGF, improved the emphysematous condition in rats.[50] Hegab *et al.* reported that twice-weekly inhalation of HGF for two weeks significantly ameliorated elastase-induced enlargement of airspaces and alveolar wall destruction (Fig. 2), and the elevated static lung compliance returned to normal levels.[51]

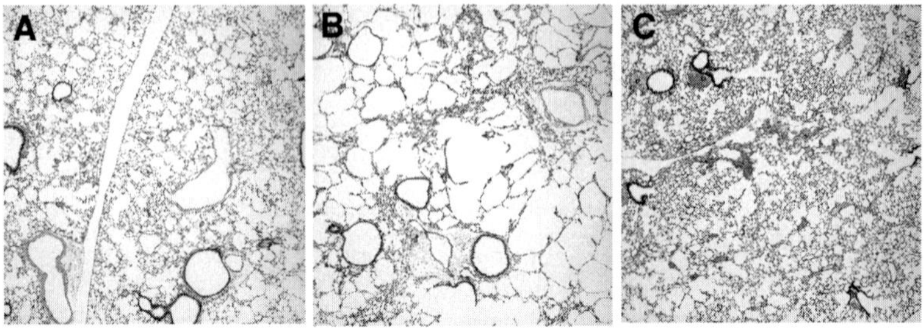

elastase	-	+	+
HGF	-	-	+

Figure 2: (A) Normal lung. (B) Elastase administration induced emphysematous changes in murine lungs. (C) HGF treatment markedly improved the emphysema. Magnification x40.

Granulocyte-colony stimulating factor (G-CSF)

G-CSF has been reported to enhance tissue regeneration and improve survival after myocardial infarction by mobilizing stem cells from bone marrow into peripheral blood.[52–54] G-CSF also induces angiogenesis in post-ischemic tissues, such as brain,[55] and hindlimb.[56] In a murine emphysema model, G-CSF treatment alone provides a significant reduction in emphysema. In G-CSF-treated mice, the alveolar mean linear intercept (Lm), used as a morphometric parameter of emphysema, showed a reduction, when compared to vehicle-treated mice. This was the same degree of reduction as observed in RA-treated mice. G-CSF increases circulating endothelial progenitor cells from the bone marrow. A combined treatment with G-CSF and RA exhibits an additive effect with increased reduction in Lm.[21] Bone marrow-derived cells were shown to partially contribute to G-CSF-induced lung regeneration. These results suggest that the lack of circulating stem cells could be a limiting factor in elderly COPD patients.

Macrophage colony-stimulating factor (M-CSF)

M-CSF is a multifunctional proinflammatory cytokine that regulates the differentiation, proliferation, and survival of monocytic progenitor cells.[57] Some reports suggest that M-CSF also induces mobilisation of endothelial progenitor cells from the bone marrow, and enhances neovascularisation.[58]

Intraperitoneal injection of M-CSF was carried out in an elastase-induced murine emphysema model, but it increased Lm and decreased alveolar surface area.[59] Alveolar macrophage expressing MMP-9 and MMP-12 increased in the lungs, suggesting M-CSF administration enhanced inflammation induced by elastase, resulting in a worsening of the emphysematous changes.

Keratinocyte growth factor (KGF)

The KGF receptor is expressed in alveolar type II epithelial cells. KGF is known to promote survival, proliferation, and migration of type II cells.[60] Intratracheal instillation of KGF induces alveolar type II hyperplasia.[61] Although pre-treatment with KGF prevented elastase-induced emphysema, post-treatment of KGF (three weeks after the elastase instillation) did not reverse alveolar enlargement.[62] These results suggest that KGF has only an anti-inflammatory effect and does not induce alveolar repair.

Adrenomedullin

Adrenomedullin is a multifunctional regulatory peptide originally isolated from a human pheochromocytoma,[63] and is reported to induce cyclic AMP production, bronchodilation, cell growth regulation, survival from apoptosis, angiogenesis and to have antimicrobial activity.[64–66] The adrenomedullin receptor is strongly expressed in the basal cells of the airway epithelium and alveolar type II epithelial cells, both of which are involved in epithelial regeneration of the lung.[67] Continuous infusion of adrenomedullin by a subcutaneous osmotic pump increases Sca-1^+ cells in peripheral blood, and regenerates alveoli and vasculature in an elastase-induced emphysema model in mice.[68]

Simvastatin

In addition to a cholesterol lowering effect, 3-hydroxy-3-methylglutaryl coenzyme A reductase inhibitors (statins) are know to have various pharmacological effects, such as an anti-inflammatory effect, improving endothelial function, among others. Statins also seem to have a beneficial effect in tissue regeneration.[69–71] Intraperitoneal injection of simvastatin was performed in an elastase-induced murine emphysema model, and a reduction of the mean linear intercept and an increase in proliferating (PCNA+) cells was observed.[72]

1.1.2. *Source of stem cells*

Stem or differentiated cells are needed for alveolar regeneration. Exogenous and endogenous progenitor cells are thought to participate in alveologenesis.

Exogenous progenitor cells

The main source of exogenous progenitor cells is the bone marrow. Rojas *et al.* have shown that injured lung cells induce proliferation and migration of bone marrow-derived stem cells.[73] Contribution of bone marrow-derived cells in alveolar regeneration after elastase-induced emphysema has been demonstrated in several reports described above. Instillation of G-CSF,[21] HGF,[48,51] and adrenomedullin[68] induces an increase in bone marrow-derived endothelial cells within the lung capillary walls during alveologenesis in a murine emphysema model. After treatment, the number of bone marrow-derived cells seen in the alveolar walls gradually decreases. It is still an open question whether bone marrow cells differentiate into alveolar cells or simply fuse with the resident stem cells.

Because most of the reagents employed in lung regeneration studies increase circulating bone marrow-derived cells, there is a possible important role for these cells. In regeneration, however, exogenous infusion of bone marrow cells had no effect on alveolar regeneration (Fig. 3). This suggests that additional mediators, such as chemokines, adhesion molecules, growth stimulators, and extracellular matrix remodelling, among others, are needed for the bone marrow-derived cells to migrate, differentiate, and induce lung regeneration.

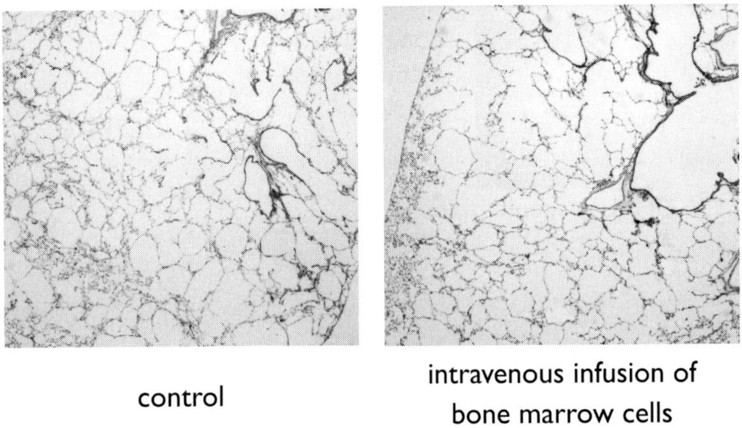

Figure 3: Administration of bone marrow-derived cells did not ameliorate emphysematous changes in mice. Bone marrow-derived cells were intravenously instilled in elastase-induced emphysema model. Representative pictures after three weeks of bone marrow cells instillation.

Endogenous progenitor cells

Based on experiments using green fluorescent protein (GFP)-chimera mice (Fig. 4), the regenerated alveoli were composed of bone marrow-derived (GFP-positive) cells and cells of non-bone marrow origin (GFP-negative).[21,48,51,68] This suggests that lung resident cells, including endogenous stem cells, contribute to alveologenesis. It is known that alveolar type II epithelial cells can repair damaged alveolar epithelium.[74] Kim *et al.* reported a new resident stem cell population, bronchioalveolar stem cells (BASCs), which differentiate into Clara cells and alveolar type II cells in adult mouse lung after injury.[75] Hegab *et al.* reported that elastase-induced lung injury alone increased cells with stem cell markers, like Sca-1, CD34 and c-Kit, within lungs before the HGF administration.[51] The numbers of Sca-1^+/SP-C^+ cells were markedly increased in response to either HGF or elastase, with a maximal increase in the elastase and HGF groups. Most were Sca-1^+ cells lung endogenous stem cells, while most of the c-Kit^+ cells were of bone marrow origin. This may suggest an important role of HGF in differentiation and regeneration, leading to lung healing.

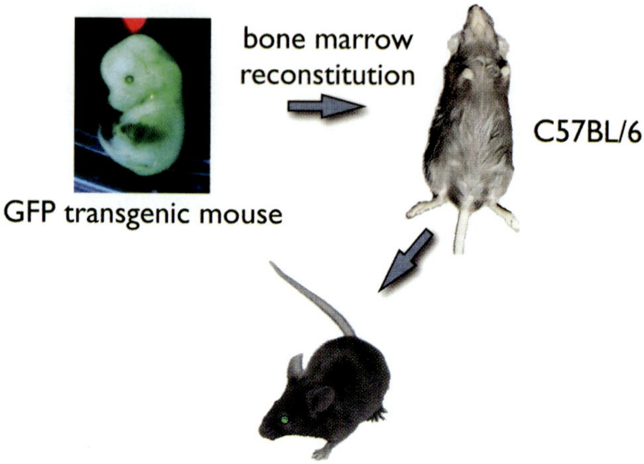

Figure 4: GFP-chimera mice. Foetal liver cells from day-13.5 GFP embryos or bone marrow cells obtained from GFP-transgenic mice were intravenously injected to recipient mice. Recipient mice were irradiated using doses of 8 Gy and 4 Gy, separated by three hours. At three weeks after transplantation, 89.5 ± 2.0% of peripheral blood mononuclear cells, 92.5 ± 2.0% of splenic lymphocytes and 85.6 ± 2.6% of bone marrow cells from recipient mice were positive for GFP.

1.2. *Compensatory lung growth*

Compensatory lung growth has been observed after pneumonectomy in children[76–78] and in experimental animal models.[77,79–89] Pneumonectomy is widely used to evaluate post-natal lung growth, because the remaining lung can be kept intact and the volume loss is easily controlled. In the pneumonectomised animal and even in aged humans, adequate gas exchange is maintained after lung resection. This physiological compensation is provided by an increase in diffusion capacity and blood flow within the remaining lung with/without compensatory lung growth.

Mechanical stresses and a variety of growth factors and hormones are thought to be the key inducers to initiate post-pneumonectomy lung growth.

1.2.1. *Mechanical stresses*

It is well known that mechanical stress changes cell function. Stretch stimulation on lung cells induces cell proliferation,[90,91] growth factor production,[92,93] and changes in gene expression. In contrast, decreased mechanical stress attenuates lung development, causing lung hypoplasia.[94] Positive airway pressure induces cell proliferation and extra-cellular matrix remodelling.[95,96] Taken together, these results indicate that mechanical stress plays an important role in lung development and growth.

Pneumonectomy induces many anatomical changes in the remaining lung and the thoracic cavity. The shift of the mediastinum toward the vacated thoracic compartment raises mechanical stress to the remaining lung, and these stresses induce gene expression such as early growth response gene-1[97,98] and cell proliferation. The remaining lung inflates more in the increased space of the thoracic cavity. This stretch signal enhances cAMP expression that plays a role in the early phase of the lung growth response.[99,100]

In addition, loss of the vascular bed by pneumonectomy increases pulmonary perfusion per unit of remaining lung tissue.[101] Chronic capillary distension and increased shear stress induce endothelial cell growth and septal remodelling. This increased blood flow and shear stress also enhance alveolar growth.[102]

1.2.2. *Growth factors and hormones for compensatory lung growth*

Retinoic acid (RA)

As in the lung emphysema model, the role of RA was extensively examined in compensatory lung growth. RA treatment enhances compensatory lung growth[20] and alveolar capillary formation after pneumonectomy.[24] However, RA does not improve lung function in mature pneumonectomised dogs.[103] Hsia *et al.* demonstrated that 55% resection of the lungs by right pneumonectomy induces compensatory growth and increased lung function.[24,84] RA enhances this compensatory lung growth in dogs. 45% resection by left pneumonectomy, however, did not induce lung growth.[104,105] These suggest that RA is a promoter of existing alveolar growth but not an initiator of alveolar growth.[105]

HGF

HGF is well known as a mediator of compensatory growth in liver.[37] HGF activates migration and proliferation of endothelial cells[106] and alveolar epithelial cells.[39,48,107,108] After pneumonectomy, HGF increases in the remaining lung, and also in liver and kidney.[41] c-Met, the HGF receptor, is transiently upregulated in alveolar type II cells. Neutralisation of endogenous HGF suppresses the compensatory DNA synthesis in lung epithelial cells, suggesting that HGF has a role in post-pneumonectomy compensatory lung growth.

Epidermal growth factor (EGF)

EGF is known to play an important role in prenatal and post-natal lung development. Alveolar type II epithelial cells express the EGF receptors, and EGF induces epithelial maturation and regeneration. In EGF receptor-deficient mice, lungs are immature and show impaired branching and deficient alveolisation.[109] Kaza *et al.* reported that EGF administration in pneumonectomised rats induces significant increases in lung volume and weight in the remaining lung.[110]

Growth hormone

The effects of growth hormone on lung volume has been reported.[111–113] Treatment with human growth hormone in children or excessive hormone in acromegalic adults induces an increase in lung volume. This increase is correlated with the standing height, suggesting that the effect of growth hormone is systemic growth,[111] including the enlargement of the thoracic cavity, rather than a direct effect on lung growth. Growth hormone did not induce lung growth in rats.[114]

Other factors

Alveolar hypoxia induces the expression of many genes, such as vascular endothelial growth factor (VEGF) and hypoxia-induced mitogenic factor (HIMF). Along with the neo-alveolisation, proper vascular growth should occur to avoid a V/Q mismatch. During lung development, VEGF is known to be deposited in the subepithelial matrix at the leading edges of branching airways, suggesting its stimulus effect of angiogenesis.[115] After pneumonectomy, alveolar type II cells express VEGF protein,[116] and

VEGF and its receptor, flk-1, increase within the lungs.[117] Sakurai *et al.* demonstrated that administration of VEGF accelerated angiogenesis and lung growth after pneumonectomy.[87] HIMF, another protein induced under hypoxic condition, upregulated in the lung during the early hyperplastic period of the compensatory lung growth. In addition, intratracheal instillation of recombinant HIMF protein induced epithelial, endothelial, and muscular proliferation in the lungs.[118]

These data suggest that alveolar hypoxia and related proteins also play a role in compensatory lung growth.[118,119]

1.2.3. *Animal models of compensatory lung growth*

The degree of compensatory lung growth differs among species and varies with age. Small animals, such as rodents, have a better capacity for compensatory growth than larger animals. Post-pneumonectomy compensation usually occurs very rapidly and completely in rodents. For example, the weight of the remaining lung doubles within 14 days after the pneumonectomy in rats[99] while 28 days are needed in rabbits,[120] and five months in dogs.[84] Age is another factor for the capacity for lung growth. In the adult dog lung, compensatory lung growth is slow and incomplete, but extensive lung resection in an immature dog stimulates a rapid and vigorous compensatory growth resulting in complete normalisation of lung function at maturity.[85] This suggests that post-pneumonectomy lung growth is maturity-dependent. Since somatic growth in rodents continues throughout their lifespan, lung can grow throughout the experiments.

It is believed that compensatory lung growth occurs only in alveoli but not in respiratory bronchioles. However, Hsia *et al.* reported that the number of respiratory bronchiole segments and branch points increases in immature lungs.[121]

1.2.4. *Cell sources for lung growth*

It is reported that circulating progenitor cells do not contribute to compensatory lung growth.[122] Recently, Nolen-Walston *et al.* evaluated the response of lung endogenous stem cells, BASCs and alveolar type II cells during compensatory lung growth after pneumonectomy in mice.[123]

They found that the number of BASCs and alveolar type II cells increased during compensatory growth, and peaked at 220% and 124% of the baseline, respectively. However, the contribution of BASCs in compensatory lung growth was 0–25%, while alveolar type II cells are necessary for regrowth, based on a cell kinetic model.

Kenzaki *et al.* implanted foetal lung tissue fragments into adult rat lungs. The implanted lung tissue was connected to the pulmonary circulation, and its alveolar spaces were opened. These changes were enhanced when the recipient lungs were partially resected. However, lung fragments obtained from adult rats did not expand after implantation.[124] These suggest that mechanical forces and premature lung cells and/or growth factors produced from premature cells are key elements for lung regrowth.

Recently, a curious case of living-donor lobar lung transplantation in a child was reported. A ten-year-old boy with idiopathic pulmonary arterial hypertension received a right single lobe transplantation from his mother (38-year-old).[125] Toyooka *et al.* followed the child's pulmonary function for six years after transplantation.[126] It is generally believed that the increase in lung volume after lung transplantation in immature recipients is the result of alveolar dilation (lung dilation) rather than alveolisation (lung growth), because the diffusing capacity of the lung for carbon monoxide adjusted for lung volume (DLco/VA) decreased after the transplantation.[127] During the follow-up for this ten-year-old boy, the patient grew 27 cm in height. At the same time, his vital capacity (VC) and forced expiratory volume in one second (FEV1.0) had nearly doubled, and the ratios of perfusion and ventilation to the transplanted lung were stable, suggesting the transplanted mature lung can grow in the growing recipient. Similar adult lung growth was reported in an animal model.[128] These reports suggest that the mature and growth-stopped lung can reinitiate quiescent pathways of growth in certain circumstances.

2. Repair Process after Acute Lung Injury

Acute lung injury/acute respiratory distress syndrome is characterised by severe alveolar damage, including apoptosis and necrosis of alveolar epithelial and endothelial cells. Newly differentiating and proliferating cells are required for lung repair and replacement of damaged cells.

Alveolar type II cells can proliferate and are thought to be the source of type I cells. Even if the type II cells are damaged, two candidate sources of newly differentiating cells can undertake the repair process: circulating bone marrow-derived progenitor cells and lung endogenous stem cells.

2.1. *Engraftment of bone marrow-derived cells*

Bone marrow-derived progenitor cells (BMDCs) have the capacity to differentiate into several phenotypes, including endothelial cells,[129] epithelial cells,[130–132] myocytes,[53,133] and neurons.[134,135] In patients with bacterial pneumonia[136] or acute lung injury,[137] the number of circulating endothelial progenitor cells (EPCs) increases, and this correlates to the outcome of the diseases. This suggests that BMDCs are released to the circulation by inflammatory stimuli, and facilitate resolution and repair of the inflammatory process.

Many studies have been performed to evaluate the role of BMDCs in lung injury models.[73,131,138–140] Several techniques have been utilised to track bone marrow cells: bone marrow reconstitution with GFP or LacZ-transgenic cells; bone marrow reconstitution with sex-mismatched donor cells; parabiotic mice. GFP-transgenic mice were surgically joined with wild type, or direct instillation of labelled progenitor cells into the lungs.

Intratracheal instillation of inflammatory stimuli, such as LPS, is known to induce a rapid and massive release of inflammatory cells from the bone marrow.[141] At the same time, the number of cells expressing Sca-1 and Flk-1, marker of bone marrow-derived circulating EPCs, also increases in peripheral blood, suggesting that circulating progenitor cells increase after LPS administration.[138] These data indicate that the signals produced by inflammatory stimuli promote the release of EPCs, and presumably all BMDCs. Therefore, LPS induces a rapid release of not only inflammatory cells but also progenitor cells from bone marrow to the circulation.

LPS administration induces severe damage to lung parenchymal cells, resulting in apoptosis and necrosis.[142] Because these dead cells are not able to divide, other cells must replace them to repair the tissue and keep organ homeostasis. Seven days after LPS-administration in GFP chimera mice, thin and flat shaped GFP-positive BMDCs appear in the alveolar walls. These cells stain positive for cytokeratin as a marker of epithelial cells or CD34 as a marker of endothelial cells.[138] This suggests that BMDCs differentiate or fuse with alveolar epithelial or pulmonary capillary endothelial cells, suggesting a possible involvement of BMDCs in lung repair. However, the frequency of BMDCs gradually and significantly decreases over a prolonged period of time (Fig. 5).[143–145] These results suggest that BMDCs initially migrate to the injured organ, and differentiate into or fuse with the parenchymal cells of the organ. However, once BMDCs reach the damaged organ, they have little or no capacity to proliferate or to develop into new cells.

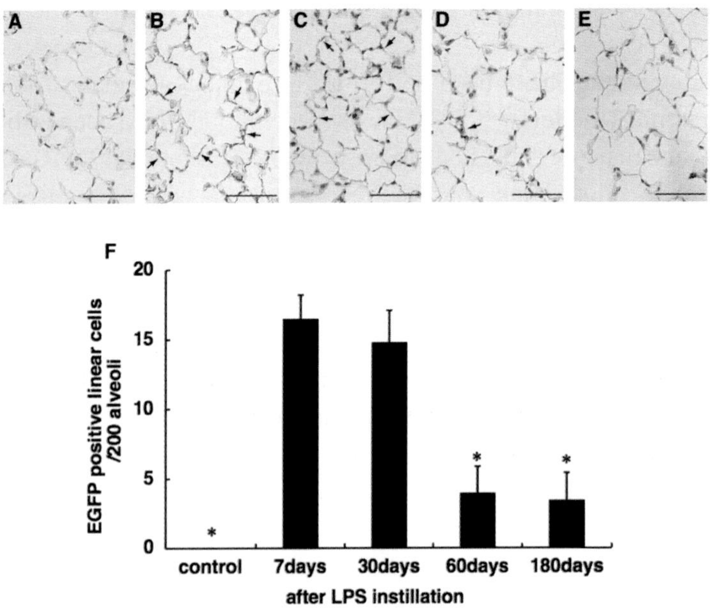

Figure 5: Presence of thin, flat bone marrow-derived cells in the alveolar walls as a function of time after LPS instillation. (A–E) Representative mouse lung sections immunostained for GFP at 0 (A), 7 (B), 30 (C), 60 (D), or 180 (E) days after intranasal LPS administration. Arrowhead indicates thin, flat GFP-positive cells. Scale bars, 50 μm. (F) The number of thin, flat GFP-positive cells in the alveolar walls of mouse lungs at the indicated time points after intranasal LPS administration. *Significantly smaller than day 7 group ($p < 0.05$).

These BMDCs appear in the lung parenchyma only when there is tissue damage.[138,146,147] This may suggest that chemokines and proper adhesion molecules expression are needed in damaged tissue. Hashimoto *et al.* reported that SDF-1/CXCR4 axis contributes to the migration of the bone marrow-derived fibrocytes to bleomycin-injured lungs.[148] It is unclear whether the same mechanisms occur in bone marrow-derived endothelial and epithelial cells.

2.2. *Possibility of immunomodulation or other beneficial effect rather than "stemness" by bone marrow cells*

Recently, many studies demonstrated the beneficial effect of administering BMDCs, EPCs, and mesenchymal stem cells (MSCs) into the injured lungs either intravenously or intratracheally. Intravenous[138] or intratracheal[149] administration of bone marrow cells or bone marrow-derived MSCs[150] ameliorated the LPS-induced lung injury in mice, while bleomycin-induced inflammation, collagen deposition, and fibrosis were reduced by intratracheal or intravenous infusion of MSCs.[73,140] Intratracheal administration of alveolar type II cells also reduced the severity of bleomycin-induced pulmonary fibrosis in rats.[151]

The percentage of engrafted BMDCs in lung parenchyma is very low. This may suggest that BMDCs and MSCs have anti-inflammatory effects rather than the capacity to differentiate into lung cells.[152,153] MSCs are known to produce many cytokines and growth factors.[154] In addition, LPS-treated lung cells co-cultured with MSCs reduced proinflammatory cytokine production,[150] suggesting that the inflammatory responses are reduced by soluble factors produced by MSCs and/or by direct contact with MSCs. MSCs also have an immunomodulatory effect on immune cells, including T cells, B cells, and natural killer cells.[155,156] Interestingly, Spees *et al.* reported that mitochondrial DNA can transfer from MSCs to other cells, and rescue mitochondrial function in the recipient cells.[157]

2.3. *Pulmonary hypertension*

The most promising bench-to-bedside cell therapy to come from the animal model is the clinical trial with autologous EPC administration for idiopathic pulmonary arterial hypertension (IPAH) (see also Chapter 12). EPC has recently been explored as a potential source for neovascularisation and increased pulmonary circulation in patients with pulmonary hypertension. EPC is known to home to the injured organ[138,158,159] and induce the release of several growth factors, such as vascular endothelial growth factor and HGF.[160] An administration of EPCs ameliorated monocrotaline-induced pulmonary arterial hypertension in rats and dogs.[161–166] A pilot study in 31 patients with IPAH found improvement in six-minute walk distance and haemodynamics, including mean pulmonary arterial pressure, pulmonary vascular resistance and cardiac output, 12 weeks after administration of autologous EPCs.[167] A phase II clinical trial in Canada known as PHACeT (Pulmonary Hypertension: Assessment of Cell Therapy) using autologous EPCs with a plasmid carrying the endothelial nitric oxide synthase gene is also underway.[168]

2.4. *The role of lung endogenous stem cells*

As in the lung regeneration and the compensatory lung growth models, endogenous stem cells contribute to lung repair. The administration of

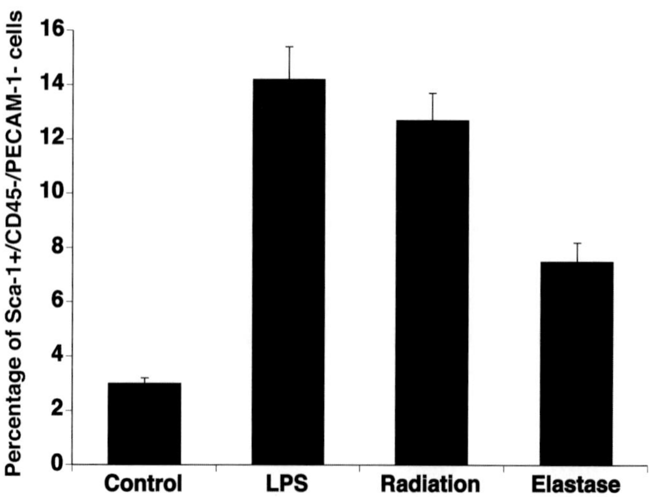

Figure 6: Lung endogenous stem cell populations ($Sca\text{-}1^+/CD45^-/PECAM\text{-}1^-$) increase after lung injury.

inflammatory stimuli, such as LPS and bleomycin, within lungs results an alveolar type I epithelium injury. Alveolar type II cells appear to proliferate and subsequently differentiate to replace the injured type I cell. A portion of the type II cell population is known to become hypertrophic. Both of these events are frequent findings in the diseased or damaged lung.

Recently reported BASCs, which have the potential to differentiate into alveolar type II cells, become hypertrophic during lung injury (Fig. 6),[51,120,169] suggesting an important role of BASCs in alveolar repair after lung injury. Hence, it is not only bone marrow-derived stem cells that migrate to the lung and participate in the healing process: endogenous lung stem cells can also proliferate in response to injury and participate in the repair process.

3. Conclusion and the Way Forward

Our knowledge of lung regeneration and repair has been expanded by the development of suitable animal models. Regenerative medicine has become one of the most promising strategies to cure or treat intractable lung dis eases. However, many challenges remain. Differences in the animal species used and changes in age are closely associated with the responses observed to the regenerative reagents or stimuli. Most of the successful experiments were performed in rodents. The rodents' lungs and thorax continue to grow throughout their entire life. Therefore, extrapolation to mature human lungs is difficult. In addition, in contrast to the murine lung, the knowledge for endogenous stem cells in human lung is more limited.

In spite of these limitations, studies in animal models are essential to understand the pathophysiology of lung regeneration and the development of novel therapeutic strategies for intractable lung diseases. Careful interpretation of animal model results must be taken into consideration. However, these studies provide valuable knowledge for the regenerative therapy of human lungs.

References

1. Pauwels, R.A. and Rabe, K.F. Burden and clinical features of chronic obstructive pulmonary disease (COPD). *The Lancet* **364**, 613–620 (2004).

and morphologic effects of elastase instillation in the rat? *Chest* **117**, 242S–244S (2000).

28. Maden, M. Retinoids have differing efficacies on alveolar regeneration in a dexamethasone-treated mouse. *Am J Respir Cell Mol Biol* **35**, 260–267 (2006).
29. Meshi, B., Vitalis, T.Z., Ionescu, D., Elliott, W.M., Liu, C., Wang, X.D., Hayashi, S., and Hogg, J.C. Emphysematous lung destruction by cigarette smoke. The effects of latent adenoviral infection on the lung inflammatory response. *Am J Respir Cell Mol Biol* **26**, 52–57 (2002).
30. Nishi, Y., Boswell, V., Ansari, T., Piprawala, F., Satchi, S., and Page, C.P. Elastase-induced changes in lung function: relationship to morphometry and effect of drugs. *Pulm Pharmacol Ther* **16**, 221–229 (2003).
31. Lucey, E.C., Goldstein, R.H., Breuer, R., Rexer, B.N., Ong, D.E., and Snider, G.L. Retinoic acid does not affect alveolar septation in adult FVB mice with elastase-induced emphysema. *Respiration* **70**, 200–205 (2003).
32. Fujita, M., Ye, Q., Ouchi, H., Nakashima, N., Hamada, N., Hagimoto, N., Kuwano, K., Mason, R.J., and Nakanishi, Y. Retinoic acid fails to reverse emphysema in adult mouse models. *Thorax* **59**, 224–230 (2004).
33. March, T.H., Bowen, L.E., Finch, G.L., Nikula, K.J., Wayne, B.J., and Hobbs, C.H. Effects of strain and treatment with inhaled all-trans-retinoic acid on cigarette smoke-induced pulmonary emphysema in mice. *COPD* **2**, 289–302 (2005).
34. Mao, J.T., Goldin, J.G., Dermand, J., Ibrahim, G., Brown, M.S., Emerick, A., McNitt-Gray, M.F., Gjertson, D.W., Estrada, F., Tashkin, D.P., and Roth, M.D. A pilot study of all-trans-retinoic acid for the treatment of human emphysema. *Am J Respir Crit Care Med* **165**, 718–723 (2002).
35. Roth, M.D., Connett, J.E., D'Armiento, J.M., Foronjy, R.F., Friedman, P.J., Goldin, J.G., Louis, T.A., Mao, J.T., Muindi, J.R., O'Connor, G.T., Ramsdell, J.W., Ries, A.L., Scharf, S.M., Schluger, N.W., Sciurba, F.C., Skeans, M.A., Walter, R.E., Wendt, C.H., and Wise, R.A. Feasibility of retinoids for the treatment of emphysema study. *Chest* **130**, 1334–1345 (2006).
36. Mao, J.T., Tashkin, D.P., Belloni, P.N., Bailey-Healy, I., Baratelli, F., and Roth, M.D. All-trans retinoic acid modulates the balance of matrix metalloproteinase-9 and tissue inhibitor of metalloproteinase-1 in patients with emphysema. *Chest* **124**, 1724–1732 (2003).
37. Nakamura, T., Nawa, K., and Ichihara, A. Partial purification and characterization of hepatocyte growth factor from serum of hepatectomized rats. *Biochem Biophys Res Commun* **122**, 1450–1459 (1984).

38. Nakamura, T., Nishizawa, T., Hagiya, M., Seki, T., Shimonishi, M., Sugimura, A., Tashiro, K., and Shimizu, S. Molecular cloning and expression of human hepatocyte growth factor. *Nature* **342**, 440–443 (1989).
39. Ohmichi, H., Matsumoto, K., and Nakamura, T. *In vivo* mitogenic action of HGF on lung epithelial cells: pulmotrophic role in lung regeneration. *Am J Physiol* **270**, L1031–L1039 (1996).
40. Yamada, T., Hisanaga, M., Nakajima, Y., Mizuno, S., Matsumoto, K., Nakamura, T., and Nakano, H. Enhanced expression of hepatocyte growth factor by pulmonary ischemia-reperfusion injury in the rat. *Am J Respir Crit Care Med* **162**, 707–715 (2000).
41. Sakamaki, Y., Matsumoto, K., Mizuno, S., Miyoshi, S., Matsuda, H., and Nakamura, T. Hepatocyte growth factor stimulates proliferation of respiratory epithelial cells during postpneumonectomy compensatory lung growth in mice. *Am J Respir Cell Mol Biol* **26**, 525–533 (2002).
42. Ohmichi, H., Koshimizu, U., Matsumoto, K., and Nakamura, T. Hepatocyte growth factor (HGF) acts as a mesenchyme-derived morphogenic factor during fetal lung development. *Development* **125**, 1315–1324 (1998).
43. Panos, R.J., Patel, R., and Bak, P.M. Intratracheal administration of hepatocyte growth factor/scatter factor stimulates rat alveolar type II cell proliferation *in vivo*. *Am J Respir Cell Mol Biol* **15**, 574–581 (1996).
44. Mason, R.J., Leslie, C.C., McCormick-Shannon, K., Deterding, R.R., Nakamura, T., Rubin, J.S., and Shannon, J.M. Hepatocyte growth factor is a growth factor for rat alveolar type II cells. *Am J Respir Cell Mol Biol* **11**, 561–567 (1994).
45. Bussolino, F., Di Renzo, M.F., Ziche, M., Bocchietto, E., Olivero, M., Naldini, L., Gaudino, G., Tamagnone, L., Coffer, A., and Comoglio, P.M. Hepatocyte growth factor is a potent angiogenic factor which stimulates endothelial cell motility and growth. *J Cell Biol* **119**, 629–641 (1992).
46. Ding, S., Merkulova-Rainon, T., Han, Z.C., and Tobelem, G. HGF receptor up-regulation contributes to the angiogenic phenotype of human endothelial cells and promotes angiogenesis *in vitro*. *Blood* **101**, 4816–4822 (2003).
47. Ono, M., Sawa, Y., Matsumoto, K., Nakamura, T., Kaneda, Y., and Matsuda, H. *In vivo* gene transfection with hepatocyte growth factor via the pulmonary artery induces angiogenesis in the rat lung. *Circulation* **106**, I264–269 (2002).
48. Ishizawa, K., Kubo, H., Yamada, M., Kobayashi, S., Suzuki, T., Mizuno, S., Nakamura, T., and Sasaki, H. Hepatocyte growth factor induces angiogenesis

in injured lungs through mobilizing endothelial progenitor cells. *Biochem Biophys Res Commun* **324**, 276–280 (2004).

49. Shigemura, N., Sawa, Y., Mizuno, S., Ono, M., Ohta, M., Nakamura, T., Kaneda, Y., and Matsuda, H. Amelioration of pulmonary emphysema by *in vivo* gene transfection with hepatocyte growth factor in rats. *Circulation* **111**, 1407–1414 (2005).
50. Shigemura, N., Okumura, M., Mizuno, S., Imanishi, Y., Nakamura, T., and Sawa, Y. Autologous transplantation of adipose tissue-derived stromal cells ameliorates pulmonary emphysema. *Am J Transplant* **6**, 2592–2600 (2006).
51. Hegab, A.E., Kubo, H., Yamaya, M., Asada, M., He, M., Fujino, N., Mizuno, S., and Nakamura, T. Intranasal HGF administration ameliorates the physiologic and morphologic changes in lung emphysema. *Mol Ther* **16**, 1417–1426 (2008).
52. Adachi, Y., Imagawa, J., Suzuki, Y., Yogo, K., Fukazawa, M., Kuromaru, O., and Saito, Y. G-CSF treatment increases side population cell infiltration after myocardial infarction in mice. *J Mol Cell Cardiol* **36**, 707–710 (2004).
53. Orlic, D., Kajstura, J., Chimenti, S., Limana, F., Jakoniuk, I., Quaini, F., Nadal-Ginard, B., Bodine, D.M., Leri, A., and Anversa, P. Mobilized bone marrow cells repair the infarcted heart, improving function and survival. *Proc Natl Acad Sci USA* **98**, 10344–10349 (2001).
54. Fukuhara, S., Tomita, S., Nakatani, T., Ohtsu, Y., Ishida, M., Yutani, C., and Kitamura, S. G-CSF promotes bone marrow cells to migrate into infarcted mice heart, and differentiate into cardiomyocytes. *Cell Transplant* **13**, 741–748 (2004).
55. Lee, S.T., Chu, K., Jung, K.H., Ko, S.Y., Kim, E.H., Sinn, D.I., Lee, Y.S., Lo, E.H., Kim, M., and Roh, J.K. Granulocyte colony-stimulating factor enhances angiogenesis after focal cerebral ischemia. *Brain Res* **1058**, 120–128 (2005).
56. Takagi, Y., Omura, T., Yoshiyama, M., Matsumoto, R., Enomoto, S., Kusuyama, T., Nishiya, D., Akioka, K., Iwao, H., Takeuchi, K., and Yoshikawa, J. Granulocyte-colony stimulating factor augments neovascularization induced by bone marrow transplantation in rat hindlimb ischemia. *J Pharmacol Sci* **99**, 45–51 (2005).
57. Stanley, E.R., Berg, K.L., Einstein, D.B., Lee, P.S., Pixley, F.J., Wang, Y., and Yeung, Y.G. Biology and action of colony — stimulating factor-1. *Mol Reprod Dev* **46**, 4–10 (1997).

58. Minamino, K., Adachi, Y., Okigaki, M., Ito, H., Togawa, Y., Fujita, K., Tomita, M., Suzuki, Y., Zhang, Y., Iwasaki, M., Nakano, K., Koike, Y., Matsubara, H., Iwasaka, T., Matsumura, M., and Ikehara, S. Macrophage colony-stimulating factor (M-CSF), as well as granulocyte colony-stimulating factor (G-CSF), accelerates neovascularization. *Stem Cells* **23**, 347–354 (2005).
59. Ishikawa, T., Aoshiba, K., Yokohori, N., and Nagai, A. Macrophage colony-stimulating factor aggravates rather than regenerates emphysematous lungs in mice. *Respiration* **73**, 538–545 (2006).
60. Ulich, T.R., Yi, E.S., Longmuir, K., Yin, S., Biltz, R., Morris, C.F., Housley, R.M., and Pierce, G.F. Keratinocyte growth factor is a growth factor for type II pneumocytes *in vivo*. *J Clin Invest* **93**, 1298–1306 (1994).
61. Fehrenbach, H., Kasper, M., Tschernig, T., Pan, T., Schuh, D., Shannon, J.M., Muller, M., and Mason, R.J. Keratinocyte growth factor-induced hyperplasia of rat alveolar type II cells *in vivo* is resolved by differentiation into type I cells and by apoptosis. *Eur Respir J* **14**, 534–544 (1999).
62. Plantier, L., Marchand-Adam, S., Antico, V.G., Boyer, L., De Coster, C., Marchal, J., Bachoual, R., Mailleux, A., Boczkowski, J., and Crestani, B. Keratinocyte growth factor protects against elastase-induced pulmonary emphysema in mice. *Am J Physiol Lung Cell Mol Physiol* **293**, L1230–L1239 (2007).
63. Kitamura, K., Kangawa, K., Kawamoto, M., Ichiki, Y., Nakamura, S., Matsuo, H., and Eto, T. Adrenomedullin: a novel hypotensive peptide isolated from human pheochromocytoma. *Biochem Biophys Res Commun* **192**, 553–560 (1993).
64. Lopez, J. and Martinez, A. Cell and molecular biology of the multifunctional peptide, adrenomedullin. *Int Rev Cytol* **221**, 1–92 (2002).
65. Kim, W., Moon, S.O., Sung, M.J., Kim, S.H., Lee, S., So, J.N., and Park, S.K. Angiogenic role of adrenomedullin through activation of Akt, mitogen-activated protein kinase, and focal adhesion kinase in endothelial cells. *FASEB J* **17**, 1937–1939 (2003).
66. Tokunaga, N., Nagaya, N., Shirai, M., Tanaka, E., Ishibashi-Ueda, H., Harada-Shiba, M., Kanda, M., Ito, T., Shimizu, W., Tabata, Y., Uematsu, M., Nishigami, K., Sano, S., Kangawa, K., and Mori, H. Adrenomedullin gene transfer induces therapeutic angiogenesis in a rabbit model of chronic hind

limb ischemia: benefits of a novel nonviral vector, gelatin. *Circulation* **109**, 526–531 (2004).

67. Martinez, A., Miller, M.J., Catt, K.J., and Cuttitta, F. Adrenomedullin receptor expression in human lung and in pulmonary tumors. *J Histochem Cytochem* **45**, 159–164 (1997).
68. Murakami, S., Nagaya, N., Itoh, T., Iwase, T., Fujisato, T., Nishioka, K., Hamada, K., Kangawa, K., and Kimura, H. Adrenomedullin regenerates alveoli and vasculature in elastase-induced pulmonary emphysema in mice. *Am J Respir Crit Care Med* **172**, 581–589 (2005).
69. Indolfi, C., Cioppa, A., Stabile, E., Di Lorenzo, E., Esposito, G., Pisani, A., Leccia, A., Cavuto, L., Stingone, A.M., Chieffo, A., Capozzolo, C., and Chiariello, M. Effects of hydroxymethylglutaryl coenzyme A reductase inhibitor simvastatin on smooth muscle cell proliferation *in vitro* and neointimal formation *in vivo* after vascular injury. *J Am Coll Cardiol* **35**, 214–221 (2000).
70. Matsuno, H., Takei, M., Hayashi, H., Nakajima, K., Ishisaki, A., and Kozawa, O. Simvastatin enhances the regeneration of endothelial cells via VEGF secretion in injured arteries. *J Cardiovasc Pharmacol* **43**, 333–340 (2004).
71. Tokunaga, T., Ikegami, T., Yoshizumi, T., Imura, S., Morine, Y., Shinohara, H., and Shimada, M. Beneficial effects of fluvastatin on liver microcirculation and regeneration after massive hepatectomy in rats. *Dig Dis Sci* **53**, 2989–2994 (2008).
72. Takahashi, S., Nakamura, H., Seki, M., Shiraishi, Y., Yamamoto, M., Furuuchi, M., Nakajima, T., Tsujimura, S., Shirahata, T., Nakamura, M., Minematsu, N., Yamasaki, M., Tateno, H., and Ishizaka, A. Reversal of elastase-induced pulmonary emphysema and promotion of alveolar epithelial cell proliferation by simvastatin in mice. *Am J Physiol Lung Cell Mol Physiol* **294**, L882–L890 (2008).
73. Rojas, M., Xu, J., Woods, C.R., Mora, A.L., Spears, W., Roman, J., and Brigham, K.L. Bone marrow-derived mesenchymal stem cells in repair of the injured lung. *Am J Respir Cell Mol Biol* **33**, 145–152 (2005).
74. Aso, Y., Yoneda, K., and Kikkawa, Y. Morphologic and biochemical study of pulmonary changes induced by bleomycin in mice. *Lab Invest* **35**, 558–568 (1976).

75. Kim, C.F., Jackson, E.L., Woolfenden, A.E., Lawrence, S., Babar, I., Vogel, S., Crowley, D., Bronson, R.T., and Jacks, T. Identification of bronchioalveolar stem cells in normal lung and lung cancer. *Cell* **121**, 823–835 (2005).
76. McBride, J.T., Wohl, M.E., Strieder, D.J., Jackson, A.C., Morton, J.R., Zwerdling, R.G., Griscom, N.T., Treves, S., Williams, A.J., and Schuster, S. Lung growth and airway function after lobectomy in infancy for congenital lobar emphysema. *J Clin Invest* **66**, 962–970 (1980).
77. Laros, C.D. and Westermann, C.J. Dilatation, compensatory growth, or both after pneumonectomy during childhood and adolescence. A thirty-year follow-up study. *J Thorac Cardiovasc Surg* **93**, 570–576 (1987).
78. Nakajima, C., Kijimoto, C., Yokoyama, Y., Miyakawa, T., Tsuchiya, Y., Kuroda, T., Nakano, M., and Saeki, M. Longitudinal follow-up of pulmonary function after lobectomy in childhood — factors affecting lung growth. *Pediatr Surg Int* **13**, 341–345 (1998).
79. Langston, C., Sachdeva, P., Cowan, M.J., Haines, J., Crystal, R.G., and Thurlbeck, W.M. Alveolar multiplication in the contralateral lung after unilateral pneumonectomy in the rabbit. *Am Rev Respir Dis* **115**, 7–13 (1977).
80. Hu, L.M., Davies, P., Adzick, N.S., Harrison, M.R., and Reid, L.M. The effects of intrauterine pneumonectomy in lambs. A morphometric study of the remaining lung at term. *Am Rev Respir Dis* **135**, 607–612 (1987).
81. Cagle, P.T. and Thurlbeck, W.M. Postpneumonectomy compensatory lung growth. *Am Rev Respir Dis* **138**, 1314–1326 (1988).
82. Rannels, D.E. and Rannels, S.R. Compensatory growth of the lung following partial pneumonectomy. *Exp Lung Res* **14**, 157–182 (1988).
83. McBride, J.T. Lung volumes after an increase in lung distension in pneumonectomized ferrets. *J Appl Physiol* **67**, 1418–1421 (1989).
84. Hsia, C.C., Herazo, L.F., Fryder-Doffey, F., and Weibel, E.R. Compensatory lung growth occurs in adult dogs after right pneumonectomy. *J Clin Invest* **94**, 405–412 (1994).
85. Takeda, S., Hsia, C.C., Wagner, E., Ramanathan, M., Estrera, A.S., and Weibel, E.R. Compensatory alveolar growth normalizes gas-exchange function in immature dogs after pneumonectomy. *J Appl Physiol* **86**, 1301–1310 (1999).
86. Mechanisms and limits of induced postnatal lung growth. *Am J Respir Crit Care Med* **170**, 319–343 (2004).

87. Sakurai, M.K., Lee, S., Arsenault, D.A., Nose, V., Wilson, J.M., Heymach, J.V., and Puder, M. Vascular endothelial growth factor accelerates compensatory lung growth after unilateral pneumonectomy. *Am J Physiol Lung Cell Mol Physiol* **292**, L742–L747 (2007).
88. Fehrenbach, H., Voswinckel, R., Michl, V., Mehling, T., Fehrenbach, A., Seeger, W., and Nyengaard, J.R. Neoalveolarisation contributes to compensatory lung growth following pneumonectomy in mice. *Eur Respir J* **31**, 515–522 (2008).
89. Burri, P.H. and Sehovic, S. The adaptive response of the rat lung after bilobectomy. *Am Rev Respir Dis* **119**, 769–777 (1979).
90. Liu, M., Skinner, S.J., Xu, J., Han, R.N., Tanswell, A.K., and Post, M. Stimulation of fetal rat lung cell proliferation *in vitro* by mechanical stretch. *Am J Physiol* **263**, L376–L383 (1992).
91. Chess, P.R., Toia, L., and Finkelstein, J.N. Mechanical strain-induced proliferation and signaling in pulmonary epithelial H441 cells. *Am J Physiol Lung Cell Mol Physiol* **279**, L43–L51 (2000).
92. Liu, M., Liu, J., Buch, S., Tanswell, A.K., and Post, M. Antisense oligonucleotides for PDGF-B and its receptor inhibit mechanical strain-induced fetal lung cell growth. *Am J Physiol* **269**, L178–L184 (1995).
93. Waters, C.M., Chang, J.Y., Glucksberg, M.R., DePaola, N., and Grotberg, J.B. Mechanical forces alter growth factor release by pleural mesothelial cells. *Am J Physiol* **272**, L552–L557 (1997).
94. Thibeault, D.W. and Haney, B. Lung volume, pulmonary vasculature, and factors affecting survival in congenital diaphragmatic hernia. *Pediatrics* **101**, 289–295 (1998).
95. Zhang, S., Garbutt, V., and McBride, J.T. Strain-induced growth of the immature lung. *J Appl Physiol* **81**, 1471–1476 (1996).
96. Berg, J.T., Fu, Z., Breen, E.C., Tran, H.C., Mathieu-Costello, O., and West, J.B. High lung inflation increases mRNA levels of ECM components and growth factors in lung parenchyma. *J Appl Physiol* **83**, 120–128 (1997).
97. Gilbert, K.A. and Rannels, D.E. Increased lung inflation induces gene expression after pneumonectomy. *Am J Physiol* **275**, L21–L29 (1998).
98. Landesberg, L.J., Ramalingam, R., Lee, K., Rosengart, T.K., and Crystal, R.G. Upregulation of transcription factors in lung in the early phase of postpneumonectomy lung growth. *Am J Physiol Lung Cell Mol Physiol* **281**, L1138–L1149 (2001).

99. Nijjar, M.S. and Thurlbeck, W.M. Alterations in enzymes related to adenosine 3′,5′-monophosphate during compensatory growth of rat lung. *Eur J Biochem* **105**, 403–407 (1980).
100. Russo, L.A., Rannels, S.R., Laslow, K.S., and Rannels, D.E. Stretch-related changes in lung cAMP after partial pneumonectomy. *Am J Physiol* **257**, E261–E268 (1989).
101. Hsia, C.C.W., Herazo, L.F., Ramanathan, M., and Johnson, R.L.J. Cardiopulmonary adaptations to pneumonectomy in dogs: IV: membrane diffusing capacity and capilary blood volume. *J Appl Physiol* **77**, 998–1005 (1994).
102. Haworth, S.G., McKenzie, S.A., and Fitzpatrick, M.L. Alveolar development after ligation of left pulmonary artery in newborn pig: clinical relevance to unilateral pulmonary artery. *Thorax* **36**, 938–943 (1981).
103. Dane, D.M., Yan, X., Tamhane, R.M., Johnson, R.L., Jr., Estrera, A.S., Hogg, D.C., Hogg, R.T., and Hsia, C.C. Retinoic acid-induced alveolar cellular growth does not improve function after right pneumonectomy. *J Appl Physiol* **96**, 1090–1096 (2004).
104. Hsia, C.C., Fryder-Doffey, F., Stalder-Nayarro, V., Johnson, R.L., Jr., Reynolds, R.C., and Weibel, E.R. Structural changes underlying compensatory increase of diffusing capacity after left pneumonectomy in adult dogs. *J Clin Invest* **92**, 758–764 (1993).
105. Yan, X., Bellotto, D.J., Dane, D.M., Elmore, R.G., Johnson, R.L., Jr., Estrera, A.S., and Hsia, C.C. Lack of response to all-trans retinoic acid supplementation in adult dogs following left pneumonectomy. *J Appl Physiol* **99**, 1681–1688 (2005).
106. Taipale, J. and Keski-Oja, J. Hepatocyte growth factor releases epithelial and endothelial cells from growth arrest induced by transforming growth factor-beta1. *J Biol Chem* **271**, 4342–4348 (1996).
107. Panos, R.J., Rubin, J.S., Csaky, K.G., Aaronson, S.A., and Mason, R.J. Keratinocyte growth factor and hepatocyte growth factor/scatter factor are heparin-binding growth factors for alveolar type II cells in fibroblast-conditioned medium. *J Clin Invest* **92**, 969–977 (1993).
108. Kim, H.J., Sammak, P.J., and Ingbar, D.H. Hepatocyte growth factor stimulates migration of type II alveolar epithelial cells on the provisional matrix proteins fibronectin and fibrinogen. *Chest* **116**, 94S–95S (1999).

109. Miettinen, P.J., Warburton, D., Bu, D., Zhao, J.S., Berger, J.E., Minoo, P., Koivisto, T., Allen, L., Dobbs, L., Werb, Z., and Derynck, R. Impaired lung branching morphogenesis in the absence of functional EGF receptor. *Dev Biol* **186**, 224–236 (1997).
110. Kaza, A.K., Laubach, V.E., Kern, J.A., Long, S.M., Fiser, S.M., Tepper, J.A., Nguyen, R.P., Shockey, K.S., Tribble, C.G., and Kron, I.L. Epidermal growth factor augments postpneumonectomy lung growth. *J Thorac Cardiovasc Surg* **120**, 916–921 (2000).
111. Brody, J.S., Fisher, A.B., Gocmen, A., and DuBois, A.B. Acromegalic pneumonomegaly: lung growth in the adult. *J Clin Invest* **49**, 1051–1060 (1970).
112. Jain, B.P., Brody, J.S., and Fisher, A.B. The small lung of hypopituitarism. *Am Rev Respir Dis* **108**, 49–55 (1973).
113. Mansell, A.L., Levison, H., and Bailey, J.D. Maturation of lung function in children with hypopituitarism. *Am Rev Respir Dis* **127**, 166–170 (1983).
114. Gaultier, C., Harf, A., and Girard, F. Lung mechanics in growing guinea-pigs treated with growth hormone. *J Dev Physiol* **8**, 315–321 (1986).
115. Healy, A.M., Morgenthau, L., Zhu, X., Farber, H.W., and Cardoso, W.V. VEGF is deposited in the subepithelial matrix at the leading edge of branching airways and stimulates neovascularization in the murine embryonic lung. *Dev Dyn* **219**, 341–352 (2000).
116. Kaner, R.J. and Crystal, R.G. Compartmentalization of vascular endothelial growth factor to the epithelial surface of the human lung. *Mol Med* **7**, 240–246 (2001).
117. Leuwerke, S.M., Kaza, A.K., Tribble, C.G., Kron, I.L., and Laubach, V.E. Inhibition of compensatory lung growth in endothelial nitric oxide synthase-deficient mice. *Am J Physiol Lung Cell Mol Physiol* **282**, L1272–L1278 (2002).
118. Li, D., Fernandez, L.G., Dodd-o, J., Langer, J., Wang, D., and Laubach, VE. Upregulation of hypoxia-induced mitogenic factor in compensatory lung growth after pneumonectomy. *Am J Respir Cell Mol Biol* **32**, 185–191 (2005).
119. Pham, I., Uchida, T., Planes, C., Ware, L.B., Kaner, R., Matthay, M.A., and Clerici, C. Hypoxia upregulates VEGF expression in alveolar epithelial cells *in vitro* and *in vivo*. *Am J Physiol Lung Cell Mol Physiol* **283**, L1133–L1142 (2002).
120. Cowan, M.J. and Crystal, R.G. Lung growth after unilateral pneumonectomy: quantitation of collagen synthesis and content. *Am Rev Respir Dis* **111**, 267–277 (1975).

121. Hsia, C.C., Zhou, X.S., Bellotto, D.J., and Hagler, H.K. Regenerative growth of respiratory bronchioles in dogs. *Am J Physiol Lung Cell Mol Physiol* **279**, L136–L142 (2000).
122. Voswinckel, R., Ziegelhoeffer, T., Heil, M., Kostin, S., Breier, G., Mehling, T., Haberberger, R., Clauss, M., Gaumann, A., Schaper, W., and Seeger, W. Circulating vascular progenitor cells do not contribute to compensatory lung growth. *Circ Res* **93**, 372–379 (2003).
123. Nolen-Walston, R.D., Kim, C.F., Mazan, M.R., Ingenito, E.P., Gruntman, A.M., Tsai, L., Boston, R., Woolfenden, A.E., Jacks, T., and Hoffman, A.M. Cellular kinetics and modeling of bronchioalveolar stem cell response during lung regeneration. *Am J Physiol Lung Cell Mol Physiol* **294**, L1158–L1165 (2008).
124. Kenzaki, K., Sakiyama, S., Kondo, K., Yoshida, M., Kawakami, Y., Takehisa, M., Takizawa, H., Miyoshi, T., Bando, Y., Tangoku, A., and Liu, M. Lung regeneration: implantation of fetal rat lung fragments into adult rat lung parenchyma. *J Thorac Cardiovasc Surg* **131**, 1148–1153 (2006).
125. Date, H., Sano, Y., Aoe, M., Matsubara, H., Kusano, K., Goto, K., Tedoriya, T., and Shimizu, N. Living-donor single-lobe lung transplantation for primary pulmonary hypertension in a child. *J Thorac Cardiovasc Surg* **123**, 1211–1213 (2002).
126. Toyooka, S., Sano, Y., Yamane, M., Oto, T., Okazaki, M., Kusano, K.F., and Date, H. Long-term follow-up of living-donor single lobe transplantation for idiopathic pulmonary arterial hypertension in a child. *J Thorac Cardiovasc Surg* **135**, 451–452 (2008).
127. Sritippayawan, S., Keens, T.G., Horn, M.V., MacLaughlin, E.F., Barr, M.L., Starnes, V.A., and Woo, M.S. Does lung growth occur when mature lobes are transplanted into children? *Pediatr Transplant* **6**, 500–504 (2002).
128. Binns, O.A., DeLima, N.F., Buchanan, S.A., Lopes, M.B., Cope, J.T., Marek, C.A., King, R.C., Laubach, V.E., Tribble, C.G., and Kron, I.L. Mature pulmonary lobar transplants grow in an immature environment. *J Thorac Cardiovasc Surg* **114**, 186–194 (1997).
129. Asahara, T., Murohara, T., Sullivan, A., Silver, M., van der Zee, R., Li, T., Witzenbichler, B., Schatteman, G., and Isner, J.M. Isolation of putative progenitor endothelial cells for angiogenesis. *Science* **275**, 964–967 (1997).
130. Krause, D.S., Theise, N.D., Collector, M.I., Henegariu, O., Hwang, S., Gardner, R., Neutzel, S., and Sharkis, S.J. Multi-organ, multi-lineage engraftment by a single bone marrow-derived stem cell. *Cell* **105**, 369–377 (2001).

131. Kotton, D.N., Ma, B.Y., Cardoso, W.V., Sanderson, E.A., Summer, R.S., Williams, M.C., and Fine, A. Bone marrow-derived cells as progenitors of lung alveolar epithelium. *Development* **128**, 5181–5188 (2001).
132. Okamoto, R., Yajima, T., Yamazaki, M., Kanai, T., Mukai, M., Okamoto, S., Ikeda, Y., Hibi, T., Inazawa, J., and Watanabe, M. Damaged epithelia regenerated by bone marrow-derived cells in the human gastrointestinal tract. *Nat Med* **8**, 1011–1017 (2002).
133. Majka, S.M., Jackson, K.A., Kienstra, K.A., Majesky, M.W., Goodell, M.A., and Hirschi, K.K. Distinct progenitor populations in skeletal muscle are bone marrow derived and exhibit different cell fates during vascular regeneration. *J Clin Invest* **111**, 71–79 (2003).
134. Brazelton, T.R., Rossi, F.M., Keshet, G.I., and Blau, H.M. From marrow to brain: expression of neuronal phenotypes in adult mice. *Science* **290**, 1775–1779 (2000).
135. Mezey, E., Chandross, K.J., Harta, G., Maki, R.A., and McKercher, S.R. Turning blood into brain: cells bearing neuronal antigens generated *in vivo* from bone marrow. *Science* **290**, 1779–1782 (2000).
136. Yamada, M., Kubo, H., Ishizawa, K., Kobayashi, S., Shinkawa, M., and Sasaki, H. Increased circulating endothelial progenitor cells in patients with bacterial pneumonia: evidence that bone marrow derived cells contribute to lung repair. *Thorax* **60**, 410–413 (2005).
137. Burnham, E.L., Taylor, W.R., Quyyumi, A.A., Rojas, M., Brigham, K.L., and Moss, M. Increased circulating endothelial progenitor cells are associated with survival in acute lung injury. *Am J Respir Crit Care Med* **172**, 854–860 (2005).
138. Yamada, M., Kubo, H., Kobayashi, S., Ishizawa, K., Numasaki, M., Ueda, S., Suzuki, T., and Sasaki, H. Bone marrow-derived progenitor cells are important for lung repair after lipopolysaccharide-induced lung injury. *J Immunol* **172**, 1266–1272 (2004).
139. Abe, S., Boyer, C., Liu, X., Wen, F.Q., Kobayashi, T., Fang, Q., Wang, X., Hashimoto, M., Sharp, J.G., and Rennard, S.I. Cells derived from the circulation contribute to the repair of lung injury. *Am J Respir Crit Care Med* **170**, 1158–1163 (2004).
140. Ortiz, L.A., Gambelli, F., McBride, C., Gaupp, D., Baddoo, M., Kaminski, N., and Phinney, D.G. Mesenchymal stem cell engraftment in lung is enhanced

in response to bleomycin exposure and ameliorates its fibrotic effects. *Proc Natl Acad Sci USA* **100**, 8407–8411 (2003).

141. Kubo, H., Graham, L., Doyle, N.A., Quinlan, W.M., Hogg, J.C., and Doerschuk, C.M. Complement fragment-induced release of neutrophils from bone marrow and sequestration within pulmonary capillaries in rabbits. *Blood* **92**, 283–290 (1998).
142. Kitamura, Y., Hashimoto, S., Mizuta, N., Kobayashi, A., Kooguchi, K., Fujiwara, I., and Nakajima, H. Fas/FasL-dependent apoptosis of alveolar cells after lipopolysaccharide-induced lung injury in mice. *Am J Respir Crit Care Med* **163**, 762–769 (2001).
143. Yamada, M., Ishizawa, K., Kobayashi, S., Suzuki, T., and Kubo, H. Bone marrow-derived progenitor cells are not a source of bronchioalveolar stem cells after LPS-induced lung injury. *Proc Am Thorac Soc* **3**, A556 (2006).
144. Serikov, V.B., Popov, B., Mikhailov, V.M., Gupta, N., and Matthay, M.A. Evidence of temporary airway epithelial repopulation and rare clonal formation by BM-derived cells following naphthalene injury in mice. *Anat Rec (Hoboken)* **290**, 1033–1045 (2007).
145. Serikov, V.B., Mikhaylov, V.M., Krasnodembskay, A.D., and Matthay, M.A. Bone marrow-derived cells participate in stromal remodeling of the lung following acute bacterial pneumonia in mice. *Lung* **186**, 179–190 (2008).
146. Herzog, E.L., Van Arnam, J., Hu, B., and Krause, D.S. Threshold of lung injury required for the appearance of marrow-derived lung epithelia. *Stem Cells* **24**, 1986–1992 (2006).
147. Bruscia, E.M., Ziegler, E.C., Price, J.E., Weiner, S., Egan, M.E., and Krause, D.S. Engraftment of donor-derived epithelial cells in multiple organs following bone marrow transplantation into newborn mice. *Stem Cells* **24**, 2299–2308 (2006).
148. Hashimoto, N., Jin, H., Liu, T., Chensue, S.W., and Phan, S.H. Bone marrow-derived progenitor cells in pulmonary fibrosis. *J Clin Invest* **113**, 243–252 (2004).
149. Gupta, N., Su, X., Popov, B., Lee, J.W., Serikov, V., and Matthay, M.A. Intrapulmonary delivery of bone marrow-derived mesenchymal stem cells improves survival and attenuates endotoxin-induced acute lung injury in mice. *J Immunol* **179**, 1855–1863 (2007).

150. Xu, J., Woods, C.R., Mora, A.L., Joodi, R., Brigham, K.L., Iyer, S., and Rojas, M. Prevention of endotoxin-induced systemic response by bone marrow-derived mesenchymal stem cells in mice. *Am J Physiol Lung Cell Mol Physiol* **293**, L131–L141 (2007).
151. Serrano-Mollar, A., Nacher, M., Gay-Jordi, G., Closa, D., Xaubet, A., and Bulbena, O. Intratracheal transplantation of alveolar type II cells reverses bleomycin-induced lung fibrosis. *Am J Respir Crit Care Med* **176**, 1261–1268 (2007).
152. Yen, C.C., Yang, S.H., Lin, C.Y., and Chen, C.M. Stem cells in the lung parenchyma and prospects for lung injury therapy. *Eur J Clin Invest* **36**, 310–319 (2006).
153. Prockop, D.J. "Stemness" does not explain the repair of many tissues by mesenchymal stem/multipotent stromal cells (MSCs). *Clin Pharmacol Ther* **82**, 241–243 (2007).
154. Haynesworth, S.E., Baber, M.A., and Caplan, A.I. Cytokine expression by human marrow-derived mesenchymal progenitor cells *in vitro*: effects of dexamethasone and IL-1 alpha. *J Cell Physiol* **166**, 585–592 (1996).
155. Rasmusson, I. Immune modulation by mesenchymal stem cells. *Exp Cell Res* **312**, 2169–2179 (2006).
156. Keating, A. Mesenchymal stromal cells. *Curr Opin Hematol* **13**, 419–425 (2006).
157. Spees, J.L., Olson, S.D., Whitney, M.J., and Prockop, D.J. Mitochondrial transfer between cells can rescue aerobic respiration. *Proc Natl Acad Sci USA* **103**, 1283–1288 (2006).
158. Werner, N., Junk, S., Laufs, U., Link, A., Walenta, K., Bohm, M., and Nickenig, G. Intravenous transfusion of endothelial progenitor cells reduces neointima formation after vascular injury. *Circ Res* **93**, e17–e24 (2003).
159. Kahler, C.M., Wechselberger, J., Hilbe, W., Gschwendtner, A., Colleselli, D., Niederegger, H., Boneberg, E.M., Spizzo, G., Wendel, A., Gunsilius, E., Patsch, J.R., and Hamacher, J. Peripheral infusion of rat bone marrow derived endothelial progenitor cells leads to homing in acute lung injury. *Respir Res* **8**, 50 (2007).
160. Rehman, J., Li, J., Orschell, C.M., and March, K.L. Peripheral blood "endothelial progenitor cells" are derived from monocyte/macrophages and secrete angiogenic growth factors. *Circulation* **107**, 1164–1169 (2003).

161. Nagaya, N., Kangawa, K., Kanda, M., Uematsu, M., Horio, T., Fukuyama, N., Hino, J., Harada-Shiba, M., Okumura, H., Tabata, Y., Mochizuki, N., Chiba, Y., Nishioka, K., Miyatake, K., Asahara, T., Hara, H., and Mori, H. Hybrid cell-gene therapy for pulmonary hypertension based on phagocytosing action of endothelial progenitor cells. *Circulation* **108**, 889–895 (2003).
162. Zhao, Y.D., Courtman, D.W., Deng, Y., Kugathasan, L., Zhang, Q., and Stewart, D.J. Rescue of monocrotaline-induced pulmonary arterial hypertension using bone marrow-derived endothelial-like progenitor cells: efficacy of combined cell and eNOS gene therapy in established disease. *Circ Res* **96**, 442–450 (2005).
163. Takahashi, M., Nakamura, T., Toba, T., Kajiwara, N., Kato, H., and Shimizu, Y. Transplantation of endothelial progenitor cells into the lung to alleviate pulmonary hypertension in dogs. *Tissue Eng* **10**, 771–779 (2004).
164. Zhao, Q., Liu, Z., Wang, Z., Yang, C., Liu, J., and Lu, J. Effect of preprocalcitonin gene-related peptide-expressing endothelial progenitor cells on pulmonary hypertension. *Ann Thorac Surg* **84**, 544–552 (2007).
165. Spees, J.L., Whitney, M.J., Sullivan, D.E., Lasky, J.A., Laboy, M., Ylostalo, J., and Prockop, D.J. Bone marrow progenitor cells contribute to repair and remodeling of the lung and heart in a rat model of progressive pulmonary hypertension. *FASEB J* **22**, 1226–1236 (2008).
166. Baber, S.R., Deng, W., Master, R.G., Bunnell, B.A., Taylor, B.K., Murthy, S.N., Hyman, A.L., and Kadowitz, P.J. Intratracheal mesenchymal stem cell administration attenuates monocrotaline-induced pulmonary hypertension and endothelial dysfunction. *Am J Physiol Heart Circ Physiol* **292**, H1120–H1128 (2007).
167. Wang, X.X., Zhang, F.R., Shang, Y.P., Zhu, J.H., Xie, X.D., Tao, Q.M., and Chen, J.Z. Transplantation of autologous endothelial progenitor cells may be beneficial in patients with idiopathic pulmonary arterial hypertension: a pilot randomized controlled trial. *J Am Coll Cardiol* **49**, 1566–1571 (2007).
168. Mucke, H.A. Pulmonary arterial hypertension: on the way to a manageable disease. *Curr Opin Investig Drugs* **9**, 957–962 (2008).
169. Kubo, H., Hegab, A.E., He, M., Ishizawa, K., and Yamada, M. Endogenous lung stem cells increased after lung injury. *Proc Am Thorac Soc* **5**, 362a–363 (2008).

Chapter 8
Fibrocytes (Reactive or Reparative)

Alberto Bellini and Sabrina Mattoli

Human fibrocytes exhibit mixed phenotypic characteristics of haematopoietic stem cells, monocytes and fibroblasts, and originate from a precursor of the monocyte lineage. They constitutively produce chemokines and growth factors that are known to modulate inflammatory reactions or promote angiogenesis and the deposition of extracellular matrix molecules. Upon exposure to transforming growth factor-β_1 and endothelin-1, fibrocytes produce large quantities of extracellular matrix components and acquire a contractile phenotype. Such differentiation of fibrocytes into myofibroblast-like cells occurs at the tissue sites during repair processes and has been found to contribute to wound healing *in vivo*. Fibrocytes and fibrocyte-derived myofibroblasts are also involved in the pathogenesis of lung disorders characterised by chronic inflammation and extensive remodelling of the bronchial wall, like asthma, or progressive fibrosis with destruction of the pulmonary architecture, like idiopathic pulmonary fibrosis. They participate in tumour-induced stromal reactions and may either promote or inhibit the metastatic progression of cancers. Prevention of excessive extracellular matrix deposition and detrimental tissue remodelling in pulmonary diseases may be achieved by inhibiting the accumulation of fibrocytes in the lungs. Moreover, *in vitro* expanded fibrocytes may serve as vehicles for the delivery of gene constructs to improve ineffective lung repair or be used in anti-cancer cell therapy.

Keywords: Fibrocytes; fibrosis; myofibroblasts; stem cells; tissue repair.

Outline

1. Introduction

Fibrocytes[1–5] are bone marrow-derived cells[6–9] that co-express haematopoietic stem cell antigens, markers of the monocyte lineage and fibroblast products.[1–6,10,11] They constitutively release chemokines and growth factors that are known to modulate ongoing inflammatory reactions or to induce the formation of new vessels and the deposition of extracellular matrix (ECM) molecules.[2,12,13] Under permissive conditions, human fibrocytes themselves produce large quantities of ECM components as well as ECM-degrading enzymes[1,3,5,11] and further differentiate into contractile cells, expressing the myofibroblast marker α-smooth muscle actin (α-SMA).[3,6,11] Converging evidence from several studies suggests that fibrocytes contribute to the new population of fibroblasts and myofibroblasts that emerge at the tissue site during wound healing,[1,3,6] in ischemic or inflammatory fibrotic processes,[8,9,11–26] and in the stromal reaction to tumour development.[15,26–34] This chapter will focus on the mechanisms and functional consequences of fibrocyte accumulation and differentiation in some lung disorders and on the possible implications for cell therapy.

2. Mechanisms of Fibrocyte Development and Differentiation

Human fibrocytes arise from a small subset of $CD14^+$ mononuclear cells,[3–5,10,25,35] which seems to represent a source of bone marrow-derived progenitors with multidifferentiation potential.[36–38] Like their precursors, human fibrocytes express the common leukocyte marker CD45, many markers of the monocyte lineage, the class I and II major histocompatibility complex (MHC) molecules, the co-stimulatory molecules CD80 and

CD86 and several CC chemokine receptors (CCRs) and CXC chemokine receptors (CXCRs), including CCR7 and CXCR4[3,5] (Table 1). The phenotypic changes occurring during the development of fibrocytes from the $CD14^+$ precursors are summarised in Table 1 and include an early upregulation of the expression of the haematopoietic stem cell/progenitor markers CD34[25] and CD105[5] on the surface of the cells and the loss of pluripotency-associated features (expression of nanog and octamer-binding factor 4A) (Table 1). The acquisition of fibroblast markers is associated with a marked downregulation of CD14 and CCR2 expression and reduced ability to proliferate.[3–5,10,25,35] The magnitude of these phenotypic changes *in vitro* largely depends on the media and supplements used in culture.[3–5,10,25,35] A cell sharing many morphological characteristics and surface markers with the fibrocyte, but expressing high levels of CD14 in conjunction with fibroblast products under certain culture conditions, is known by the name of monocyte-derived mesenchymal progenitor,[38] and may represent a more immature progenitor along the pathway that leads from the $CD14^+$ multipotent precursor to a mature mesenchymal cell.

Several factors affect the development of fibrocytes from the $CD14^+$ precursors and the further differentiation of fibrocytes into mature mesenchymal cells[3,5,10–12,25,35] (Fig. 1). Direct contact between peripheral blood $CD14^+$ mononuclear cells and T lymphocytes in co-cultures[3] and stimulation of peripheral blood $CD14^+$ mononuclear cells with transforming growth factor-β_1 (TGF-β_1) increase the yield of fibrocytes.[3,4] Platelet-derived growth factor (PDGF)[25] and cytokines produced by helper type 2 T lymphocytes (Th2 cells), such as interleukin (IL)-4 and IL-13,[35] also promote the differentiation of the $CD14^+$ precursors into fibrocytes. By contrast, exposure of the $CD14^+$ precursors to immune complexes or to serum amyloid P (SAP) inhibits the development of fibrocytes.[10] SAP is a constitutive serum protein that belongs to the pentraxin family of autacoids. It may be present in elevated concentration at tissue sites during the early phase of inflammatory reactions, as a consequence of the extravasations of serum proteins, but it is likely cleared or inactivated at a later stage.[5,10] Two cytokines normally associated with inflammatory responses mediated by helper type 1 T lymphocytes (Th1 cells), interferon (IFN)-γ and IL-12, inhibit the differentiation of fibrocytes from $CD14^+$ mononuclear cells.[35] IL-1β does not affect the development of fibrocytes from the $CD14^+$ precursors, but it reduces the release of collagens from

Table 1: Phenotypic changes occurring during the development of fibrocytes and their differentiation into fibroblast- and myofibroblast-like cells *in vitro*.

Markers	Fibrocyte Precursors	Fibrocytes	Fibroblast/ Myofibroblast-Like Cells
Stem cell markers			
• Flk1	+/–	–	–
• Oct-4A	++	+/–	–
• Nanog	++	–	–
Hematopoietic progenitor markers			
• CD34	+/–	+	+/–
• CD105	+/–	+	NA
Leukocyte markers			
• CD45	++	++	+, +/–
Markers of cells of the monocyte lineage			
• CD11b	++	++	+, +/–
• CD13	++	++	+, +/–
• CD14	++	+/–	–
• CD31	+	+/–	–
MHC and co-stimulatory molecules			
• MHC class I	++	++	NA
• MHC class II	++	++	NA
• CD80	+	+	NA
• CD86	++	++	+/–
Chemokine receptors			
• CCR2	++	+/–	–
• CCR5	+	+	NA
• CCR7	+	+	+
• CXCR4	++	+	+
Mesenchymal products			
• Fibronectin	–	+	++
• Proteoglycans	–	+	++
• Collagens	–	+	++
• α-SMA	–	–	+

The fibrocyte precursor was isolated from a specific $CD14^{+}$ light-density fraction of peripheral blood mononuclear cells (containing immature cells) using a validated technique, based on a modification of the standard discontinuous Percoll gradient method. The symbols represent no expression (–), marginal or no expression (+/–) and increasing level of expression (+, ++). Abbreviations: α-SMA, α-smooth muscle actin; CCR, CC chemokine receptor; CXCR, CXC chemokine receptor; MHC, major histocompatibility complex; NA, not assessed; OCT-4A, octamer-binding factor 4A.

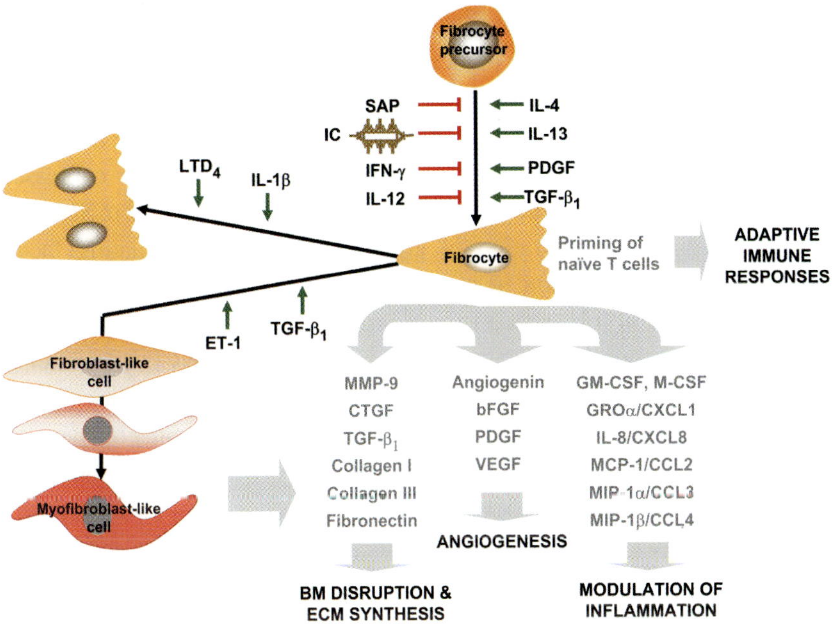

Figure 1: Signals that may promote (green arrows) or inhibit (red lines with blunt ends) the accumulation of fibrocytes at the tissue site and their effects on inflammatory responses and tissue remodelling. BM, basement membrane; CCL, CC chemokine ligand; CXCL, CXC chemokine ligand; CTGF, connective tissue growth factor; ET-1, endothelin-1; ECM, extracellular matrix; GM-CSF, granulocyte-macrophage colony-stimulating factor; IFN-γ, interferon-γ; GRO, growth-related oncogene; IC, immune complexes; IL, interleukin; LTD_4, leukotriene D_4; MCP-1, monocyte chemoattractant protein-1; M-CSF, monocyte-macrophage colony-stimulating factor; MIP, macrophage inflammatory protein; MMP-9, matrix metalloproteinase-9; PDGF, platelet-derived growth factor; SAP, serum amyloid P; TGF-β_1, transforming growth factor-β_1; VEGF, vascular endothelial growth factor. (Designed with the use of the ScienceSlides 2005 software, VisiScience Corporation, Chapel Hill, NC, USA.)

mature fibrocytes.[2] Both this cytokine[2] and leukotriene D_4[39] induce a weak proliferative response in mature fibrocytes (Fig. 1).

The maturation of fibrocytes from their $CD14^+$ precursors may conceivably occur in the bone marrow, in the peripheral blood or at the tissue site, depending on the balance between factors that promote or inhibit such maturation process in these compartments. However, at least five studies suggest that the development of fibrocytes from $CD14^+$ mononuclear cells, as well as their further differentiation into myofibroblasts, predominantly occurs in peripheral organs or in the arterial walls. In one of these studies,[9] fibrotic cardiomyopathy was induced in mice by multiple daily episodes of transient coronary artery occlusion. The development of the fibrotic changes was associated with the recruitment of $CD45^+$ cells of the monocyte lineage from the peripheral blood, some of which subsequently expressed CD34, cardiac fibroblast markers and the myofibroblast marker α-SMA. In another study,[40] administration of granulocyte colony-stimulating factor after myocardial infarction in mice, enhanced the accumulation of bone marrow-derived fibroblasts and myofibroblasts in the ischemic area, and these cells were found to arise from the mobilised subset of $CD14^+$ mononuclear cells. In a third study,[22] the possible role of fibrocytes in the pathogenesis of chronic pulmonary hypertension was evaluated by using an animal model of hypoxia-induced pulmonary vascular remodelling. Circulating cells of the monocyte/macrophage lineage from the chronically hypoxic rats were selectively labelled *in vivo* and the labelled cells were subsequently identified in the remodelled adventitial layer of the pulmonary arteries. Many of the labelled $CD14^+$ cells taken up from the circulation progressively acquired the ability to synthesise new collagen while retaining expression of CD45 and CD11b. A substantial proportion of the labelled cells also showed α-SMA immunoreactivity. In a fourth study,[25] the investigators labelled peripheral blood leukocytes to evaluate the possibility that circulating cells contributed to vascular remodelling in an ovine model of carotid artery intimal hyperplasia. A small fraction of the labelled circulating leukocytes that infiltrated the intima *in vivo* progressively acquired the fibrocyte phenotype and showed α-SMA immunoreactivity during the ongoing remodelling process. Finally, in another study in humans,[41] cells of the monocyte lineage were found to transform into fibroblasts and myofibroblasts in the fibrous cap of atherosclerotic plaques.

Several groups have used various murine models of organ injury to evaluate the dependence of fibrocyte accumulation on the functional chemokine receptors expressed by these cells, and have demonstrated a significant involvement of CCR2,[20,23] CCR5,[42] CCR7[24] and CXCR4.[19] While CCR5, CCR7 and CXCR4 are expressed to various extents by both fibrocytes and their $CD14^+$ precursors, the expression of CCR2 is markedly downregulated during the development of fibrocytes from the $CD14^+$ mononuclear cells (Table 1). In this respect, there seems to be no major differences between murine and human cells.[3,5,43] Therefore, taken

specimens from patients with IPF/UIP, fibrocytes were often present in proximity of fibroblastic foci and there was a significant correlation between the density of fibrocytes and the number of fibroblastic foci in the lungs. Fibrocytes and fibrocyte-derived myofibroblasts actively synthesised collagen at the tissue site, as demonstrated by the expression of prolyl-4-hydroxylase, an enzyme critically involved in collagen synthesis. Fibrocytes and fibrocyte-derived myofibroblasts showed CXCR4 immunoreactivity and there were elevated levels of the CXCR4 ligand CXCL12 both in the plasma and in the bronchioalveolar lavage fluid of patients with IPF/UIP. The results of the latter study suggest that fibrocytes may contribute to the fibrotic process in this disease and that the CXCR4/CXCL12 axis may play an important role in the intrapulmonary recruitment of fibrocytes or their precursors.

4.3. *Cancer-induced stromal reaction and lung metastasis*

Immature monocytes released from the bone marrow are actively recruited into solid tumours from the peripheral blood and represent a prominent component of the leukocyte infiltrate, particularly in malignant lesions.[86] A substantial proportion of the mesenchymal cells that contribute to the development of cancer-induced stroma in mice also appear to originate from bone marrow-derived precursors.[47,87–90] $CD45^+CD34^-$ α-SMA^+ cells of bone marrow origin have been identified as an important component of cancer-induced stroma in the advanced stage of development of an implanted carcinoma in bone marrow chimeric mice,[47] and such cells show the phenotype of fibrocytes undergoing further differentiation into mature myofibroblasts.

In addition, the CCR5/CCL3 axis seems to regulate the development of the lung metastasis of invasive carcinomas in mice, and deficiency of the CCR5 or CCL3 gene markedly reduces the number of metastasis foci and the intratumoral accumulation of macrophages and fibrocytes.[90] The ability of fibrocytes to release ECM-degrading enzymes and growth factors that promote tumour invasion and angiogenesis also supports the hypothesis that these cells may potentially contribute to the metastatic progression of invasive tumours.[90] Nonetheless, cells co-expressing monocyte markers and collagen I have been shown to encapsulate the peritoneal implants of a mammary carcinoma in mice and prevent tumour growth, particularly in the presence of T lymphocytes.[91]

Concerning the potential involvement of human fibrocytes in cancer development, several studies have demonstrated the presence of $CD34^+$ fibroblasts in the stroma of various human benign and invasive tumours.[15,26–34] $CD34^+\alpha$-SMA^- fibroblasts, resembling fibrocytes, are usually found in the stroma encapsulating benign tumours while $CD34^-\alpha$-SMA^+ fibroblasts are particularly frequent around invasive carcinoma. These data would suggest that human fibrocytes are not an important component of cancer-induced stroma. However, cancerous cells and the cells at the interface of malignant lesions produce high amounts of TGF-β_1,[34,89] and hypoxic regions of cancers contain elevated levels of ET-1.[86] Therefore, the apparent loss of $CD34^+$ fibrocyte-like cells and the concomitant increase in the number of α-SMA^+ $CD34^-$ cells in the stroma surrounding malignant lesions[26,30,32] may reflect an increased differentiation of $CD34^+$ fibrocytes into mature $CD34^-$ myofibroblasts as a result of the exposure to elevated concentrations of TGF-β_1 and ET-1 in the micro-environment. More studies are needed to understand the functional role of fibrocytes and fibrocyte-derived myofibroblasts in cancer-induced stromal reaction.

5. Conclusions and Implications for Therapy

The numerous studies conducted over the last decade have revealed that human fibrocytes may serve as an important source of fibroblast- and myofibroblast-like cells during normal or aberrant reparative processes and in fibrotic disorders or stromal reactions to tumour development. Signals that may promote or inhibit the maturation of fibrocytes and their differentiation into lesional fibroblasts and myofibroblasts in those conditions have been at least in part defined (Fig. 1) and may serve as therapeutic targets. Inhibitors of IL-4 and IL-13, TGF-β antagonists and ET-1 receptor antagonists are currently under development for the treatment of asthma[53] or IPF/UIP.[93] In the murine model of bleomycin-induced pulmonary fibrosis, the administration of exogenous SAP, either before or following the intra-tracheal injection of bleomycin, markedly reduces the accumulation of fibrocytes in the lungs, and this effect is associated with reduced deposition of collagen in the interstitium and reduced loss of functional parenchyma.[94] SAP is a protein normally present in human serum and may represent a promising therapeutic option for human fibrotic disorders, particularly IPF/UIP. Fibrocytes express the chemokine receptors CCR5, CCR7 and CXCR4 and migrate *in vitro* in response to the corresponding ligands CCL3-CCL5, CCL19-CCL21 and CXCL12. Inhibition of signalling through these receptors has resulted in the prevention of injury-induced fibrotic reactions in various animal models of pulmonary or renal fibrosis.[19,24,43] Unfortunately, such receptors are expressed by many cell types, and some of these cells play a critical role in adaptive immune responses, host protection against infections, killing of intracellular pathogens and tumour resistance.[50–52,95] Therefore, the selection of these receptors as therapeutic targets for inhibiting or facilitating fibrocyte accumulation and differentiation at tissue sites may not represent a sufficiently safe approach.

Fibrocytes can be easily developed from circulating $CD14^+$ mononuclear cells and expanded *ex vivo*. Tracking experiments in murine models of wound healing[3] and chronic allergic asthma[11] have demonstrated that intravenously injected fibrocytes rapidly migrate to wounded or chronically inflamed sites and localise to areas of ongoing ECM deposition.[3,11] Given this propensity of intravenously injected fibrocytes to migrate into inflamed tissues, the use of autologous fibrocytes for cell therapy seems to be an attractive approach for certain conditions that do not require many repeated interventions. In particular, expanded autologous fibrocytes may be used for accelerating the healing of acute tissue injuries, such as extensive burns or lung tissue damage resulting from the inhalation of toxic agents. Expanded autologous cells may also serve as vehicles for the delivery of therapeutic gene constructs to improve ineffective lung repair processes. Fibrocytes are present in the stroma encapsulating benign tumours but also participate in the cancer-induced stromal reactions. It is still unclear if these cells inhibit or promote cancer progression. However, expanded fibrocytes could potentially be used in anti-cancer cell therapy to encapsulate tumours and prevent their invasion, as well as for the delivery of cytotoxic compounds.

24. Sakai, N., Wada, T., Yokoyama, H., *et al.* Secondary lymphoid tissue chemokine (SLC/CCL21)/CCR7 signaling regulates fibrocytes in renal fibrosis. *Proc Natl Acad Sci USA* **103**, 14098–14103 (2006).
25. Varcoe, R.L., Mikhail, M., Guiffre, A.K., *et al.* The role of the fibrocyte in intimal hyperplasia. *J Thromb Haemost* **4**, 1125–1133 (2006).
26. Nimphius, W., Moll, R., Olbert, P., *et al.* CD34$^+$ fibrocytes in chronic cystitis and noninvasive and invasive urothelial carcinomas of the urinary bladder. *Wirchows Archiv* **450**, 179–185 (2007).
27. Barth, P.J., Ramaswamy, A., and Moll, R. CD34(+) fibrocytes in normal cervical stroma, cervical intraepithelial neoplasia III, and invasive squamous cell carcinoma of the cervix uteri. *Virchows Arch* **441**, 564–568 (2002).
28. Barth, P.J., Ebrahimsade, S., Ramaswamy, A., *et al.* CD34$^+$ fibrocytes in invasive ductal carcinoma, ductal carcinoma *in situ* and benign breast lesions. *Virchows Arch* **440**, 298–303 (2002).
29. Ramaswamy, A., Moll, R., and Barth, P.J. CD34(+) fibrocytes in tubular carcinomas and radial scars of the breast. *Virchows Arch* **443**, 536–540 (2003).
30. Chauhan, H., Abraham, A., Phillips, J.R.A., *et al.* There is more than one kind of myofibroblast: analysis of CD34 expression in benign, *in situ*, and invasive breast cancer. *J Clin Pathol* **56**, 271–276 (2003).
31. Kuroda, N., Toi, M., Nakayama, H., *et al.* The distribution and role of myofibroblasts and CD34-positive stromal cells in normal pancreas and various pancreatic lesions. *Histol Histopathol* **19**, 59–67 (2004).
32. Barth, P.J., Schenck zu Schweinsberg, T., Ramaswamy, A., *et al.* CD34$^+$ fibrocytes, alpha-smooth muscle antigen-positive myofibroblasts, and CD117 expression in the stroma of invasive squamous cell carcinomas of the oral cavity, pharynx, and larynx. *Virchows Arch* **444**, 231–234 (2004).
33. Yazhou, C., Wenlv, S., Weidong, Z., *et al.* Clinicopathological significance of stromal myofibroblasts in invasive ductal carcinoma of the breast. *Tumour Biol* **25**, 290–295 (2004).
34. Kojc, N., Zidar, N., Vodopivec, B., *et al.* Expression of CD34, alpha-smooth muscle actin, and transforming growth factor beta1 in squamous intraepithelial lesions and squamous cell carcinoma of the larynx and hypopharynx. *Human Pathol* **36**, 16–21 (2005).
35. Shao, D.D., Suresh, R., Vakil, V., *et al.* Pivotal advance: Th-1 cytokines inhibit, and Th-2 cytokines promote fibrocyte differentiation. *J Leukoc Biol* **83**, 1323–1333 (2008).

36. Zhao, Y., Glesne, D., and Huberman, E. A human peripheral blood monocyte-derived subset acts as pluripotent stem cells. *Proc Natl Acad Sci USA* **100**, 2426–2431 (2003).
37. Romagnani, P., Annunziato, F., Liotta, F., *et al.* $CD14^{+}CD34^{low}$ cells with stem cell phenotypic and functional features are the major source of circulating endothelial progenitors. *Circ Res* **97**, 314–322 (2005).
38. Kuwana, M., Okazaki, Y., Kodama, H., *et al.* Human circulating $CD14^{+}$ monocytes as a source of progenitors that exhibit mesenchymal cell differentiation. *J Leukoc Biol* **74**, 833–845 (2003).
39. Vannella, K.M., McMillan, T.R., Charbeneau, R.P., *et al.* Cysteinyl leukotrienes are autocrine and paracrine regulators of fibrocyte function. *J Immunol* **179**, 7883–7890 (2007).
40. Fujita, J., Mori, M., Kawada, H., *et al.* Administration of granulocyte colony-stimulating factor after myocardial infarction enhances the recruitment of hematopoietic stem cell-derived myofibroblasts and contributes to cardiac repair. *Stem Cells* **25**, 2750–2759 (2007).
41. Medbury, H., Tarran, S.L., Guiffre, A.K., *et al.* Monocytes contribute to the atherosclerotic cap by transformation into fibrocytes. *Int Angiol* **27**, 114–123 (2008).
42. Ishida, Y., Kimura, A., Kondo, T., *et al.* Essential roles of the CC chemokine ligand 3-CC chemokine receptor 5 axis in bleomycin-induced pulmonary fibrosis through regulation of macrophage and fibrocyte infiltration. *Am J Pathol* **170**, 843–854 (2007).
43. Bellini, A. and Mattoli, S. The role of the fibrocyte, a bone marrow-derived mesenchymal progenitor, in reactive and reparative fibroses. *Lab Invest* **87**, 858–870 (2007).
44. Hong, K.M., Burdick, M.D., Phillips, R.J., *et al.* Characterization of human fibrocytes as circulating adipocyte progenitors and the formation of human adipose tissue in SCID mice. *FASEB J* **19**, 2029–2031 (2005).
45. Wang, J., Jiao, H., Stewart, T., *et al.* Accelerated wound healing in leukocyte-specific, protein 1-deficient mouse is associated with increased infiltration of leukocytes and fibrocytes. *J Leukoc Biol* **82**, 1554–1563 (2007).
46. Naik-Mathuria, B., Pilling, D., Crawford, J.R., *et al.* Serum amyloid P inhibits dermal wound healing. *Wound Repair Reg* **16**, 266–273 (2008).
47. Ishii, G., Sangai, T., Sugiyama, K., *et al.* *In vivo* characterization of bone marrow-derived fibroblasts recruited into fibrotic lesions. *Stem Cells* **23**, 699–706 (2005).

48. Chesney, J., Bacher, M., Bender, A., *et al.* The peripheral blood fibrocyte is a potent antigen-presenting cell capable of priming naïve T cells *in situ. Proc Natl Acad Sci USA* **94**, 6307–6312 (1997).
49. Balmelli, C., Ruggli, N., McCullough, K., *et al.* Fibrocytes are potent stimulators of anti-virus cytotoxic T cells. *J Leukoc Biol* **77**, 923–933 (2005).
50. Rollins, B.J. Chemokines. *Blood* **90**, 909–928 (1997).
51. Mantovani, A. The chemokine system: redundancy for robust outputs. *Immunol Today* **20**, 254–257 (1999).
52. Mantovani, A., Sica, A., Sozzani, S., *et al.* The chemokine system in diverse forms of macrophage activation and polarization. *Trends Immunol* **25**, 677–686 (2004).
53. Barnes, P.J. The cytokine network in asthma and chronic obstructive pulmonary disease. *J Clin Invest* **118**, 3546–3556 (2008).
54. Hartert, T.V. and Peebles, R.S., Jr. Epidemiology of asthma: the year in review. *Curr Opin Pulm Med* **6**, 4–9 (2000).
55. Kiley, J., Smith, R., and Noel, P. Asthma phenotypes. *Curr Opin Pulm Med* **13**, 19–23 (2007).
56. Laitinen, L.A., Heino, M., Laitinen, A., *et al.* Damage of the airway epithelium and bronchial reactivity in patients with asthma. *Am Rev Respir Dis* **131**, 599–606 (1985).
57. Shahana, S., Bjornsson, E., Ludviksdottir, D., *et al.* Ultrastructure of bronchial biopsies from patients with allergic and non-allergic asthma. *Respir Med* **99**, 429–443 (2005).
58. Mattoli, S. Allergen-induced generation of mediators in the mucosa. *Environ Health Perspect* **109** (Suppl 4), 553–557 (2001).
59. Marini, M., Avoni, E., Hollemborg, J., *et al.* Cytokine mRNA profile and cell activation in bronchoalveolar lavage fluid from nonatopic patients with symptomatic asthma. *Chest* **102**, 661–669 (1992).
60. Ackerman, V., Marini, M., Vittori, E., *et al.* Detection of cytokines and their cell sources in bronchial biopsy specimens from asthmatic patients: relationship to atopic status, symptoms, and level of airway hyperresponsiveness. *Chest* **105**, 687–696 (1994).
61. Elias, J.A., Zhu, Z., Chupp, G., *et al.* Airway remodeling in asthma. *J Clin Invest* **104**, 1001–1006 (1999).
62. Bousquet, J., Jeffery, P.K., Busse, W.W., *et al.* Asthma. From bronchoconstriction to airway inflammation and remodeling. *Am J Respir Crit Care Med* **161**, 1720–1745 (2000).

63. Brewster, C.E.P., Howarth, P.H., Djukanovic, R., *et al.* Myofibroblasts and subepithelial fibrosis in bronchial asthma. *Am J Respir Cell Mol Biol* **3**, 507–511 (1990).
64. Stewart, A.G., Tomlinson, P.R., and Wilson, J. Airway wall remodeling in asthma: a novel target for the development of anti-inflammatory drugs. *Trend Pharmacol Sci* **14**, 275–279 (1993).
65. Gabbrielli, S., Di Lollo, S., Stanflin, N., *et al.* Myofibroblasts and elastic and collagen fiber hyperplasia in the bronchial mucosa: a possible basis for the progressive irreversibility of airflow obstruction in asthma. *Pathologica* **86**, 157–160 (1994).
66. Fish, J.E. and Peters, S.P. Airway remodeling and persistent airway obstruction in asthma. *J Allergy Clin Immunol* **104**, 509–516 (1999).
67. Rose, C.E., Jr., Sung, S.J., and Fu, S.M. Significant involvement of CCL2 (MCP-1) in inflammatory disorders of the lung. *Microcirculation* **10**, 273–288 (2003).
68. Mattoli, S. and Schmidt, M. Role in asthmatic lung disease. Chapter 6. In: Bucala, R. (ed.), *Fibrocytes: New Insights into Tissue Repair and Systemic Fibroses* (Singapore: World Scientific Publishing Co. Pte. Ltd., 2007), pp. 105–123.
69. O'Byrne, P.M., Dolovich, J., and Hargreave, F.E. Late asthmatic responses. *Am Rev Respir Dis* **136**, 740–751 (1987).
70. Lemanske, R.F. and Kaliner, M.A. Late phase allergic reactions. In: Middleton, E., Jr., Reed, C.E., Ellis, E.F., Adkinson, N.F., Yuninger, J.W., and Busse, W.W. (eds.), *Allergy: Principles and Practice*, 4th edition, (St. Louis: Mosby Yearbook, 1993), pp. 320–361.
71. Gizycki, M.J., Adelroth, E., Rogers, A., *et al.* Myofibroblast involvement in the allergen-induced late response in mild atopic asthma. *Am J Respir Cell Mol Biol* **16**, 664–673 (1997).
72. Kariyawasam, H.H., Aizen, M., Barkans, J., *et al.* Remodeling and airway hyperesponsiveness but not cellular inflammation persist after allergen challenge in asthma. *Am J Respir Crit Care Med* **175**, 896–904 (2007).
73. Saunders, R., Siddiqui, S., Kaur, D., *et al.* Fibrocyte localization to the airway smooth muscle is a feature of asthma. *J Allergy Clin Immunol* [published online ahead of print] (2009) doi:10.1016/j.jaci.2008.10.048.
74. Wang, C.-H., Huang, C.-D., Lin, H.-C., *et al.* Increased circulating fibrocytes in asthma with chronic airflow obstruction. *Am J Respir Crit Care Med* **178**, 583–591 (2008).

75. American Thoracic Society, European Respiratory Society. American Thoracic Society/European Respiratory Society international multidisciplinary consensus classification of the idiopathic interstitial pneumonias. *Am J Respir Crit Care Med* **165**, 277–304 (2002).
76. Gross, T.J. and Hunninghake, G.W. Idiopatic pulmonary fibrosis. *N Engl J Med* **345**, 517–525 (2001).
77. Selman, M., King, T.E., and Pardo, A. Idiopatic pulmonary fibrosis: prevailing and evolving hypotheses about its pathogenesis and implications for therapy. *Ann Intern Med* **134**, 136–151 (2001).
78. Thannickal, V.J., Toews, G.B., White, E.S., *et al.* Mechanisms of pulmonary fibrosis. *Annu Rev Med* **55**, 395–417 (2004).
79. Mason, R.J., Schwarz, M.I., Hunninghake, G.W., *et al.* NHLBI workshop summary, pharmacological therapy for idiopatic pulmonary fibrosis. Past, present and future. *Am J Respir Crit Care Med* **160**, 1771–1777 (1999).
80. Perez, A., Rogers, R.M., and Dauber, J.H. The prognosis of idiopatic pulmonary fibrosis. *Am J Respir Cell Mol Biol* **29**(3 Suppl), S19–S26 (2003).
81. Katzenstein, A.A. and Myers, J.L. Idiopatic pulmonary fibrosis: clinical relevance of pathologic classification. *Am J Respir Crit Care Med* **157**, 1301–1315 (1998).
82. King, T.E., Jr., Schwarz, M.I., Brown, K., *et al.* Idiopatic pulmonary fibrosis: relationship between histopathologic features and mortality. *Am J Respir Crit Care Med* **164**, 1025–1032 (2001).
83. Choi, E.S., Pierce, E.M., Jakubzick, C., *et al.* Focal interstitial CC chemokine receptor 7 (CCR7) expression in idiopatic interstitial pneumonia. *J Clin Pathol* **59**, 28–39 (2006).
84. Mehrad, B., Burdick, M.D., Zisman, D.A., *et al.* Circulating peripheral blood fibrocytes in human fibrotic interstitial lung disease. *Biochem Biophys Res Commun* **353**, 104–108 (2007).
85. Andersson-Sjöland, A., de Alba, C.G., Nihlberg, K., *et al.* Fibrocytes are a potential source of lung fibroblasts in idiopatic pulmonary fibrosis. *Int J Biochem Cell Biol* **40**, 2129–2140 (2008).
86. Murdoch, C., Giannoudis, A., and Lewis, C.E. Mechanisms regulating the recruitment of macrophages into hypoxic areas of tumors and other ischemic tissues. *Blood* **104**, 2224–2234 (2004).
87. Ishii, G., Sangai, T., Oda, T., *et al.* Bone-marrow-derived myofibroblasts contribute to the cancer-induced stromal reaction. *Biochem Biophys Res Commun* **309**, 232–240 (2003).

88. Direkze, N.C., Hodivala-Dilke, K., Jeffery, R., *et al.* Bone marrow contribution to tumor-associated myofibroblasts and fibroblasts. *Cancer Res* **64**, 8492–8495 (2004).
89. Sangai, T., Ishii, G., Kodama, S., *et al.* Effect of differences in cancer cells and tumor growth sites on recruiting bone marrow-derived endothelial cells and myofibroblasts in cancer-induced stroma. *Int J Cancer* **115**, 885–892 (2005).
90. Wu, Y., Li, Y.-Y., Matsushima, K., *et al.* CCL3-CCR5 axis regulates intratumoral accumulation of leukocytes and fibroblasts and promotes angiogenesis in murine lung metastasis process. *J Immunol* **181**, 6384–6393 (2008).
91. Vaage, J. and Harlos, J.P. Collagen production by macrophages in tumour encapsulation and dormancy. *Br J Cancer* **63**, 758–762 (1991).
92. Bridle, K.R., Crawford, D.H., Powell, L.W., *et al.* Role of myofibroblasts in tumour encapsulation of hepatocellular carcinoma in haemochromatosis. *Liver* **21**, 96–104 (2001).
93. Gharaee-Kermani, M., Gyetko, M.R., Hu, B., *et al.* New insights into the pathogenesis and treatment of idiopatic pulmonary fibrosis: a potential role for stem cells in the lung parenchyma and implications for therapy. *Pharm Res* **24**, 819–841 (2007).
94. Pilling, D., Roife, D., Wang, M., *et al.* Reduction of bleomycin-induced pulmonary fibrosis by serum amyloid P. *J Immunol* **179**, 4035–4044 (2007).
95. Olson, T. and Ley, K. Chemokines and chemokine receptors in leukocyte trafficking. *Am J Physiol Regul Integr Comp Physiol* **283**, R7–R28 (2002).

Chapter 9
Three-Dimensional Pulmonary Constructs

Christine M. Finck, Blair Roszell and Peter I. Lelkes

In this chapter we discuss how to engineer 3D pulmonary tissue constructs *in vitro* using primary isolates of foetal mouse distal lung cells. When cultured in hydrogel-based 3D constructs, the mixed cell population, comprised epithelial, mesenchymal and endothelial cells, organised into alveolar forming unit (AFU)-like sacculated structures, which, in terms of morphology and cytodifferentiation, were reminiscent of native distal lung. By using a unique, serum-free medium supplemented with a cocktail of tissue-specific growth factors, we were able to induce concomitant alveolisation and neovascularisation when culturing the cells in the hydrogels, but not in scaffolds composed of synthetic polymers. Our data suggest that our *in vitro* model is capable of recapitulating the parallel morphogenesis of epithelial and endothelial pulmonary tissue components, which may occur through dynamic paracrine interactions. These results also stress the importance of the complex input from co-cultures, tissue-specific growth factors and integrin signalling for successful tissue engineering *in vitro*. In a mouse model *in vivo*, incorporation of the primary lung cell isolates into Matrigel plugs, implanted either subcutaneously (s.c.), or under the kidney capsule, leads to the formation of sacculated AFUs in close proximity to patent capillaries. Effective functional vascularisation, however, was only observed upon addition of angiogenic growth factors to the scaffolds and their controlled release over time. Use of a fluorescent cell tracker confirmed that the neovessels in the constructs comprised endothelial cells from both the host and the grafts. These data demonstrate that it is feasible to generate vascularised pulmonary tissue constructs *in vivo* with proper epithelial differentiation, and that the degree of vascularisation may be manipulated by incorporating the release of an angiogenic factor within the construct.

Keywords: Mouse; distal lung; foetal pulmonary cells; tissue engineering; collagen; alveoli; endothelial cells; fibroblast growth factor; capillaries.

Outline

1. Introduction

Regeneration and replacement of damaged tissues remains a challenge to both medicine and bioengineering. One goal for tissue engineers is to develop biomimetic substitutes that can restore functionality to engineered tissue. Such a regenerated tissue will allow for maintenance of cell and tissue architecture with normal mechanical and physiologic properties.[1] Some measure of success has been achieved with "simple" tissues such as cartilage, blood vessels, currently in clinical trials, and skin that is being sold commercially.[2,3] Using appropriate combinations of cells, scaffolds and culture conditions, researchers were recently able to tissue-engineer a functional bladder.[3] Machiarini *et al*. recently reported the successful implantation of a tissue-engineered trachea that was prepared by seeding a decellularised donor with autologous cells and mesenchymal stem cell-derived chondrocytes obtained from the recipient patient.[4]

While the ultimate goal remains full *in situ* regeneration of whole organs, such as an entire limb, recent advances in biotechnology might advance our capability to engineer more complex organ growth *in vitro* and then to utilise these engineered tissues for clinical purposes, such as high-fidelity 3D tissue models for drug/toxicology screening, or for studying basic mechanisms of tissue assembly in health and disease. The ability to grow functional three-dimensional lung tissue from cultured cells for transplantation would be invaluable in ameliorating the morbidity and/or preventing mortality associated with neonatal pulmonary hypoplasia or bronchopulmonary dysplasia, both of which are of significant concern in perinatal medicine. For example, pulmonary hypoplasia is found in as many as 15–20% of all neonatal autopsies, contributing to more than 2,800 deaths annually.[5] In adults, chronic obstructive pulmonary disease (COPD) is the fourth highest cause of death in the USA.[6] Therefore, restoring functionality to poorly developed neonatal lungs or damaged adult lungs would be life-saving.

2. Culture of Distal Lung Cells *In Vitro*

The focus of this chapter is on the distal lung, the formation of which is a complex developmental process involving distinct inductive cues from (and heterotypic cross-talk between) cells derived from the different germ lines. Foetal development and differentiation of the distal lung encompasses a number of different growth factors, such as FGFs, VEGF and HGF, and distinct signalling pathways, such as wnt, shh, FGF and VEGF.[7–9] Similarly, multiple steps are required to generate a functionally and anatomically correct 3D distal lung tissue model *in vitro*, with increasing difficulty at each successive iterative step. The cellular composition of the distal airways is multifaceted, consisting of pneumocytes (type I and type II), endothelial cells, and mesenchymal cells, in addition to some 40 other cell subtypes. Hence, expansion of the primary cell isolates, including maintaining the diverse cellular phenotypes and re-establishing appropriate cell-to-cell interactions, will be essential to the long-term goal of tissue-engineering a high-fidelity 3D model of the distal lung (see also Chapter 6).

3. Two-Dimensional Models

Previous attempts at culturing purified populations of (neonatal and adult) type II alveolar epithelial cells, mostly in two-dimensional monolayers, provide a basic understanding of pulmonary epithelial transport properties and/or function, but yield a less than optimal model for lung tissue engineering.[10,11] As a caveat, culturing methods for alveolar type II cells differ depending on the species and age of the donor. Differences in methods of primary culture have been reported for human and rat foetal cells;[12,13] reagents used to prevent fibroblast overgrowth in human tissue isolates may not be effective when used in rat pulmonary cell cultures.[13] If the need arises to expand the primary isolates, the cells should be cultured in a defined medium in which foetal calf serum is replaced with a specific cocktail of adhesion and growth factors to maintain phenotypic diversity, avoid epithelial dedifferentiation and prevent overgrowth of the cultures with mesenchymal cells.

Dame Julia Polak and collaborators have sought to alleviate this issue by establishing methods to drive embryonic human and mouse embryonic stem (ES) cells into the distal lung cell phenotype[14,15] and then culturing the ES-derived pulmonary cells on 3D synthetic scaffolds.[16] At present these approaches are hampered mainly by the very low yield and the lack of functional (and morphologically correct) assembly of the alveolar-like structures. Cortiella and co-workers[17] have used adult ovine somatic lung progenitor cells to generate a mixed population of distal lung progenitor cells, which were grown on synthetic scaffolds made of Pluronic 127 and yielded "identifiable pulmonary structures" upon subcutaneous implantation, but not *in vitro*. Another, more recent approach to distal airway engineering is to utilise mixed populations of foetal pulmonary cells,[18,19] based on the well-documented fact that pure populations of type II pneumocytes isolated from either neonatal or adult animals lose many of their special features, such as the ability to produce surfactant protein.[20,21] Morphological change and lamellar body loss can be demonstrated after several days of primary culture and suggests that alveolar type II cells in 2D monoculture rapidly dedifferentiate.[22] Attempts to solve this problem include altering the physical environment by providing an air–liquid interface,[23] or, alternatively, using cells from mature foetuses in combination with plating the cells on physiologic (i.e. less stiff) substrates such as a gelatin disc, Matrigel™ or collagen.[24–26] Rather than attempting to establish purified monocultures of pulmonary (epithelial) cells, we took the approach that preserving heterotypic cell populations will facilitate the maintenance of distinct cellular phenotypes present in the primary

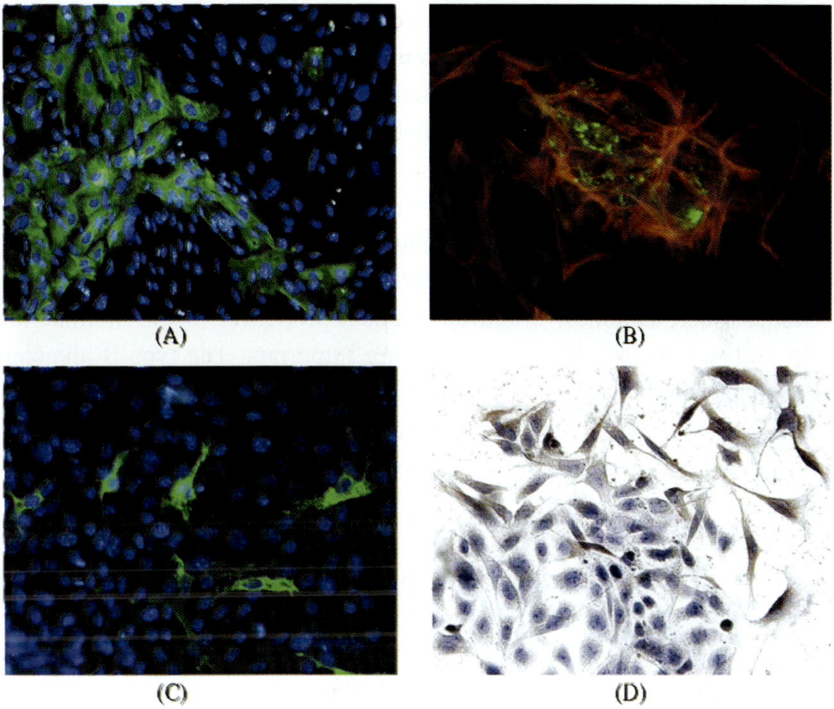

Figure 1: Immunohistochemical analysis of primary foetal pulmonary cell isolates following four days of *in vitro* culture on Matrigel-coated tissue culture chamber slides with SF-ITS medium (except C with SF-GF medium). (A) Cytokeratin (green) staining of epithelial cells, bisbenzimide (blue) nuclear counterstain (200×). (B) Prosurfactant protein C (green) staining, indicative of AE2 cells, rhodamine phalloidin (red) cytoskeletal counter-staining (400×). (C) GSL Iso B4 (green) staining of endothelial cells, bisbenzimide (blue) nuclear counterstaining (400×). (D) Vimentin (brown, DAB horse radish peroxidase) staining of mesenchymal cells, haematoxylin (blue) nuclear counterstaining (400×).[18]

isolates. Indeed, mixed populations of foetal pulmonary cells (FPC) isolated from E17-day mouse embryos and co-cultured on Matrigel™-coated 2D surfaces maintained epithelial, mesenchymal, and endothelial differentiation (Fig. 1).

4. Three-Dimensional Growth

It has been shown that maintaining alveolar cell shape in culture preserves the differentiated state, as evidenced by surfactant synthesis.[27,28] Thus, establishing 3D heterogeneous co-cultures using biomimetic scaffolds (see below) in combination with optimised media may be a valid approach to preventing lung cell dedifferentiation. Successful 3D organ growth requires an environment that facilitates cellular adhesion and provides tissue-specific growth factors within a 3D scaffold. Ideally, once the organ-specific morphology is established, degradation of the scaffold over time would be desirable to leave behind functional 3D lung constructs. In order to promote proper cell differentiation, the physicochemical characteristics, such as adhesive and mechanical properties of the biomaterial onto/into which the cells are seeded, are important. The type II alveolar cell phenotype is attenuated when seeded on plastic in a planar configuration or on bovine corneal endothelial cell matrix, human amniotic membrane, fibronectin, and type IV collagen.[29–31] When these type II cells are grown in collagen gels or on complex matrices, this transition is not as prominent.[29] Cross-linked type I collagen films support some respiratory epithelial cell growth; however, collagen gels may be far superior.[1] The basement membrane in the lung is composed of type IV collagen, laminin, entactin, perlecan, SPARC, and fibronectin.[32] The closest available product to extracellular matrix (ECM) is Matrigel™, a reconstituted basement membrane derived from the Engelbreth-Holm-Swarm (EHS) mouse sarcoma. Matrigel™ is composed of ECM proteins laminin, collagen IV, nidogen/entactin and proteoglycan, as well as a large number of growth factors, such as TGFßs, FGFs, EGF, PDGF and IGF.[33] Pulmonary type II cells grown on Matrigel™ retain their differentiated form when compared with cultured cells on a plastic surface.[26,27] Mondrinos *et al.*[18] demonstrated that foetal pulmonary cells (FPC) cultured in 3D Matrigel™ hydrogels assembled into cyst-like structures that displayed branching morphogenesis (Fig. 2A), in addition to maintenance of epithelial differentiation (Figs. 2C and 2D). Thus these structures histologically resemble distal mouse foetal (E17) lung in the canalicular stage (Fig. 2B).

The drawback of these gelatinous matrices is that they lack defined 3D architecture and have poor mechanical properties. The natural ECM of the lung displays a fibrous morphology.[34] For this reason, we and others have investigated the use of fibrous scaffolds fabricated by electrospinning of natural ECM proteins, such as collagen or elastin, to generate 3D matrices for lung cell growth.[17,18,35] The small diameters of the

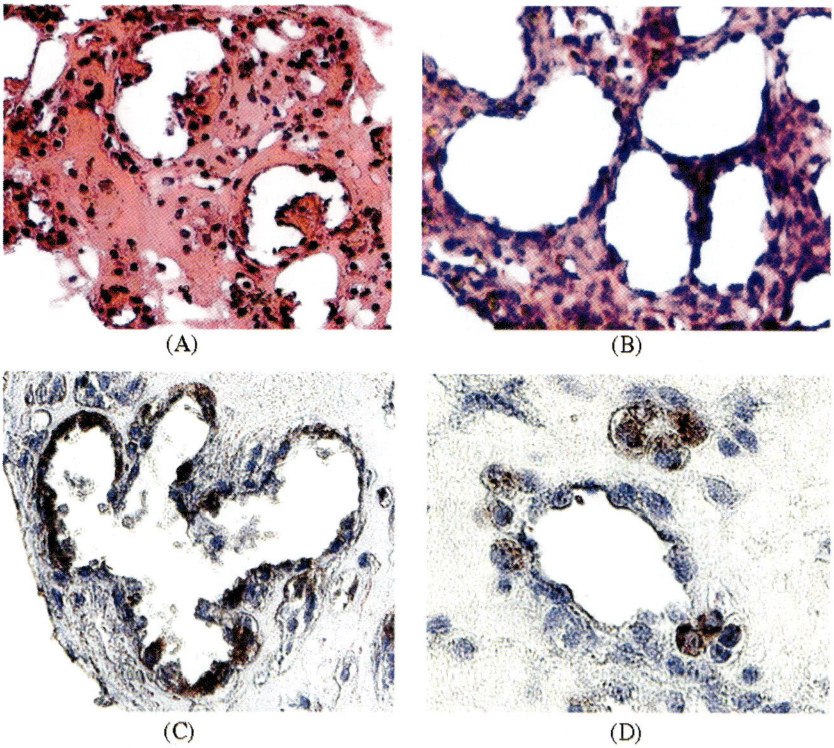

Figure 2: Histological and phenotypic analysis of foetal pulmonary cells cultured in Matrigel hydrogel for one week with SF-GF medium. (A) H&E staining of foetal pulmonary cell construct (400×). (B) H&E staining of embryonic day-17 mouse lung for histological comparison (400×). (C) Cytokeratin (red, AEC horse radish peroxidase) immunostaining of foetal pulmonary cell constructs demonstrating epithelial lining of branching AFU structures (original magnification 400×). (D) Prosurfactant protein C (red, AEC horse radish peroxidase) staining of foetal pulmonary constructs demonstrating presence of type II alveolar epithelial cells (original magnification 400×).[18]

electrospun fibres have a large surface-to-volume ratio, which enables absorption of liquids while facilitating cellular adhesion and cell–cell contact.[36] Recent results suggest that nano-topography is beneficial for maintaining the function and differentiation of some mesenchymal cell types.[37] In line with this notion, culture of FPC on fibrous or porous scaffolds made of biodegradable synthetic polymers, such as PLGA or PLLA (Fig. 3A), did not result in histotypic alveolar morphogenesis.

vasculogenesis. In studies of two-dimensional lung culture, addition of FGF10 can induce generalised budding.[39]

The formation of histotypic 3D branching structures internally lined with epithelial cells from cultured FPCs in a complex matrix like Matrigel™ is not surprising. Importantly, these structures formed in both a defined medium as well as in a medium containing foetal bovine serum (FBS), suggesting that morphogenic/differentiative effects of the plethora of growth factors contained in Matrigel™ may override the well-known tendency of FBS for preferentially stabilizing mesenchymal differentiation. Culturing the heterotypic FPS isolates in a better defined 3D environment, such as a hydrogel made of type I collagen in conjunction with a serum-free, defined medium (SF), reveals the pivotal role that specific growth/differentiation factors (GF) play in formation of the distal alveolar-like structures *in vitro*.[19] In emulating the involvement of multiple growth factors in the development of the distal lung, we chose to explore in our 3D *in vitro* model a cocktail of three members of the fibroblast growth factor family, namely FGFs 2, 7, and 10. Culturing FPC in 3D collagen gels in a FBS-containing medium results in the induction of a round/elliptical cyst (Fig. 4A), while in the absence of FBS, a defined medium supplemented with ITS (Insulin-Transferrin-Selenium) and a cocktail of fibroblast growth family members (FGFs 2/7/10), the cells organise into a elaborate branching structure (Fig. 4B).

The cysts formed in the presence of FBS (Fig. 4A), or in serum free-medium supplemented only with ITS, are "solid" amorphous epithelial aggregates, i.e. filled with cytokeratin-positive cells (Fig. 4C). By contrast, in the presence of optimised medium (SF-GF), the branching alveolar-like structures are hollow and lined with cytokeratin-positive epithelial cells (Fig. 4D).

Shown in Figs. 5–7 is the effect of each of the FGFs, alone or in the various combinations, on both the epithelial and endothelial component in the 3D constructs. The spherical (epithelial) cysts formed in the basal media + ITS (Fig. 5A) are somewhat dilated in the presence of either FGF 7 or FGF 2 (Figs 5C and 5D), while FGF 10 alone induces significant branching (Fig. 5B). FGF 7/10 seem to enhance the sacculation (Fig. 5E), while addition of all three FGFs (2/7/10) leads to a complex, sacculated 3D structure, including the formation of clefts (Fig. 5F), as seen *in situ* in the canalicular stage of alveolar differentiation in foetal distal lungs.

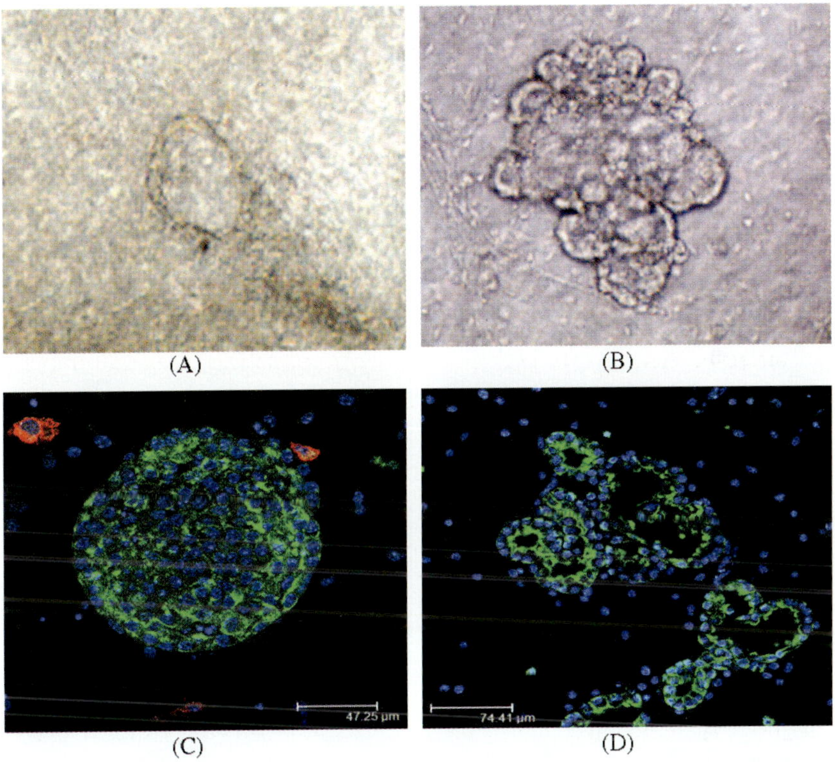

Figure 4: Morphology of FPC cultured for seven days in type I collagen hydrogels in medium supplemented with FBS (A and C) or a cocktail of growth factors (ITS + FGF2, 7, 10, B and D). The cells inside the solid cyst (C) as well as the ones lining the luminal faces of the hollow branching structures (D) are of epithelial nature, as revealed by immunofluorescent staining for cytokeratin (green). Nuclear counterstaining with bisbenzimide (blue). (A) and (B). Phase contrast images (original magnification – 100×). (C) and (D): Single optical slices made by confocal microscopy.[19]

In addition to studying the role of FGF supplementation on the *in vitro* morphogenesis of the 3D constructs, we also evaluated whether the same FGFs, alone or in combination, might affect the differentiation of epithelial cells lining the lumen of the alveolar-like cysts. Specifically, the samples were immunostained for prosurfactant protein C (pro-SpC), a marker for AE II cells. As seen in Fig. 6, all constructs stain positive for pro-SpC, indicating that when cultured in permissive 3D hydrogels, the AEII retain their cytodifferentiation and organise in a histotypic fashion, i.e. they will line the inner face of alveolar-forming units (AFUs). Semi-quantitative

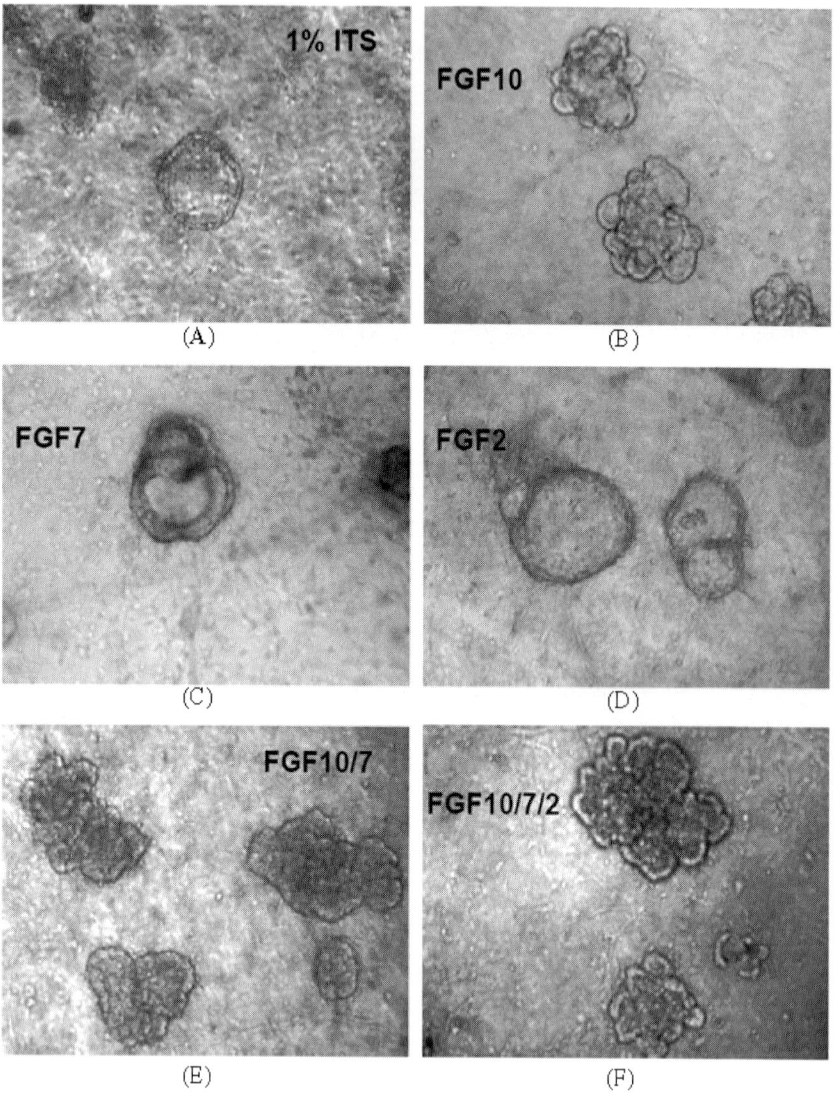

Figure 5: FGF supplementation modulates epithelial morphology in 3D constructs resembling alveolar forming units (AFUs). Representative phase contrast micrographs of AFUs following seven days of culture in the presence of FGF10, FGF7 and FGF2 alone and in combination (100x magnification in all panels). (A) 1% ITS, (B) FGF10, (C) FGF7, (D) FGF2, (E) FGF10/7, and (F) FGF 10/7/2.[19]

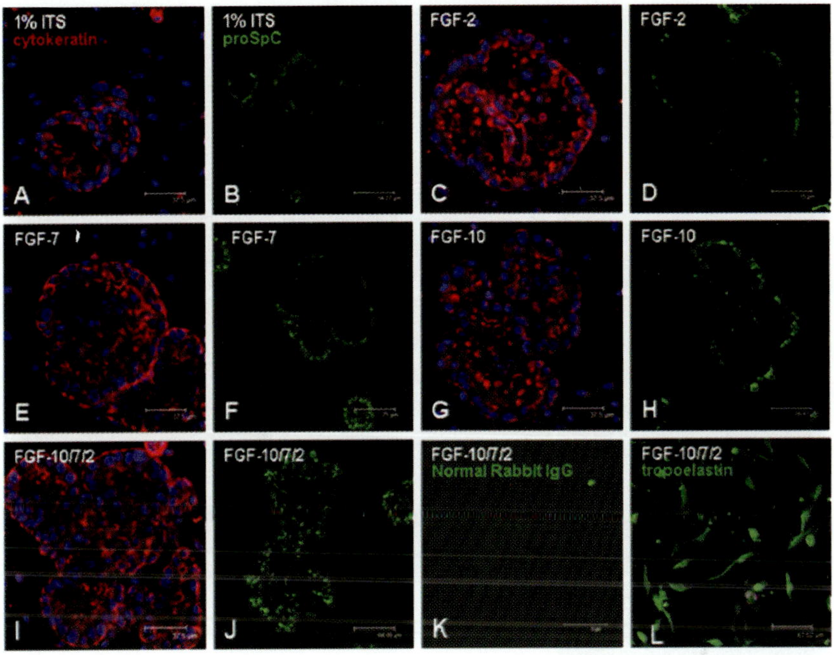

Figure 6: Confocal microscopic visualisation of epithelial morphogenesis and cytodifferentiation. Optical sections through AFUs stained for cytokeratin (red) to visualise epithelial cells and counterstained with DAPI (blue) for nuclei. Cytodifferentiation was assessed by prosurfactant protein C (proSpC, green) staining (B, D, F, H and J). (A) 1% ITS, scale bar = 37.5 μm; (B) 1% ITS, scale bar = 54 μm; (C) FGF2, scale bar = 37.5 μm; (D) FGF2, scale bar = 75 μm; (E) FGF7, scale bar = 37.5 μm; (F) FGF7, scale bar = 75 μm; (G) FGF10, scale bar = 37.5 μm; (H) FGF10, scale bar = 50 μm; (I) FGF10/7/2, scale bar = 37.5 μm; (J) FGF10/7/2, scale bar = 64 μm; (K) FGF10/7/2, normal rabbit IgG, scale bar = 75 μm; and (L) FGF10/7/2, tropoelastin staining of fibroblastic cells, scale bar = 48 μm.[19]

evaluation of the images shown in Fig. 6 indicates that the strongest staining (largest number of SpC positive and highest level of SpC expression) is observed in the samples exposed to the combination of all three FGFs (Fig. 6J). In addition to epithelial cysts, the interstitial spaces between the AFUs abound, especially in the case of FGF 2/7/10 supplementation with tropoelastin-positive mesenchymal cells (Fig. 6L).

Since primary FPC isolates contain abundant endothelial cells (EC; see Fig. 1), we also investigated the role of FGF supplementation in maintaining an endothelial phenotype and inducing vascular morphogenesis,

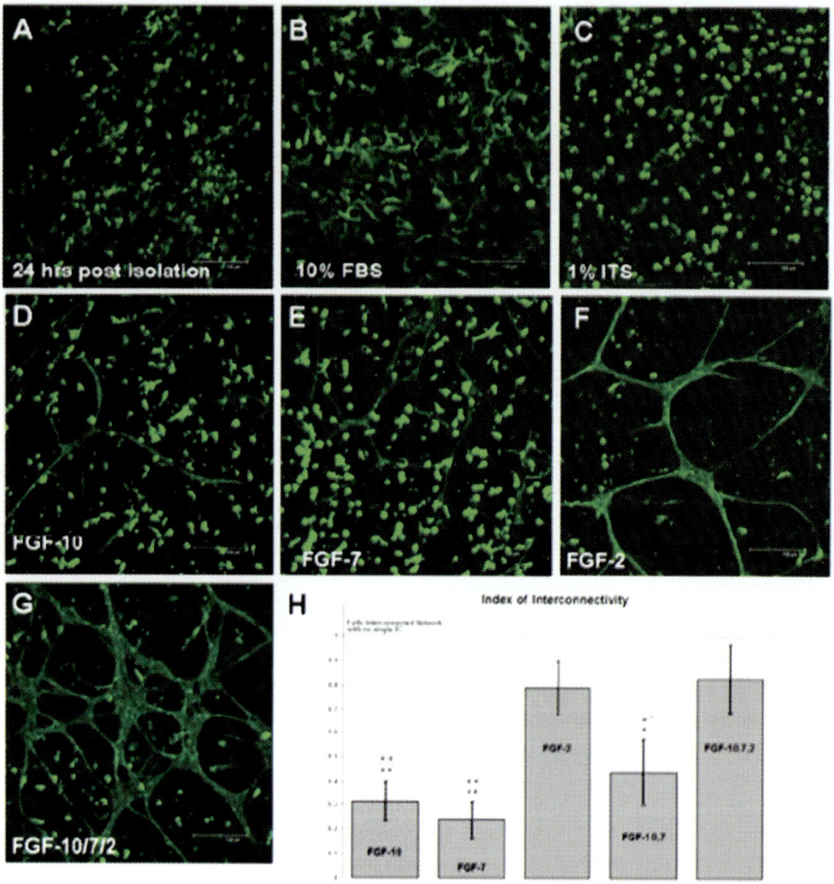

Figure 7: Modulation of vascular morphogenesis by FGF following seven days of *in vitro* culture. The images shown in A–G are 3D z-projections of confocal images (75–100 micron thick segments) of constructs stained with isolectinB4 for endothelial cells (green). (A) Distribution of endothelial cells within constructs 24 hours post-isolation, prior to application of growth factors (day 0 of experiment); (B–G) from seven-day cultures; (B) 10% FBS; (C) 1% ITS; (D) FGF10; (E) FGF7; (F) FGF2; and (G) FGF10/7/2 (scale bar = 150 microns in all images). (H) Quantitative image analysis of isolectinB4 staining of endothelial network formation across FGF supplementation conditions.[19]

specifically the formation of capillary-like networks, using Griffonia simplifolica lectin (GSL)-isoform 4 as a specific marker for murine endothelial cells.[19]

As seen in Fig. 7A, 24 hours post-isolation, the cultures contain large numbers of GSL-4 positive endothelial cells. Upon seven days in 3D culture in collagen gels, the cells cultured in FBS appear mesenchymal/fibroblast-like (flattened, elongated, disorganised; Fig. 7B), while most of the ECs maintained in 1% ITS remain rounded (Fig. 7C). FGF 10 and 7 induce a low degree of endothelial "tubular" morphogenesis (Figs. 7D and 7E, respectively). It will be of interest to study the differential sensitivity of individual EC to these growth factors/morphogens (endothelial cell heterogeneity). By contrast, in the presence of FGF2, essentially all EC assemble into a network of thin capillary-like structures (Fig. 7F). The visual impression can easily be substantiated by quantitative image analysis, confirming that FGF2 and the combination of the three FGFs generate an essentially similar degree of interconnectivity.[19] However, the combination of FGFs 2/7/10 yields a much more robust network of thickened capillary-like structures. Taken together, these data demonstrate the differential susceptibility of epithelial and endothelial cells to the morphogenic actions of distinct members of the fibroblast growth factor family.

6. Importance of Endothelial–Epithelial Interactions in Heterogeneous Co-Culture

Heterogeneous cultures of cells are capable of maintaining epithelial differentiation and growth.[49] In the lung, there is evidence that type II cell–fibroblast interactions maintain type II cell differentiation.[28] For example, direct contacts between alveolar type II cells and fibroblasts appear to have a trophic effect on cultured alveolar type II cells, as evidenced by increasing levels of mRNA for surfactant protein A (SpA).[49] Type II cell proliferation is *enhanced* in culture by an ECM with fibroblasts beneath it,[46] while in another co-culture system, alveolar type II cells *inhibited* fibroblast proliferation, an effect postulated to be mediated through prostaglandin E2.[50] Rat lung alveolar type II cells appear to release a factor that stimulates type I collagen secretion by fibroblasts, thus demonstrating the communication that occurs between mesenchymal and epithelial cells.[10] The production of laminin α5 has been shown *in vitro* to enhance the development of distal airways.[51] This laminin is only produced with epithelial–mesenchymal cell contact.[51] Finally, the pulmonary mesenchyme can act instructively on the tracheal epithelium to induce a program of type II cell differentiation.[52]

There is also evidence that mitogenic factors are stored in the endothelial extracellular matrix. Organs such as the liver require vasculogenic endothelial cells and nascent vessels for early stages of organogenesis.[53] Alveolar cells grown on an endothelial matrix showed the greatest adherence, cell division and spreading to reach confluence compared with cells on an epithelial matrix or collagen.[54] The results suggest that some component(s) of the endothelial matrix at the alveolar basement membrane facilitates epithelial cell growth.

Endothelial–epithelial interactions appear to be reciprocal; Mondrinos *et al.*[19] showed that foetal pulmonary cells stimulated with epithelial-specific growth factors respond by initiating a program of vascular network formation characterised by tight interactions with alveolar epithelial buds[19] (Fig. 8). Under sub-optimal conditions (SF medium supplemented only with ITS), endothelial cells migrate to the vicinity of epithelial cell clusters and seem to undergo morphogenic changes when in close apposition/direct contact with the epithelial clusters (Fig. 8A).

By contrast, under optimised conditions (SF-GF medium), EC capillary-like structures enrobe the sacculated epithelial structures and penetrate the clefts (Fig. 8B), thus closely representing true "alveolar forming units". In addition, the capillary-like vascular networks surrounding the epithelial alveolar-like structures exhibit true lumen (Fig. 8C) and expand by "classical" sprouting angiogenesis (Fig. 8D).

Taken together, it seems clear that the cross-talk between the epithelial and endothelial components of the alveolar forming units also occurs *in vitro*, similar to what has been reported *in vivo*. This bidirectional signalling appears to be pivotal for establishing the proper morphology (and function) of the emerging alveolar-like structures *in vitro*, thus closely mimicking the formation of the blood–air interface that provides for the gas exchange across the alveoli and capillaries. The capability of recapitulating the formation of the distal airways in an *in vitro* model opens new avenues for studying the mechanisms that lead to the normal and pathologic development of these structures. Such tissue constructs representing high-fidelity models of normal and pathological states of distal lung could accelerate high-throughput drug discovery and/or toxicity assessments. Moreover, in honing the ability to generate morphologically correct alveolar forming units, this 3D model might serve as the basis for

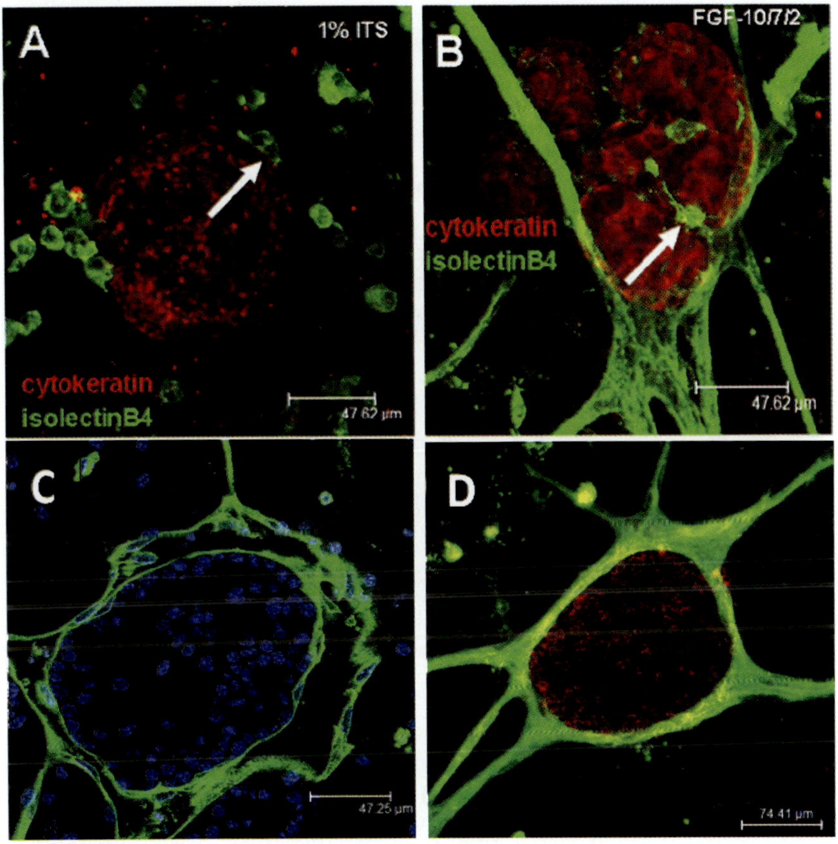

Figure 8: Visualisation of epithelial–endothelial interaction; 75–100 one-micron thick reconstructed confocal segments of AFU structures in seven-day constructs maintained in either 1% ITS (A) or FGF10/7/2 (B) with epithelial cells cytokeratin-labelled (red) and endothelial cells isolectinB4-labelled (green). Note the polarised appearance of the single EC tightly interfaced with the AFU in (A) (arrow). By contrast the AFU in (B), *viz.* in the FGF10/7/2 supplemented medium, is enrobed by interconnected EC (arrow). (C) Formation of lumen containing endothelial cell lined microvasculature surrounding the cytokeratin positive "epithelial cyst" single confocal optical section. (D) Sprouting angiogenesis of EC surrounding epithelial cell containing 75–100 reconstituted segments, as above. Scale bar = 48 μm (A, B and C), 75 μm (D).[19]

a tissue engineering-based approach for augmenting/repairing/replacing diseased lungs in neonates and/or adults. For that, however, one has to show first that the *in vitro* generated 3D models can also integrate into host tissue.

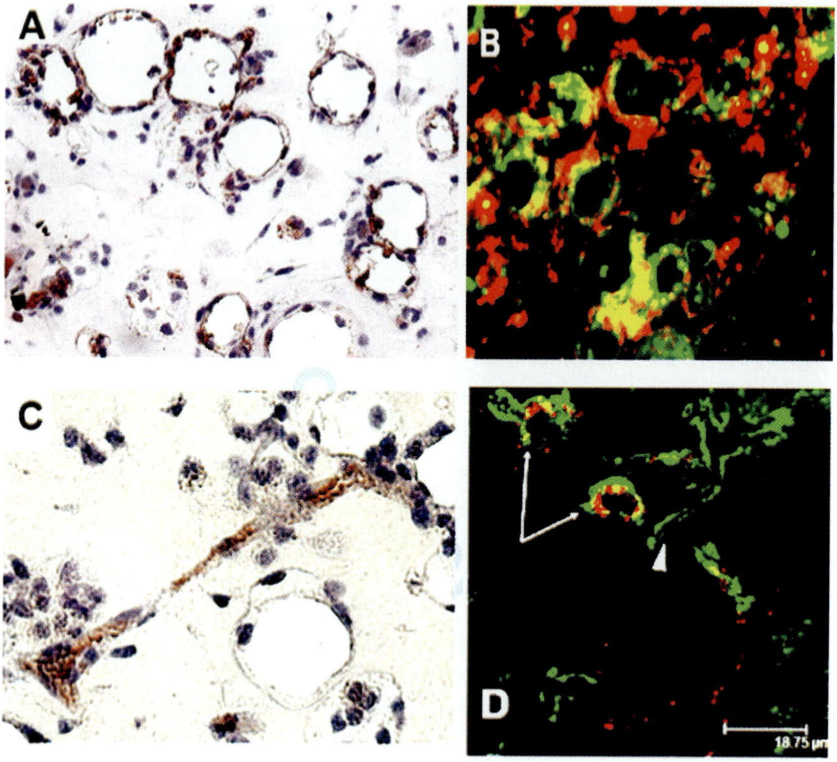

Figure 9: Immunohistochemical staining of alveolar constructs generated for seven days *in vivo*. (A) Immunoperoxidase staining for cytokeratin (AEC substrate chromogen reaction product appears in red) confirms the epithelial phenotype of the abundant ductal structures observed within Matrigel + FPC + FGF2 constructs (magnification = 400×). (B) Prosurfactant protein C immunostaining (green) confirms the distal lung differentiation status of the cells lining the ductal epithelium, the graft origin of which is highlighted by CMTPX CellTracker™ labelling (orange) prior to admixing in Matrigel™ (magnification = 400×). (C) Immunoperoxidase staining for von Willebrand factor (vWF, AEC substrate chromogen reaction product appears in red) illustrates the presence of vWF-positive endothelial cells (ECs) within the constructs (original magnification 400×). (D) Confocal optical section (250 nm thickness) of an isolectinB4 whole mount stained MG + FPC + FGF2 construct, identifying endothelial cells (green), along with CMTPX labelling to identify ECs derived from engrafted FPC (orange) reveals mixed graft (arrows) and host (arrowhead) origin of endothelial structures (scale bar = 18.75 microns).[55]

As a first step toward this goal, we have implanted *in vitro* generated 3D distal lung constructs subcutaneously and under the kidney capsule in a murine model, and demonstrated both continued maintenance of the differentiated phenotypes of the transplanted EC and AEII cells and rapid vascularisation.[55] This finding is in line with the notion that transplanted endothelial cells help maintain alveolar differentiation and remain differentiated themselves. In addition to the direct contribution of endothelial cells to epithelial growth, endothelial cells can also function to connect a tissue-engineered structure to the host vasculature upon transplantation. The generation of vascularised bone,[56] hepatic tissue,[57] skeletal muscle,[58,59] and cardiac muscle[60] by *in vivo* implantation of parenchymal cells demonstrates that neovascularisation may be mediated by angiogenesis from the host blood supply. Other studies have shown an enhancement of graft-host neovascularisation when endothelial cells are engrafted along with parenchymal cells.[61] In line with this notion, we also demonstrated (Fig. 9) that donor-derived endothelial cells can contribute to implanting vascularisation alongside host-derived vessels.[61] Somewhat troubling at this stage is the finding that we have to provide a long-term source of exogenous growth factors (specifically FGF2) to maintain a healthy level of vascularisation of the implants. The solution might be simply to increase the number of cells in our constructs, which would then serve as the missing "natural" source for producing the required angiogenic growth factors. However this brings up the issue of cell sourcing, which is a perennial problem in tissue engineering. One possible solution is to optimise the targeted differentiation of (embryonic) stem cells toward multiple pulmonary lineages.

7. Conclusions and Future Challenges

Three-dimensional pulmonary growth *in vitro* is a complex, multifaceted process. Successful generation of histotypic alveolar forming units requires the use of a heterogeneous co-culture of progenitor cells, possibly and preferably foetal pulmonary cells, in conjunction with optimised three-dimensional culture conditions and the prudent choice of a cocktail of defined growth factors that promote concomitant epithelial and endothelial differentiation. At the same time, and in keeping with other systems,[58] these culture conditions have to allow for the presence of, but avoid overgrowth with, mesenchymal cells (fibroblasts), which contribute important cues for both epithelial and endothelial cells differentiation.[62]

The ability to generate complex distal lung tissues *in vitro* raises the spectre of being able to repair diseased neonatal and/or adult lungs and aid in advancing drug discovery for a plethora of pulmonary diseases in the not-too-distant future. At this stage, however, there are numerous challenges that need to be addressed. In terms of the basic understanding of how a lung develops, our model will allow detailed dissection of critical steps that are necessary for the "normal" assembly and development of the distal lung. As a logical next step, our 3D model will also allow us to study the onset and progression of numerous pulmonary diseases, such as Pulmonary Hypertension (PH), Bronchopulmonary Dysplasia (BPD), and Chronic Obstructive Pulmonary Disease (COPD). In terms of clinical applications, methods still need to be developed for functionally engrafting *in vitro* generated pieces (or even entire lobes) of the lung. Alternatively, this model, like others, might provide an interim solution until the development of true regenerative medicine, in which we will be able to apply what we have learned from this model, the concepts and the tools, to developing cell/agent-based therapies, which will allow the diseased and dysfunctional lung to repair and heal itself.

References

1. Sigurdson, L., Carney, D.E., Hou, Y., Hall, L., III, Hard, R., Hicks, W., Jr., Bright, F.V., and Gardella, J.A., Jr. A comparative study of primary and immortalized cell adhesion characteristics to modified polymer surfaces: toward the goal of effective re-epithelialization. *J Biomed Mater Res* **59**(2), 357–365 (2002).
2. Grikscheit, T.C. and Vacanti, J.P. The history and current status of tissue engineering: the future of pediatric surgery. *J Pediatr Surg* **37**(3), 277–288 (2002).
3. Atala, A., Bauer, S.B., Soker, S., Yoo, J.J., and Retik, A.B. Tissue-engineered autologous bladders for patients needing cystoplasty. *Lancet* **367**(9518), 1241–1246 (2006).
4. Macchiarini, P., Jungebluth, P., Go, T., Asnaghi, M.A., Rees, L.E., Cogan, T.A., Dodson, A., Martorell, J., Bellini, S., Parnigotto, P.P., Dickinson, S.C., Hollander, A.P., Mantero, S., Conconi, M.T., and Birchall, M.A. Clinical transplantation of a tissue-engineered airway. *Lancet* **372**(9655), 2023–2030 (2008).
5. Murphy, S.L. Deaths: final data for 1998. *Natl Vital Stat Rep* **48**(11), 1–105 (2000).

6. Andrade, C.F., Wong, A.P., Waddell, T.K., Keshavjee, S., and Liu, M. Cell-based tissue engineering for lung regeneration. *Am J Physiol Lung Cell Mol Physiol* **292**(2), L510–L518 (2007).
7. Cardoso, W.V. and Lu, J. Regulation of early lung morphogenesis: questions, facts and controversies. *Development* **133**(9), 1611–1624 (2006).
8. Voelkel, N.F., Vandivier, R.W., and Tuder, R.M. Vascular endothelial growth factor in the lung. *Am J Physiol Lung Cell Mol Physiol* **290**(2), L209–L221 (2006).
9. Kumar, V.H., Lakshminrusimha, S., El Abiad, M.T., Chess, P.R., and Ryan, R.M. Growth factors in lung development. *Adv Clin Chem* **40**, 261–316 (2005).
10. Shannon, J.M., Pan, T., Nielsen, L.D., Edeen, K.E., and Mason, R.J. Lung fibroblasts improve differentiation of rat type II cells in primary culture. *Am J Respir Cell Mol Biol* **24**(3), 235–244 (2001).
11. Rice, W.R., Conkright, J.J., Na, C.L., Ikegami, M., Shannon, J.M., and Weaver, T.E. Maintenance of the mouse type II cell phenotype *in vitro*. *Am J Physiol Lung Cell Mol Physiol* **283**(2), L256–L264 (2002).
12. Sugahara, K., Tokumine, J., Teruya, K., and Oshiro, T. Alveolar epithelial cells: differentiation and lung injury. *Respirology* **11**(Suppl), S28–S31 (2006).
13. Yasunaga, M., Tada, S., Torikai-Nishikawa, S., Nakano, Y., Okada, M., Jakt, L.M., Nishikawa, S., Chiba, T., Era, T., and Nishikawa, S. Induction and monitoring of definitive and visceral endoderm differentiation of mouse ES cells. *Nat Biotechnol* **23**(12), 1542–1550 (2005).
14. Ali, N.N., Edgar, A.J., Samadikuchaksaraei, A., Timson, C.M., Romanska, H.M., Polak, J.M., and Bishop, A.E. Derivation of type II alveolar epithelial cells from murine embryonic stem cells. *Tissue Eng* **8**(4), 541–550 (2002).
15. Samadikuchaksaraei, A., Cohen, S., Isaac, K., Rippon, H.J., Polak, J.M., Bielby, R.C., and Bishop, A.E. Derivation of distal airway epithelium from human embryonic stem cells. *Tissue Eng* **12**(4), 867–875 (2006).
16. Lane, S., Rippon, H.J., and Bishop, A.E. Stem cells in lung repair and regeneration. *Regen Med* **2**(4), 407–415 (2007).
17. Cortiella, J., Nichols, J.E., Kojima, K., Bonassar, L.J., Dargon, P., Roy, A.K., Vacant, M.P., Niles, J.A., and Vacanti, C.A. Tissue-engineered lung: an *in vivo* and *in vitro* comparison of polyglycolic acid and pluronic F-127 hydrogel/somatic lung progenitor cell constructs to support tissue growth. *Tissue Eng* **12**(5), 1213–1225 (2006).

18. Mondrinos, M.J., Koutzaki, S., Jiwanmall, E., Li, M., Dechadarevian, J.P., Lelkes, P.I., and Finck, C.M. Engineering three-dimensional pulmonary tissue constructs. *Tissue Eng* **12**(4), 717–728 (2006).
19. Mondrinos, M.J., Koutzaki, S., Lelkes, P.I., and Finck, C.M. A tissue-engineered model of fetal distal lung tissue. *Am J Physiol Lung Cell Mol Physiol* **293**(3), L639–L650 (2007).
20. Dobbs, L.G. Isolation and culture of alveolar type II cells. *Am J Physiol* **258**(4 Pt 1), L134–L147 (1990).
21. Nandkumar, M.A., Yamato, M., Kushida, A., Konno, C., Hirose, M., Kikuchi, A., and Okano, T. Two-dimensional cell sheet manipulation of heterotypically co-cultured lung cells utilizing temperature-responsive culture dishes results in long-term maintenance of differentiated epithelial cell functions. *Biomaterials* **23**(4), 1121–1130 (2002).
22. Eckenhoff, R.G., Rannels, S.R., and Fisher, A.B. Secretory granule calcium loss after isolation of rat alveolar type II cells. *Am J Physiol* **260**(2 Pt 1), L129–L135 (1991).
23. Kohsa, K., Yamada, H., and Sugihara, H. Influence of air exposure treatment on alveolar type II epithelial cells cultured on extracellular matrix. *Cell Struct Funct* **21**(1), 81–89 (1996).
24. Douglas, W.H. and Teel, R.W. An organotypic *in vitro* model system for studying pulmonary surfactant production by type II alveolar pneumonocytes. *Am Rev Respir Dis* **113**(1), 17–23 (1976).
25. Douglas, W.H., Moorman, G.W., and Teel, R.W. The formation of histotypic structures from monodisperse fetal rat lung cells cultured on a three-dimensional substrate. *In Vitro* **12**(5), 373–381 (1976).
26. Blau, H., Guzowski, D.E., Siddiqi, Z.A., Scarpelli, E.M., and Bienkowski, R.S. Fetal type 2 pneumocytes form alveolar-like structures and maintain long-term differentiation on extracellular matrix. *J Cell Physiol* **136**(2), 203–214 (1988).
27. Rannels, S.R., Grove, R.N., and Rannels, D.E. Matrix-derived soluble components influence type II pneumocytes in primary culture. *Am J Physiol* **256**(3 Pt 1), C621–C629 (1989).
28. Shannon, J.M., Jennings, S.D., and Nielsen, L.D. Modulation of alveolar type II cell differentiated function *in vitro*. *Am J Physiol* **262**(4 Pt 1), L427–L436 (1992).
29. Paine, R., III, Christensen, P., Toews, G.B., and Simon, R.H. Regulation of alveolar epithelial cell ICAM-1 expression by cell shape and cell-cell interactions. *Am J Physiol* **266**(4 Pt 1), L476–L484 (1994).

30. Robinson, P.C., Voelker, D.R., and Mason, R.J. Isolation and culture of human alveolar type II epithelial cells. Characterization of their phospholipid secretion. *Am Rev Respir Dis* **130**(6), 1156–1160 (1984).
31. Lwebuga-Mukasa, J.S., Ingbar, D.H., and Madri, J.A. Repopulation of a human alveolar mtrix by adult rat type II pneumocytes *in vitro*. A novel system for type II pneumocyte culture. *Exp Cell Res* **162**(2), 423–435 (1986).
32. Yurchenco, P.D., Amenta, P.S., and Patton, B.L. Basement membrane assembly, stability and activities observed through a developmental lens. *Matrix Biol* **22**(7), 521–238 (2004).
33. Kleinman, H.K. and Martin, G.R. Matrigel: basement membrane matrix with biological activity. *Semin Cancer Biol* **15**(5), 378–386 (2005).
34. Suki, B., Ito, S., Stamenovic, D., Lutchen, K.R., and Ingenito, E.P. Biomechanics of the lung parenchyma: critical roles of collagen and mechanical forces. *J Appl Physiol* **98**(5), 1892–1899 (2005).
35. Van Vranken, B.E., Rippon, H.J., Samadikuchaksaraei, A., Trounson, A.O., and Bishop, A.E. The differentiation of distal lung epithelium from embryonic stem cells. *Curr Protoc Stem Cell Biol* Chapter 1 Unit 1G.1 (2007).
36. Li, W.J., Laurencin, C.T., Caterson, E.J., Tuan, R.S., and Ko, F.K. Electrospun nanofibrous structure: a novel scaffold for tissue engineering. *J Biomed Mater Res* **60**(4), 613–621 (2002).
37. Thapa, A., Miller, D.C., Webster, T.J., and Haberstroh, K.M. Nano-structured polymers enhance bladder smooth muscle cell function. *Biomaterials* **24**(17), 2915–2926 (2003).
38. Adamson, I.Y. and Young, L. Alveolar type II cell growth on a pulmonary endothelial extracellular matrix. *Am J Physiol* **270**(6 Pt 1), L1017–L1022 (1996).
39. Kaplan, F. Molecular determinants of fetal lung organogenesis. *Mol Genet Metab* **71**(1–2), 321–341 (2000).
40. Hislop, A.A. Airway and blood vessel interaction during lung development. *J Anat* **201**(4), 325–334 (2002).
41. Shannon, J.M., Gebb, S.A., and Nielsen, L.D. Induction of alveolar type II cell differentiation in embryonic tracheal epithelium in mesenchyme-free culture. *Development* **126**(8), 1675–1688 (1999).
42. Sugahara, K., Mason, R.J., and Shannon, J.M. Effects of soluble factors and extracellular matrix on DNA synthesis and surfactant gene expression in primary cultures of rat alveolar type II cells. *Cell Tissue Res* **291**(2), 295–303 (1998).

43. Miettinen, P.J., Warburton, D., Bu, D., Zhao, J.S., Berger, J.E., Minoo, P., Koivisto, T., Allen, L., Dobbs, L., Werb, Z., and Derynck, R. Impaired lung branching morphogenesis in the absence of functional EGF receptor. *Dev Biol* **186**(2), 224–236 (1997).
44. Andreadis, S.T., Hamoen, K.E., Yarmush, M.L., and Morgan, J.R. Keratinocyte growth factor induces hyperproliferation and delays differentiation in a skin equivalent model system. *FASEB J* **15**(6), 898–906 (2001).
45. Mason, I.J., Fuller-Pace, F., Smith, R., and Dickson, C. FGF-7 (keratinocyte growth factor) expression during mouse development suggests roles in myogenesis, forebrain regionalisation and epithelial-mesenchymal interactions. *Mech Dev* **45**(1), 15–30 (1994).
46. Adamson, I.Y., Young, L., and Bakowska, J. Enhanced alveolar type II cell growth on a pulmonary extracellular matrix over fibroblasts. *Am J Physiol* **272**(3 Pt 1), L413–L417 (1997).
47. Sugahara, K., Rubin, J.S., Mason, R.J., Aronsen, E.L., and Shannon, J.M. Keratinocyte growth factor increases MRNAs for SP-A and SP-B in adult rat alveolar type II cells in culture. *Am J Physiol* **269**(3 Pt 1), L344–L350 (1995).
48. Sekine, K., Ohuchi, H., Fujiwara, M., Yamasaki, M., Yoshizawa, T., Sato, T., Yagishita, N., Matsui, D., Koga, Y., Itoh, N., and Kato, S. Fgf10 is essential for limb and lung formation. *Nat Genet* **21**(1), 138–141 (1999).
49. Griffin, M., Bhandari, R., Hamilton, G., Chan, Y.C., and Powell, J.T. Alveolar type II Cell-fibroblast interactions, synthesis and secretion of surfactant and type I collagen. *J Cell Sci* **105**(Pt 2), 423–432 (1993).
50. Pan, T., Mason, R.J., Westcott, J.Y., and Shannon, J.M. Rat alveolar type II cells inhibit lung fibroblast proliferation *in vitro*. *Am J Respir Cell Mol Biol* **25**(3), 353–361 (2001).
51. Schuger, L., Skubitz, A.P., Zhang, J., Sorokin, L., and He, L. Laminin alpha1 chain synthesis in the mouse developing lung: requirement for epithelial-mesenchymal contact and possible role in bronchial smooth muscle development. *J Cell Biol* **139**(2), 553–562 (1997).
52. Shannon, J.M. Induction of alveolar type II cell differentiation in fetal tracheal epithelium by grafted distal lung mesenchyme. *Dev Biol* **166**(2), 600–614 (1994).
53. Matsumoto, K., Yoshitomi, H., Rossant, J., and Zaret, K.S. Liver organogenesis promoted by endothelial cells prior to vascular function. *Science* **294**(5542), 559–563 (2001).

54. Adamson, I.Y., King, G.M., and Young, L. Influence of extracellular matrix and collagen components on alveolar type 2 cell morphology and function. *In Vitro Cell Dev Biol* **25**(6), 494–502 (1989).
55. Crisanti, M.C., Koutzaki, S.H., Mondrinos, M.J., Lelkes, P.I., and Finck, C.M. Novel methods for delivery of cell-based therapies. *J Surg Res* **146**(1), 3–10 (2008).
56. Kim, W.S. and Kim, H.K. Tissue engineered vascularized bone formation using *in vivo* implanted osteoblast-polyglycolic acid scaffold. *J Korean Med Sci* **20**(3), 479–482 (2005).
57. Yokoyama, T., Ohashi, K., Kuge, H., Kanehiro, H., Iwata, H., Yamato, M., and Nakajima, Y. *In vivo* engineering of metabolically active hepatic tissues in a neovascularized subcutaneous cavity. *Am J Transplant* **6**(1), 50–59 (2006).
58. Levenberg, S., Rouwkema, J., Macdonald, M., Garfein, E.S., Kohane, D.S., Darland, D.C., Marini, R., van Blitterswijk, C.A., Mulligan, R.C., D'Amore, P.A., and Langer, R. Engineering vascularized skeletal muscle tissue. *Nat Biotechnol* **23**(7), 879–884 (2005).
59. Messina, A., Bortolotto, S.K., Cassell, O.C., Kelly, J., Abberton, K.M., and Morrison, W.A. Generation of a vascularized organoid using skeletal muscle as the inductive source. *FASEB J* **19**(11), 1570–1572 (2005).
60. Birla, R.K., Borschel, G.H., Dennis, R.G. and Brown, D.L. Myocardial engineering *in vivo*: formation and characterization of contractile, vascularized three-dimensional cardiac tissue. *Tissue Eng* **11**(5–6), 803–813 (2005).
61. Mondrinos, M.J., Koutzaki, S.H., Poblete, H.M., Crisanti, M.C., Lelkes, P.I., and Finck, C.M. *In vivo* pulmonary tissue engineering: contribution of donor-derived endothelial cells to construct vascularization. *Tissue Eng A* **14**(3), 361–368 (2008).
62. Mondrinos, M. *In Vitro Tissue Engineered Pulmonary Constructs.* (Drexel University Library, Thesis, 2009).

Chapter 10

Stem Cells as Vehicles for Gene Therapy in Lung Repair

Shirley H. J. Mei and Duncan J. Stewart

There has been considerable research on harnessing the unique properties of stem and progenitor cells to develop more effective therapeutic approaches for respiratory diseases with limited treatment options. Cell-based gene therapy, combined with genetic engineering to enhance regenerative cell function, can be a powerful strategy able to provide additional benefits, and may overcome many of the limitations of cell or gene therapy alone. This dual strategy not only allows the direct targeting of cells to the lung for regenerative interventions, but also provides a site-specific source to release therapeutic proteins to the lung microvasculature or alveoli. This review discusses relevant literature on the use of stem/progenitor cells-based gene therapy for the treatment of acute and chronic pulmonary disorders, including acute lung injury, pulmonary arterial hypertension, emphysema, cystic fibrosis, and cancer therapy for lung metastasis. Different types of stem and progenitor cells such as mesenchymal stem cells or endothelial progenitor cells, which have been used in these therapeutic approaches, will also be covered.

Keywords: Stem/progenitor cells; gene therapy; acute lung injury; mesenchymal stem cells; endothelial progenitor cells; pulmonary arterial hypertension; cystic fibrosis.

Outline

1. Introduction

Gene therapy holds tremendous promise for lung diseases but has proven difficult to translate into effective clinical treatment strategies,[1] in part because of the difficulties in achieving sufficient transfection efficiency *in vivo*, as well as the challenges in targeting gene transfer selectively to the lung. Intratracheal delivery of plasmid or viral DNA vectors has the advantage of selective gene transfer directly to the lung. However, the transfection efficiency of these methods is limited by physical and immunological barriers presented by the epithelium of the lung,[1,2] which may be exacerbated by excessive edema and mucin production in the context of acute lung injury or cystic fibrosis.[1] To some extent, this can be overcome by the use of adenoviral vectors, for example, which can produce efficient transfection into lung tissues; however, inherent inflammatory responses even to second and third generation viral vectors has created a major obstacle for its use in clinical trials.[3]

The combination of cell and gene therapies can be an attractive approach to overcome certain limitations associated with each individual strategy.[1] This dual strategy not only allows the direct targeting of the lung for clinical intervention, but also provides a site-specific source for the release of therapeutic proteins and/or other cellular products of interest by the retained cells.[4–7] In cell-based gene therapy, cells from a given source, either autologous (isolated from patient's own tissue) or allogeneic (isolated from an unrelated donor), are expanded *in vitro* and genetically manipulated to express the therapeutic gene of interest before re-introduction into the patient (Fig. 1). This process permits the cells to be

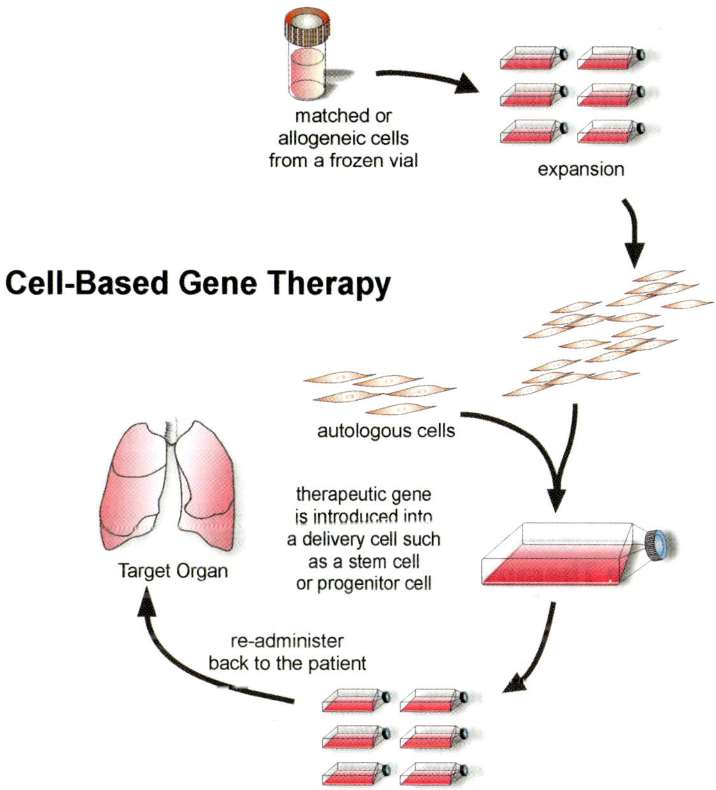

Figure 1: Cell-based gene therapy for pulmonary diseases. In cell-based gene therapy, cells from a given source are isolated and culturally expanded *ex vivo*. Culturally expanded cells are then genetically manipulated to express a therapeutic gene of interest before re-administration back into the patient.

genetically engineered *ex vivo* to overexpress a therapeutic transgene product. Cells can be used to target transgene expression to the lung, even sub-selectively to specific regions of the lung. For example, cells larger than circulating blood elements will be efficiently filtered by the lung after intravenous or intrapulmonary arterial administration, and will lodge in the distal arteriolar bed, where they can locally express the transgene to maximise the desired therapeutic effect.[4–7] Similarly, cells can be administered by intratracheal instillation to be distributed to the distal airways.[8,9] Finally, cells can be injected directly into the lung tissue.[10]

first reported by Asahara *et al.* in 1997,[46] who identified EPCs by the surface antigen expression of CD34, expressed on all haematopoietic stem cells, and Flk-1 (vascular endothelial growth factor receptor 2, VEGFR2), expressed by both early haematopoietic stem cells and endothelial cells. However, there is considerable confusion surrounding the definition of EPCs, and currently there is no agreement on markers that can be used for the identification of "true" EPCs (as is reviewed in detail in Chapter 12). Nevertheless, regardless of their origin or method of selection, EPCs possess many unique properties that make these cells attractive for cell therapy, or cell-based gene therapy, of lung diseases.

EPCs can be found to localise within pulmonary vasculature after systemic delivery[11] or home to sites of tumours[47] in the context of metastatic lung disease; properties that allow EPCs to be used as a tool to target both vascular diseases and cancer. In cancer research, genetic modification of EPCs to express the suicide gene has been used as a therapy to reduce lung tumour metastasis.[48] Many researchers have also utilised the neoangiogenic potential of EPCs in pulmonary vascular disease, such as the case for pulmonary arterial hypertension.[11] Again, these and other applications will be discussed in more detail below and in other chapters.

2.3. *Other cell types*

Many reports have now described the existence of resident lung and progenitor cells from various parts of the lung, including proximal airway, bronchiolar, or alveolar epithelial regions[18,49] (as is discussed in greater detail in Chapter 4). Krause *et al.* were the first to show that bone marrow-derived cells can differentiate into epithelial cells in lung and other organs.[50] Other studies have shown that mouse embryonic stem cells can be induced to become cells resembling type II alveolar[51] and tracheal epithelial[52] cells. Isolation of adult stem cells from umbilical cord blood,[53,54] amniotic fluid,[55] adipose[56] and even placental[57] tissues have been reported. To our knowledge, genetic modification of these cells for pulmonary disease has not been described. Nevertheless, these cells might hold enormous potentials to be utilised for cell-based therapies for lung repair in the near future.

3. Pulmonary Diseases Targeted by Stem/Progenitor Cells-Based Gene Therapy

3.1. *Acute lung injury*

Acute respiratory distress syndrome (ARDS) is a clinically important complication of severe acute lung injury (ALI) in humans, and is a significant cause of morbidity and mortality in critically ill patients.[58–61] Infectious etiologies, such as sepsis and pneumonia, are leading causes of ALI/ARDS.[58,59] In humans, ALI/ARDS is characterised by a severe acute inflammatory response in the lungs and neutrophilic alveolitis.[58] It is a relatively common and potentially lethal clinical syndrome, with an incidence as high as 80 cases per 100,000 population per year.[62] Despite decades of research, few effective therapeutic strategies for clinical ARDS have emerged, and current specific options for treatment are limited.[62–66] ARDS continues to be an important contributor to prolonged mechanical ventilation in the intensive care unit, and ARDS-associated mortality remains high at 30–50% despite optimal supportive care.[58,62,64,65]

To increase the potential efficacy of a cell-based gene therapy strategy, it is crucial to choose an appropriate transgene with biological activities that are complementary to the potential therapeutic effects of cells *per se*. In the context of ALI, angiopoietin 1 (ANGPT1) may represent a good example of such complementarity. Although this system is better known for its effects on vascular development, acting in concert with VEGF to promote neovasculature stabilisation and maturation during embryonic development, ANGPT1 is highly expressed in nearly all post-natal vascular beds, and is thought to act as a homeostatic factor, maintaining quiescence of vascular endothelium in the adult. ANGPT1 exerts potent anti-inflammatory, anti-permeable, and endothelial-protective actions by acting on the endothelial-selective receptor tyrosine kinase, Tie2, whereas angiopoietin 2 (ANGPT2) acts as a context-dependent inhibitor of the Tie2 receptor and has opposite effects. Circulating levels of ANGPT2 were found to be elevated in the plasma of patients suffering from sepsis,[67,68] and the administration of ANGPT2 led to a disruption of the endothelial barrier both *in vitro* and in a mouse model *in vivo*.[68] Therefore, we proposed a therapeutic approach to restore the balance between ANGPT1 and ANGPT2 expression, which may be useful in modulating endothelial activation and inflammation in response to ALI. Some of the early features of ARDS can be reproduced by administration of the bacterial endotoxin, lipopolysaccharide (LPS). Using a gene therapy approach, Witzenbichler *et al.* reported the first study that employed systemic pre-treatment with an adenoviral construct containing ANGPT1, two days prior to intraperitoneal injection of LPS.[69] Twelve hours after injury, they found adenoviral ANGPT1 pre-treatment reduced adhesion molecule expression in the lungs and prolonged survival in mice with endotoxic shock. A very similar study recently published by Huang *et al.* also used an adenoviral construct to overexpress ANGPT1 to examine its therapeutic efficacy as a pre-treatment (five days prior) or post-treatment (three hours post) for LPS-induced septic shock in mice.[70] Their results indicated that the pre-treatment strategy improved survival and decreased pulmonary vascular permeability, whereas post-treatment with adenoviral ANGPT1 showed no benefit.

Using an intratracheal instillation model of LPS, our laboratory provided the first demonstration of successful cell-based gene therapy for ALI.[71] In this study, pre-treatment with syngeneic somatic cells (i.e. fibroblasts), transfected with human ANGPT1, reduced airspace inflammation by 60% and markedly decreased the expression of endothelial-selective adhesion molecules.[71] The protective effect of ANGPT1 was further validated by using a transgenic model. Mice that overexpressed ANGPT1, targeted to the vascular endothelium, demonstrated reduced sensitivity to LPS and showed improved survival.[71] These results clearly support the potential utility of ANGPT1 for cell-based gene therapy in the treatment of ALI. However, fibroblasts are unlikely to provide any additional benefits above and beyond the effects of the therapeutic transgene.

We reasoned that MSCs, with their immunomodulatory and regenerative properties, might have therapeutic actions in their own right to modulate lung injury. Therefore, bone marrow-derived MSCs, with or without ANGPT1 transfection, were administered via the intrajugular vein to mice 30 minutes after LPS-induced ALI.[7] Administration of MSCs to the pulmonary circulation partially prevented LPS-induced lung inflammation three days after lung injury. Moreover, treatment with plasmid ANGPT1-transfected MSCs resulted in a further and near complete improvement in both alveolar inflammation and permeability (Fig. 2).

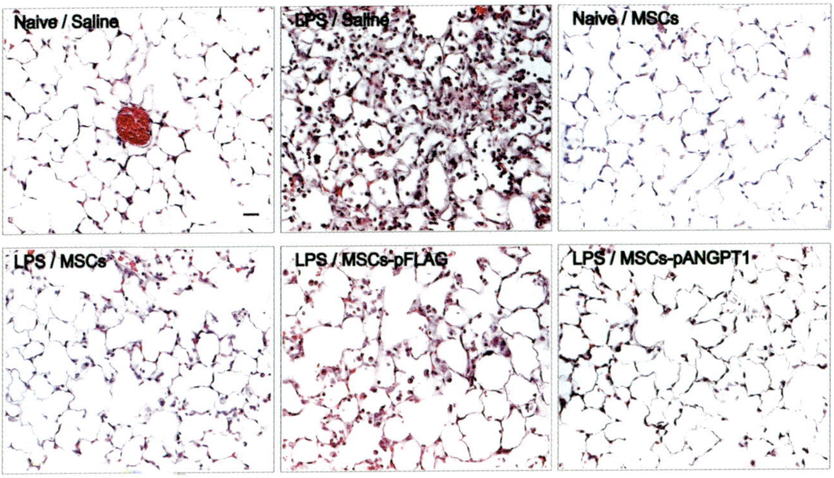

Figure 2: Histological evaluation of therapeutic potential of MSCs and MSCs-pANGPT1 on LPS-induced lung injury in mice. Representative images of haematoxylin and eosin-stained lung sections from six experimental groups. Lungs were fixed with 4% paraformaldehyde, embedded in paraffin, and then cut into 5 μm thick sections before being stained. Photomicrographs were obtained with a Nikon Eclipse E800 microscope with a 40× objective. Scale bar = 20 μm. Figure taken from Mei *et al.* (2007).[7]

These data demonstrated that bone marrow-derived MSCs can prevent and rescue ALI when administered into the pulmonary circulation of the LPS-injured mouse lung, possibly by virtue of their immunosuppressive/immunomodulatory actions. Moreover, this effect was enhanced when these cells were engineered to overexpress the vasculoprotective factor ANGPT1, a specific inhibitor of EC inflammation and permeability (Fig. 3). A similar study by Yu and colleagues also used the therapeutic combination of MSCs and ANGPT1 in a LPS-nebulisation model (20 minutes of exposure per day for seven days).[41] MSCs alone, lentiviral ANGPT1 alone, or lentiviral ANGPT1 transduced MSCs were intravenously administrated to mice two hours after the first dose of LPS inhalation. They observed a significant effect from MSCs transduced with lentiviral ANGPT1, but little effect in mice treated with either MSCs or lentiviral ANGPT1 alone. In addition, several other reports have shown benefits with unmodified MSCs[9,72] for the treatment of LPS-induced ALI in mice or polymicrobial sepsis.[73] These studies demonstrate the feasibility

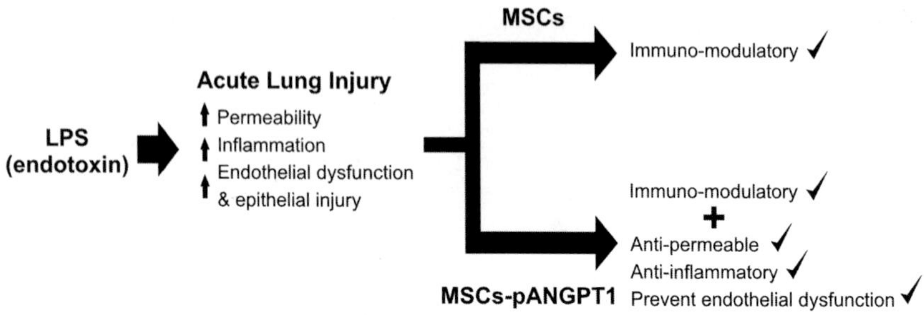

Figure 3: Proposed mechanism for the therapeutic activity of MSCs and MSCs-pANGPT1 on LPS-induced ALI. LPS functions as a potent pro-inflammatory microbial product to cause acute pulmonary inflammation that progresses to ALI, characterised by pulmonary vascular leak resulting from pulmonary endothelial dysfunction and alveolar epithelial injury. By virtue of their immuno-modulatory action, MSCs effectively decrease LPS-induced inflammation but do not completely protect against LPS-induced ALI. The combination of MSCs and ANGPT1 (MSC-pANGPT1) not only suppresses inflammation but also prevents LPS-induced ALI by further stabilisation of pulmonary endothelial cell function.

and effectiveness of cell and cell-based gene therapy for experimental ALI, and may provide a basis for the development of an innovative approach for the prevention and treatment of ALI/ARDS.

3.2. *Pulmonary arterial hypertension*

Pulmonary arterial hypertension (PAH) is a disease characterised by severe remodelling changes to the pulmonary vasculature that results in narrowing and occlusion of the small pulmonary arteries, particularly at the levels of the pre-capillary (or intra-acinar) arterioles.[74] These abnormalities cause a progressive increase in pulmonary vascular resistance, ultimately leading to right heart failure and death. Even with recent advances in therapeutic options, the prognosis for many PAH patients remains poor, with an average survival after diagnosis of only two to five years.[75] Although the precise pathogenesis remains unclear, there is a general agreement that PAH results from injury to the pulmonary vascular endothelium, leading to narrowing and ultimately occlusion or degeneration of pre-capillary arterioles.[76]

It is now widely accepted that EPCs play a critical role in neovascularisation[77] and the repair of endothelial injury, in part by participating in post-natal vasculogenesis.[78] Therefore, it is not surprising that EPCs have been considered as an attractive cell type for the therapy for PAH, as is discussed in greater detail in Chapter 12. In particular, Zhao *et al.* studied bone marrow-derived EPCs, isolated by differential culture (i.e. early outgrowth EPCs) that had been engineered to overexpress endothelial nitric oxide synthase (eNOS) in the treatment of monocrotaline (MCT)-induced pulmonary hypertension in a rat model.[11] A product of the enzyme eNOS, nitric oxide, is a known vasodilator and mediator for angiogenesis, both of which have potential therapeutic relevance for PAH.[79] They showed that the delivery of EPCs three days after MCT-induced lung injury resulted in a near complete prevention of PAH in rats. These cells were also found to incorporate into arteriolar endothelium in some animals. To provide further evidence that this therapy can be employed clinically, the effects of EPCs, with or without eNOS gene transfer, were studied in the treatment model of established PAH. Cell therapy was delayed until three weeks after MCT injury, at which time severe PAH was already apparent, and therapeutic efficacy was evaluated two weeks later. In contrast to the prevention model, EPCs alone were only partially effective, preventing a further increase in right ventricular systemic pressure (RVSP) from day 21 to day 35. However, in those animals that had received eNOS-transfected EPCs, there was a significant reduction in the measurement of RVSP to levels not different from normal (sham-treated) controls. Furthermore, these animals also showed improved survival and micro-vascular perfusion.[11] A separate study by Nagaya *et al.* demonstrated that similar results could be obtained by using plasmid DNA-gelatin complex to engineer human umbilical cord blood-derived EPCs to express adrenomedulin, a potent vasodilator.[53] Based on these encouraging preclinical data, the first clinical trial using EPC-based eNOS gene therapy for PAH has been initiated in Toronto and Montreal, and details are provided in Chapter 12.

MSCs have also been used for cell-based eNOS gene therapy of experimental PAH.[12] Treatment with MSCs alone or MSCs transduced with eNOS one week after MCT injury lowered RVSP compared to animals not receiving cell therapy; however, only eNOS expressing MSCs significantly improved survival. Taken together, these studies strongly suggest that stem/progenitor cell-based gene therapy is both feasible and effective in treating experimental PAH.

3.3. *Emphysema*

Emphysema is a form of chronic obstructive lung disease (COPD), often caused by long-term exposure to a toxic substance such as tobacco smoke.[80] People with genetic deficiency of $alpha_1$-antitrypsin (AAT) have a marked predisposition for developing emphysema.[81] The current standard of care for patients with AAT deficiency and lung dysfunction is a weekly infusion of AAT protein derived from human plasma,[82] a therapy that is associated with high cost and risk of infection. Therefore, there has been great interest in developing alternative treatment strategies, in particular, gene therapy to correct the AAT deficiency. However, the results of gene therapy alone have been disappointing, mainly due to the failure to achieve a durable increase in the expression AAT *in vivo*.[83,84]

A study by Kotton and colleagues provided "proof of concept" using haematopoietic stem cells (HSCs) transduced with lentiviral vector encoding the human AAT gene.[85] The lentiviral approach has the potential to produce stable transfection of self-renewing, circulating HSCs. The authors transplanted a small number of transduced HSCs into irradiated mice recipients, which resulted in a sustained increase of AAT *in vivo* for 31 weeks. Even though the authors concluded that their initial study did not result in production of sufficient AAT protein to reach the theoretical threshold for protective levels (800 μg/mL), this report suggests that more effective strategies to increase transgene protein expression in the lung, such as cell-based gene therapy, may still be effective in the treatment of severe AAT deficiency.

Even in the absence of a genetic cause, stem cells and cell-based gene therapies may be of value in the treatment of common forms of emphysema resulting mostly from tobacco smoke exposure.[56] Using surgically joined parabiotic mice, Abe *et al.* demonstrated that type I alveolar epithelial cells and lung fibroblasts can be derived from circulating stem and progenitor cells and contribute to the repair of lung injury in a radiation/elastase-induced emphysema model.[86] As well, a clinical trial using allogeneic MSCs injected intravenously in patients with acute myocardial infarction may provide a potential indication of MSCs' efficacy in improving lung function.[87] In this study, pulmonary function was evaluated as part of a safety analysis, since a significant proportion of the cells were retained initially within the lungs. Of interest, patients receiving MSCs showed a significant improvement in forced expiratory volume in one second (FEV1), one of the primary indicators of lung function. Based on these intriguing results, a clinical trial using MSC therapy for patients with chronic obstructive pulmonary diseases is being planned (see also Chapters 1 and 11).

3.4. *Cystic fibrosis*

Cystic fibrosis (CF) is one of the most common genetic diseases in populations of European descent,[88] and is caused by a recessive mutation in the cystic fibrosis transmembrane conductance regulator (CFTR) gene that encodes a transmembrane chloride channel expressed on the epithelium. Cystic fibrosis is a devastating disease affecting multiple organs, but it is the associated pulmonary complications that account for most of the morbidity and mortality. Therefore, most efforts have been focused on treating the respiratory system of CF patients, including numerous gene therapy approaches targeting CF lungs.[89] Since the CFTR gene was first cloned, there have been 25 phase I/II clinical trials carried out with either viral and nonviral gene transfer.[83] Most of these expensive and time-consuming trials were unsuccessful, largely due to low transfection efficiency into human airway epithelial cells.

Cell-based gene therapy has been an emerging concept in the development of new strategies to treat CF,[90] which might overcome some of the limitations of the early gene therapy approaches. This approach is particularly useful for treatment of the later stages of CF disease, in which the damaged epithelium and accumulation of excess mucus can impair the efficiency of traditional viral vector-based gene therapy. Using an *in vitro* system, Wang and Bunnell first showed that human MSCs have the potential to differentiate into airway epithelia after MSCs have been co-cultured with human airway epithelial cells.[91] They went on to demonstrate further that MSCs from cystic fibrosis patients can be isolated and the genetic abnormality can be corrected by gene transfer *ex vivo*, using a viral vector expressing the functional CFTR gene. These "corrected" MSCs, after co-culturing with human airway epithelial cells, can contribute to chloride secretion upon cAMP stimulation, a critical determinant for cell therapy in cystic fibrosis. Bruscia *et al.* used bone marrow transplantation to deliver Cftr-positive bone marrow-derived cells to Cftr-deficient mice after receiving different doses of irradiation.[92] They provided evidence to show that in some of the transplanted mice, even with low engraftment frequency, there was detectable *in vivo* chloride channel activity in the nasal epithelium and gastrointestinal track. A separate study by Loi *et al.* used two different populations of Cftr-expressed cells (cultured marrow stromal cells and whole bone marrow cells) into naphthalene-injured, Cftr-deficient mice.[44] Naphthalene was used in this study to injure airway epithelial cells and increase recruitment of cells to the lungs. Even though they demonstrated that adult marrow cells can be recruited to lungs and induced to express the Cftr gene, the efficiency of this phenomenon was so low that the authors concluded these cells were unlikely to contribute any therapeutic benefit. Nevertheless, these reports provide support in principle for the concept of cell-based gene therapy for CF, and should encourage more experimental work to develop more efficient strategies that could be applied to the treatment of CF.

3.5. *Cancer therapy for lung metastasis*

Despite the considerable research effort that has been invested into developing therapy against aggressive or metastatic tumours, the prognosis for patients with these tumours remains poor.[93] The lung is one of the most frequently targeted organs by metastasizing cancer cells, and most of the patients with lung metastasis usually have an average survival rate of less than five years.[94] Recently, strategies that use stem/progenitor cells to target gene delivery to tumours are being intensively explored as a means to overcome toxicity and the lack of efficacy observed in existing gene therapy approaches. Stem or progenitor cells administered systemically have been demonstrated to lodge in or home to primary and metastastic tumours, a necessary property for the purpose of delivering therapeutic genes. This inherent tumour tropism, particularly exhibited by neural stem/progenitor cells[95] or MSCs,[42,96] has been exploited by researchers to genetically modify these cells to express prodrug-activating enzymes, apoptosis inducers, cell cycle modulators, anti-angiogenesis factors, and immune-enhancing agents.[93]

Taking advantage of this unique characteristic to migrate to the site of tumour tissues, Kanehira *et al.* transduced mouse MSCs with adenoviral vector encoding for NK4, an antagonist of hepatocyte growth factor.[42] Hepatocyte growth factor is a known inducer of tumour growth and promoter for angiogenesis, both of which are requirements for the metastasis of tumour cells.[97] These authors found that overexpression of NK4 by genetically modified MSCs inhibited lung tumour progression and metastasis, and prolonged mice survival without causing adverse side effects that can often be seen with the systemic administration of viral vectors.[42] Another study by Xin *et al.* used intravenous injection of MSCs expressing CX3CL1, an immuno-stimulatory chemokine, to tumour-bearing mice.[43] Their study also showed that treatment resulted in a strong inhibitory effect on lung metastases, with prolonged survival of the tumour-bearing mice. Two separate studies by Ren *et al.* used MSCs to express interferon-beta[98] or interferon alpha,[99] and both strategies reduced tumour size and significantly prolonged survival. The lungs of treated mice showed increased apoptosis and decreased proliferation in tumour cells, as well as decreases in blood vessel counts.

Endothelial progenitor cells (EPCs) have also been used as cellular vehicles for cancer gene therapy due to their ability to home to both the tumour site and the tumour's supporting vasculature.[47] Since most tumours depend on the development of new vessels to grow and metastasise, a therapy to targeting disruption of these vessels may be particularly effective in eliminating tumour metastasis. Wei *et al.* showed that embryonic EPCs, which have been genetically modified with a suicide gene, can efficiently target lung metastases[48] by interfering with the development of the tumour circulation.

4. Conclusion and the Way Forward

Acute and chronic pulmonary diseases can be treated safely and effectively by delivering *ex vivo* transfected stem or progenitor cells to the circulation in animal models. Compared to using either the cells or gene therapy alone, the combination of stem/progenitor cells and gene therapy has been proven to provide synergistically an often enhanced therapeutic effect. While certain strategies (such as pulmonary hypertension) are in the process of being translated from bench to bedside, more preclinical studies will be required before other experimental therapies are ready for clinical testing.

Specifically, a number of issues and challenges will need to be addressed before cell-based strategies can be used in therapy under an acute setting, such as in the case of ALI. Although a growing number of reports have suggested MSCs may be "immune-privileged", it will be important to explore whether the use of allogeneic MSCs will be equally effective in any of these pulmonary disorders. Other issues related to the potential tumourgenicity of MSCs *in vivo*[100] and the determination of proper cell dosing during treatment to avoid pulmonary embolism, should also be carefully considered. Finally, more efficient methods to recruit stem or progenitor cells to the lung, as well as development of better protocols to obtain sufficient number of stem or progenitor cells to achieve therapeutic efficacy will also need to be addressed.

References

1. Weiss, D.J. Delivery of gene transfer vectors to lung: obstacles and the role of adjunct techniques for airway administration. *Mol Ther* **6**(2), 148–152 (2002).
2. Davis, P.B. and Cooper, M.J. Vectors for airway gene delivery. *AAPS J* **9**(1), E11–E17 (2007).
3. Thomas, C.E., Ehrhardt, A., and Kay, M.A. Progress and problems with the use of viral vectors for gene therapy. *Nat Rev Genet* **4**(5), 346–358 (2003).
4. Zhao Y.D., Campbell, A.I., Robb, M., Ng, D., and Stewart, D.J. Protective role of angiopoietin-1 in experimental pulmonary hypertension. *Circ Res* **92**(9), 984–991 (2003).
5. Campbell, A.I., Zhao, Y., Sandhu, R., and Stewart, D.J. Cell-based gene transfer of vascular endothelial growth factor attenuates monocrotaline-induced pulmonary hypertension. *Circulation* **104**(18), 2242–2248 (2001).
6. Campbell, A.I., Kuliszewski, M.A., and Stewart, D.J. Cell-based gene transfer to the pulmonary vasculature: endothelial nitric oxide synthase overexpression inhibits monocrotaline-induced pulmonary hypertension. *Am J Respir Cell Mol Biol* **21**(5), 567–575 (1999).

7. Mei, S.H.J., McCarter, S.D., Deng, Y., Parker, C.H., Liles, W.C., and Stewart, D.J. Prevention of LPS-induced acute lung injury in mice by mesenchymal stem cells overexpressing angiopoietin 1. *PLoS Med* **4**(9), e269 (2007).
8. Zhang, D., Wu, M., Nelson, D.E., Pasula, R., and Martin, W.J., 2nd. Alpha-1-antitrypsin expression in the lung is increased by airway delivery of gene-transfected macrophages. *Gene Ther* **10**(26), 2148–2152 (2003).
9. Gupta, N., Su, X., Popov, B., Lee, J.W., Serikov, V., and Matthay, M.A. Intrapulmonary delivery of bone marrow-derived mesenchymal stem cells improves survival and attenuates endotoxin-induced acute lung injury in mice. *J Immunol* **179**(3), 1855–1863 (2007).
10. Takahashi, M., Nakamura, T., Toba, T., Kajiwara, N., Kato, H., and Shimizu, Y. Transplantation of endothelial progenitor cells into the lung to alleviate pulmonary hypertension in dogs. *Tissue Eng* **10**(5–6), 771–779 (2004).
11. Zhao, Y.D., Courtman, D.W., Deng, Y., Kugathasan, L., Zhang, Q., and Stewart, D.J. Rescue of monocrotaline-induced pulmonary arterial hypertension using bone marrow-derived endothelial-like progenitor cells: efficacy of combined cell and eNOS gene therapy in established disease. *Circ Res* **96**(4), 442–450 (2005).
12. Kanki-Horimoto, S., Horimoto, H., Mieno, S., Kishida, K., Watanabe, F., Furuya, E., and Katsumata, T. Implantation of mesenchymal stem cells overexpressing endothelial nitric oxide synthase improves right ventricular impairments caused by pulmonary hypertension. *Circulation* **114** (Suppl 1), I181–I185 (2006).
13. Koc, O.N. and Lazarus, H.M. Mesenchymal stem cells: heading into the clinic. *Bone Marrow Transplant* **27**(3), 235–239 (2001).
14. Barry, F.P. and Murphy, J.M. Mesenchymal stem cells: clinical applications and biological characterization. *Int J Biochem Cell Biol* **36**(4), 568–584 (2004).
15. Horwitz, E.M. Stem cell plasticity: the growing potential of cellular therapy. *Arch Med Res* **34**(6), 600–606 (2003).
16. Dominici, M., Hofmann, T.J., and Horwitz, E.M. Bone marrow mesenchymal cells: biological properties and clinical applications. *J Biol Regul Homeost Agents* **15**(1), 28–37 (2001).

17. Baksh, D., Song, L., and Tuan, R.S. Adult mesenchymal stem cells: characterization, differentiation, and application in cell and gene therapy. *J Cell Mol Med* **8**(3), 301–316 (2004).
18. Stripp, B.R. and Shapiro, S.D. Stem cells in lung disease, repair, and the potential for therapeutic interventions: state-of-the-art and future challenges. *Am J Respir Cell Mol Biol* **34**(5), 517–518 (2006).
19. Prockop, D.J., Gregory, C.A., and Spees, J.L. One strategy for cell and gene therapy: harnessing the power of adult stem cells to repair tissues. *Proc Natl Acad Sci USA* **100** (Suppl 1), 11917–11923 (2003).
20. Horwitz, E.M., Le Blanc, K., Dominici, M., Mueller, I., Slaper-Cortenbach, I., Marini, F.C., Deans, R.J., Krause, D.S., and Keating, A. Clarification of the nomenclature for MSC: the International Society for Cellular Therapy position statement. *Cytotherapy* **7**(5), 393–395 (2005).
21. Prockop, D.J. Marrow stromal cells as stem cells for nonhematopoietic tissues. *Science* **276**(5309), 71–74.
22. Dominici, M., Le Blanc, K., Mueller, I., Slaper-Cortenbach, I., Marini, F., Krause, D., Deans, R., Keating, A., Prockop, D., and Horwitz, E. Minimal criteria for defining multipotent mesenchymal stromal cells. The International Society for Cellular Therapy position statement. *Cytotherapy* **8**(4), 315–317 (2006).
23. Korbling, M. and Estrov, Z. Adult stem cells for tissue repair — a new therapeutic concept? *N Engl J Med* **349**(6), 570–582 (2003).
24. Chen, S.L., Fang, W.W., Ye, F., Liu, Y.H., Qian, J., Shan, S.J., Zhang, J.J., Chunhua, R.Z., Liao, L.M., Lin, S., and Sun, J.P. Effect on left ventricular function of intracoronary transplantation of autologous bone marrow mesenchymal stem cell in patients with acute myocardial infarction. *Am J Cardiol* **94**(1), 92–95 (2004).
25. Ringden, O., Uzunel, M., Rasmusson, I., Remberger, M., Sundberg, B., Lonnies, H., Marschall, H.U., Dlugosz, A., Szakos, A., Hassan, Z., Omazic, B., Aschan, J., Barkholt, L., and Le Blanc, K. Mesenchymal stem cells for treatment of therapy-resistant graft-versus-host disease. *Transplantation* **81**(10), 1390–1397 (2006).
26. Grove, J.E., Lutzko, C., Priller, J., Henegariu, O., Theise, N.D., Kohn, D.B., and Krause, D.S. Marrow-derived cells as vehicles for delivery of gene therapy to pulmonary epithelium. *Am J Respir Cell Mol Biol* **27**(6), 645–651 (2002).

27. Ortiz, L.A., Gambelli, F., McBride, C., Gaupp, D., Baddoo, M., Kaminski, N., and Phinney, D.G. Mesenchymal stem cell engraftment in lung is enhanced in response to bleomycin exposure and ameliorates its fibrotic effects. *Proc Natl Acad Sci USA* **100**(14), 8407–8411 (2003).
28. Kotton, D.N., Ma, B.Y., Cardoso, W.V., Sanderson, E.A., Summer, R.S., Williams, M.C., and Fine, A. Bone marrow-derived cells as progenitors of lung alveolar epithelium. *Development* **128**(24), 5181–5188 (2001).
29. Giordano, A., Galderisi, U., and Marino, I.R. From the laboratory bench to the patient's bedside: an update on clinical trials with mesenchymal stem cells. *J Cell Physiol* **211**(1), 27–35 (2007).
30. Koc, O.N., Day, J., Nieder, M., Gerson, S.L., Lazarus, H.M., and Krivit, W. Allogeneic mesenchymal stem cell infusion for treatment of metachromatic leukodystrophy (MLD) and Hurler syndrome (MPS-IH). *Bone Marrow Transplant* **30**(4), 215–222 (2002).
31. Horwitz, E.M., Gordon, P.L., Koo, W.K., Marx, J.C., Neel, M.D., McNall, RY, Muul L., and Hofmann, T. Isolated allogeneic bone marrow-derived mesenchymal cells engraft and stimulate growth in children with osteogenesis imperfecta: implications for cell therapy of bone. *Proc Natl Acad Sci USA* **99**(13), 8932 8937 (2002).
32. Krampera, M., Cosmi, L., Angeli, R., Pasini, A., Liotta, F., Andreini, A., Santarlasci, V., Mazzinghi, B., Pizzolo, G., Vinante, F., Romagnani, P., Maggi, E., Romagnani, S., and Annunziato, F. Role for interferon-gamma in the immunomodulatory activity of human bone marrow mesenchymal stem cells. *Stem Cells* **24**(2), 386–398 (2006).
33. Frank, M.H. and Sayegh, M.H. Immunomodulatory functions of mesenchymal stem cells. *Lancet* **363**(9419), 1411–1412 (2004).
34. Le Blanc, K. Immunomodulatory effects of fetal and adult mesenchymal stem cells. *Cytotherapy* **5**(6), 485–489 (2003).
35. Di Nicola, M., Carlo-Stella, C., Magni, M., Milanesi, M., Longoni, P.D., Matteucci, P., Grisanti, S., and Gianni, A.M. Human bone marrow stromal cells suppress T-lymphocyte proliferation induced by cellular or nonspecific mitogenic stimuli. *Blood* **99**(10), 3838–3843 (2002).
36. Ryan, J.M., Barry, F.P., Murphy, J.M., and Mahon, B.P. Mesenchymal stem cells avoid allogeneic rejection. *J Inflamm (Lond)* **2**, 8 (2005).
37. Gotherstrom, C., Ringden, O., Tammik, C., Zetterberg, E., Westgren, M., and Le Blanc, K. Immunologic properties of human fetal mesenchymal stem cells. *Am J Obstet Gynecol* **190**(1), 239–245 (2004).

38. Majumdar, M.K., Keane-Moore, M., Buyaner, D., Hardy, W.B., Moorman, M.A., McIntosh, K.R., and Mosca, J.D. Characterization and functionality of cell surface molecules on human mesenchymal stem cells. *J Biomed Sci* **10**(2), 228–241 (2003).
39. Devine, S.M. and Hoffman, R. Role of mesenchymal stem cells in hematopoietic stem cell transplantation. *Curr Opin Hematol* **7**(6), 358–363 (2000).
40. Reiser, J., Zhang, X.Y., Hemenway, C.S., Mondal, D., Pradhan, L., and La Russa, V.F. Potential of mesenchymal stem cells in gene therapy approaches for inherited and acquired diseases. *Expert Opin Biol Ther* **5**(12), 1571–1584 (2005).
41. Xu, J., Qu, J., Cao, L., Sai, Y., Chen, C., He, L., and Yu, L. Mesenchymal stem cell-based angiopoietin-1 gene therapy for acute lung injury induced by lipopolysaccharide in mice. *J Pathol* **214**(4), 472–481 (2008).
42. Kanehira, M., Xin, H., Hoshino, K., Maemondo, M., Mizuguchi, H., Hayakawa, T., Matsumoto, K., Nakamura, T., Nukiwa, T., and Saijo, Y. Targeted delivery of NK4 to multiple lung tumors by bone marrow-derived mesenchymal stem cells. *Cancer Gene Ther* **14**(11), 894–903 (2007).
43. Xin, H., Kanehira, M., Mizuguchi, H., Hayakawa, T., Kikuchi, T., Nukiwa, T., and Saijo, Y. Targeted delivery of CX3CL1 to multiple lung tumors by mesenchymal stem cells. *Stem Cells* **25**(7), 1618–1626.
44. Loi, R., Beckett, T., Goncz, K.K., Suratt, B.T., and Weiss, D.J. Limited restoration of cystic fibrosis lung epithelium *in vivo* with adult bone marrow-derived cells. *Am J Respir Crit Care Med* **173**(2), 171–179 (2006).
45. Ward, M.R., Stewart, D.J., and Kutryk, M.J. Endothelial progenitor cell therapy for the treatment of coronary disease, acute MI, and pulmonary arterial hypertension: current perspectives. *Catheter Cardiovasc Interv* **70**(7), 983–998 (2007).
46. Asahara, T., Murohara, T., Sullivan, A., Silver, M., van der Zee, R., Li, T., Witzenbichler, B., Schatteman, G., and Isner, J.M. Isolation of putative progenitor endothelial cells for angiogenesis. *Science* **275**(5302), 964–967 (1997).
47. Debatin, K.M., Wei, J., and Beltinger, C. Endothelial progenitor cells for cancer gene therapy. *Gene Ther* **15**(10), 780–786 (2008).
48. Wei, J., Blum, S., Unger, M., Jarmy, G., Lamparter, M., Geishauser, A., Vlastos, G.A., Chan, G., Fischer, K.D., Rattat, D., Debatin, K.M., Hatzopoulos, A.K., and Beltinger, C. Embryonic endothelial progenitor

cells armed with a suicide gene target hypoxic lung metastases after intravenous delivery. *Cancer Cell* **5**(5), 477–488 (2004).
49. Gomperts, B.N. and Strieter, RM. Stem cells and chronic lung disease. *Annu Rev Med* **58**, 285–298 (2007).
50. Krause, D.S., Theise, N.D., Collector, M.I., Henegariu, O., Hwang, S., Gardner, R., Neutzel, S., and Sharkis, S.J. Multi-organ, multi-lineage engraftment by a single bone marrow-derived stem cell. *Cell* **105**(3), 369–377 (2001).
51. Rippon, H.J., Polak, J.M., Qin, M., and Bishop, A.E. Derivation of distal lung epithelial progenitors from murine embryonic stem cells using a novel three-step differentiation protocol. *Stem Cells* **24**(5), 1389–1398 (2006).
52. Coraux, C., Nawrocki-Raby, B., Hinnrasky, J., Kileztky, C., Gaillard, D., Dani, C., and Puchelle, E. Embryonic stem cells generate airway epithelial tissue. *Am J Respir Cell Mol Biol* **32**(2), 87–92 (2005).
53. Nagaya, N., Kangawa, K., Kanda, M., Uematsu, M., Horio, T., Fukuyama, N., Hino, J., Harada-Shiba, M., Okumura, H., Tabata, Y., Mochizuki, N., Chiba, Y., Nishioka, K., Miyatake, K., Asahara, T., Hara, H., and Mori, H. Hybrid cell-gene therapy for pulmonary hypertension based on phagocytosing action of endothelial progenitor cells. *Circulation* **108**(7), 889–895 (2003).
54. Sueblinvong, V., Loi, R., Eisenhauer, P.L., Bernstein, I.M., Suratt, B.T., Spees, J.L., and Weiss, D.J. Derivation of lung epithelium from human cord blood-derived mesenchymal stem cells. *Am J Respir Crit Care Med* **177**(7), 701–711.
55. Kunisaki, S.M., Freedman, D.A., and Fauza, D.O. Fetal tracheal reconstruction with cartilaginous grafts engineered from mesenchymal amniocytes. *J Pediatr Surg* **41**(4), 675–682; discussion 675–682 (2006).
56. Shigemura, N., Okumura, M., Mizuno, S., Imanishi, Y., Nakamura, T., and Sawa, Y. Autologous transplantation of adipose tissue-derived stromal cells ameliorates pulmonary emphysema. *Am J Transplant* **6**(11), 2592–2600 (2006).
57. Fukuchi, Y., Nakajima, H., Sugiyama, D., Hirose, I., Kitamura, T., and Tsuji, K. Human placenta-derived cells have mesenchymal stem/progenitor cell potential. *Stem Cells* **22**(5), 649–658 (2004).
58. Ware, L.B. and Matthay, M.A. The acute respiratory distress syndrome. *N Engl J Med* **342**(18), 1334–1349 (2000).

59. Goss, C.H., Brower, R.G., Hudson, L.D., and Rubenfeld, G.D. Incidence of acute lung injury in the United States. *Crit Care Med* **31**(6), 1607–1611 (2003).
60. Mendez, J.L. and Hubmayr, R.D. New insights into the pathology of acute respiratory failure. *Curr Opin Crit Care* **11**(1), 29–36 (2005).
61. Rubenfeld, G.D., Caldwell, E., Peabody, E., Weaver, J., Martin, D.P., Neff, M., Stern, E.J., and Hudson, L.D. Incidence and outcomes of acute lung injury. *N Engl J Med* **353**(16), 1685–1693 (2005).
62. Slutsky, A.S. and Hudson, L.D. PEEP or no PEEP — lung recruitment may be the solution. *N Engl J Med* **354**(17), 1839–1841 (2006).
63. Crimi, E. and Slutsky, A.S. Inflammation and the acute respiratory distress syndrome. *Best Pract Res Clin Anaesthesiol* **18**(3), 477–492 (2004).
64. Ventilation with lower tidal volumes as compared with traditional tidal volumes for acute lung injury and the acute respiratory distress syndrome. The Acute Respiratory Distress Syndrome Network. *N Engl J Med* **342**(18), 1301–1308.
65. Matthay, M.A., Zimmerman, G.A., Esmon, C., Bhattacharya, J., Coller, B., Doerschuk, C.M., Floros, J., Gimbrone, M.A., Jr., Hoffman, E., Hubmayr, R.D., Leppert, M., Matalon, S., Munford, R., Parsons, P., Slutsky, A.S., Tracey, K.J., Ward, P., Gail, D.B., and Harabin, AL. Future research directions in acute lung injury: summary of a National Heart, Lung, and Blood Institute working group. *Am J Respir Crit Care Med* **167**(7), 1027–1035 (2003).
66. Mehta, D., Bhattacharya, J., Matthay, M.A., and Malik, A.B. Integrated control of lung fluid balance. *Am J Physiol Lung Cell Mol Physiol* **287**(6), L1081–L1090 (2004).
67. Orfanos, S.E., Kotanidou, A., Glynos, C., Athanasiou, C., Tsigkos, S., Dimopoulou, I., Sotiropoulou, C., Zakynthinos, S., Armaganidis, A., Papapetropoulos, A., and Roussos, C. Angiopoietin-2 is increased in severe sepsis: correlation with inflammatory mediators. *Crit Care Med* **35**(1), 199–206 (2007).
68. Parikh, S.M., Mammoto, T., Schultz, A., Yuan, H.T., Christiani, D., Karumanchi, S.A., and Sukhatme, V.P. Excess circulating angiopoietin-2 may contribute to pulmonary vascular leak in sepsis in humans. *PLoS Med* **3**(3), e46 (2006).
69. Witzenbichler, B., Westermann, D., Knueppel, S., Schultheiss, H.P., and Tschope, C. Protective role of angiopoietin-1 in endotoxic shock. *Circulation* **111**(1), 97–105.

70. Huang, Y.Q., Sauthoff, H., Herscovici, P., Pipiya, T., Cheng, J., Heitner, S., Szentirmai, O., Carter, B., and Hay, J.G. Angiopoietin-1 increases survival and reduces the development of lung edema induced by endotoxin administration in a murine model of acute lung injury. *Crit Care Med* **36**(1), 262–267 (2008).
71. McCarter, S.D., Mei, S.H., Lai, P.F., Zhang, Q., Parker, C.H., Suen, R.S., Hood R.D., Zhao, Y.D., Deng, Y., Han, R.N., Dumont, D.J., and Stewart, D.J. Cell-based angiopoietin-1 gene therapy for acute lung injury. *Am J Respir Crit Care Med* **175**(10), 1014–1026 (2007).
72. Xu, J., Woods, C.R., Mora, A.L., Joodi, R., Brigham, K.L., Iyer, S., and Rojas, M. Prevention of endotoxin-induced systemic response by bone marrow-derived mesenchymal stem cells in mice. *Am J Physiol Lung Cell Mol Physiol* **293**(1), L131–L141 (2007).
73. Nemeth, K., Leelahavanichkul, A., Yuen, P.S., Mayer, B., Parmelee, A., Doi, K., Robey, P.G., Leelahavanichkul, K., Koller, B.H., Brown, J.M., Hu, X., Jelinek, I., Star, R.A., and Mezey, E. Bone marrow stromal cells attenuate sepsis via prostaglandin E(2)-dependent reprogramming of host macrophages to increase their interleukin-10 production. *Nat Med* **15**(1), 42–49 (2009).
74. Farber, H.W. and Loscalzo, J. Pulmonary arterial hypertension. *N Engl J Med* **351**(16), 1655–1665 (2004).
75. Humbert, M., Sitbon, O., and Simonneau, G. Treatment of pulmonary arterial hypertension. *N Engl J Med* **351**(14), 1425–1436 (2004).
76. Puri, A., McGoon, M.D., and Kushwaha, S.S. Pulmonary arterial hypertension: current therapeutic strategies. *Nat Clin Pract Cardiovasc Med* **4**(6), 319–329 (2007).
77. Rafii, S. and Lyden, D. Therapeutic stem and progenitor cell transplantation for organ vascularization and regeneration. *Nat Med* **9**(6), 702–712 (2003).
78. Asahara, T. and Kawamoto, A. Endothelial progenitor cells for postnatal vasculogenesis. *Am J Physiol Cell Physiol* **287**(3), C572–C579.
79. Cooke, J.P. and Losordo, D.W. Nitric oxide and angiogenesis. *Circulation* **105**(18), 2133–2135 (2002).
80. Hogg, J.C. and Timens, W. The pathology of chronic obstructive pulmonary disease. *Annu Rev Pathol* **4**, 435–459 (2008).
81. Carrell, R.W. and Lomas, D.A. Alpha1-antitrypsin deficiency — a model for conformational diseases. *N Engl J Med* **346**(1), 45–53 (2002).

82. Kohnlein, T. and Welte, T. Alpha-1 antitrypsin deficiency: pathogenesis, clinical presentation, diagnosis, and treatment. *Am J Med* **121**(1), 3–9 (2008).
83. Alexander, B.L., Ali, R.R., Alton, E.W., Bainbridge, J.W., Braun, S., Cheng, S.H., Flotte, T.R., Gaspar, H.B., Grez, M., Griesenbach, U., Kaplitt, M.G., Ott, M.G., Seger, R., Simons, M., Thrasher, A.J., Thrasher, A.Z., and Yla-Herttuala, S. Progress and prospects: gene therapy clinical trials (part 1). *Gene Ther* **14**(20), 1439–1447 (2007).
84. Wood, A.M. and Stockley, R.A. Alpha one antitrypsin deficiency: from gene to treatment. *Respiration* **74**(5), 481–492 (2007).
85. Wilson, A.A., Kwok, L.W., Hovav, A.H., Ohle, S.J., Little, F.F., Fine, A., and Kotton, D.N. Sustained expression of alpha1-antitrypsin after transplantation of manipulated hematopoietic stem cells. *Am J Respir Cell Mol Biol* **39**(2), 133–141 (2008).
86. Abe, S., Boyer, C., Liu, X., Wen, F.Q., Kobayashi, T., Fang, Q., Wang, X., Hashimoto, M., Sharp, J.G., and Rennard, S.I. Cells derived from the circulation contribute to the repair of lung injury. *Am J Respir Crit Care Med* **170**(11), 1158–1163 (2004).
87. Schuleri, K.H., Amado, L.C., Boyle, A.J., Centola, M., Saliaris, A.P., Gutman, M.R., Hatzistergos, K.E., Oskouei, B.N., Zimmet, J.M., Young, R.G., Heldman, A.W., Lardo, A.C., and Hare, J.M. Early improvement in cardiac tissue perfusion due to mesenchymal stem cells. *Am J Physiol Heart Circ Physiol* **294**(5), H2002–H2011 (2008).
88. Rowe, S.M., Miller, S., and Sorscher, E.J. Cystic fibrosis. *N Engl J Med* **352**(19), 1992–2001 (2005).
89. Kolb, M., Martin, G., Medina, M., Ask, K., and Gauldie, J. Gene therapy for pulmonary diseases. *Chest* **130**(3), 879–884 (2006).
90. Flotte, T.R., Ng, P., Dylla, D.E., McCray, P.B., Jr., Wang, G., Kolls, J.K., and Hu, J. Viral vector-mediated and cell-based therapies for treatment of cystic fibrosis. *Mol Ther* **15**(2), 229–241 (2007).
91. Wang, G., Bunnell, B.A., Painter, R.G., Quiniones, B.C., Tom, S., Lanson, N.A., Jr., Spees, J.L., Bertucci, D., Peister, A., Weiss, D.J., Valentine, V.G., Prockop, D.J., and Kolls, J.K. Adult stem cells from bone marrow stroma differentiate into airway epithelial cells: potential therapy for cystic fibrosis. *Proc Natl Acad Sci USA* **102**(1), 186–191 (2005).

92. Bruscia, E.M., Price, J.E., Cheng, E.C., Weiner, S., Caputo, C., Ferreira, E.C., Egan, M.E., and Krause, D.S. Assessment of cystic fibrosis transmembrane conductance regulator (CFTR) activity in CFTR-null mice after bone marrow transplantation. *Proc Natl Acad Sci USA* **103**(8), 2965–2970 (2006).
93. Aboody, K.S., Najbauer, J., and Danks, M.K. Stem and progenitor cell-mediated tumor selective gene therapy. *Gene Ther* **15**(10), 739–752 (2008).
94. Spira, A. and Ettinger, D.S. Multidisciplinary management of lung cancer. *N Engl J Med* **350**(4), 379–392 (2004).
95. Benedetti, S., Pirola, B., Pollo, B., Magrassi, L., Bruzzone, M.G., Rigamonti, D., Galli, R., Selleri, S., Di Meco, F., De Fraja, C., Vescovi, A., Cattaneo, E., and Finocchiaro, G. Gene therapy of experimental brain tumors using neural progenitor cells. *Nat Med* **6**(4), 447–450 (2000).
96. Ozawa, K., Sato, K., Oh, I., Ozaki, K., Uchibori, R., Obara, Y., Kikuchi, Y., Ito, T., Okada, T., Urabe, M., Mizukami, H., and Kume, A. Cell and gene therapy using mesenchymal stem cells (MSCs). *J Autoimmun* **30**(3), 121–127 (2008).
97. Bellusci, S., Moens, G., Gaudino, G., Comoglio, P., Nakamura, T., Thiery, J.P., and Jouanneau, J. Creation of an hepatocyte growth factor/scatter factor autocrine loop in carcinoma cells induces invasive properties associated with increased tumorigenicity. *Oncogene* **9**(4), 1091–1099 (1994).
98. Ren, C., Kumar, S., Chanda, D., Kallman, L., Chen, J., Mountz, J.D., and Ponnazhagan, S. Cancer gene therapy using mesenchymal stem cells expressing interferon-beta in a mouse prostate cancer lung metastasis model. *Gene Ther* **15**(21), 1446–1453 (2008).
99. Ren, C., Kumar, S., Chanda, D., Chen, J., Mountz, J.D., and Ponnazhagan, S. Therapeutic potential of mesenchymal stem cells producing interferon-alpha in a mouse melanoma lung metastasis model. *Stem Cells* **26**(9), 2332–2338 (2008).
100. Tolar, J., Nauta, A.J., Osborn, M.J., Panoskaltsis Mortari, A., McElmurry, R.T., Bell, S., Xia, L., Zhou, N., Riddle, M., Schroeder, T.M., Westendorf, J.J., McIvor, R.S., Hogendoorn, P.C., Szuhai, K., Oseth, L., Hirsch, B., Yant, S.R., Kay, M.A., Peister, A., Prockop, D.J., Fibbe, W.E., and Blazar, B.R. Sarcoma derived from cultured mesenchymal stem cells. *Stem Cells* **25**(2), 371–379 (2007).

Chapter 11
Repair of the Lung Epithelium in Cystic Fibrosis

Viranuj Sueblinvong and Daniel J. Weiss

Cystic fibrosis (CF) remains a devastating and incurable disease. One of the first targets for lung gene therapy and studies of lung gene transfer, early promising results were achieved *in vitro* and in small animal models. However, studies in primate models and in patients were discouraging, despite a large number of clinical trials, and gene therapy approaches for CF have generally fallen out of favour. Newer approaches with cell-based therapy, utilizing either embryonic stem cells or extra embryonic-derived adult stem cells (bone marrow or cord blood), have been investigated recently and may provide viable future therapeutic options. In parallel, further understanding of the role of endogenous progenitor cells in CF lungs may also provide both mechanistic understanding and potential therapeutic approaches for CF lung disease. In this chapter, endogenous lung progenitors in CF and the potential use of cell therapy-based approaches for CF will be considered.

Keywords: Cystic fibrosis; stem cell; cell therapy; lung epithelium.

Outline

1. Introduction and Pathophysiology of Cystic Fibrosis

Cystic Fibrosis (CF) is an autosomal recessive disease caused by mutations in the gene encoding for the cystic fibrosis transmembrane conductance regulator (CFTR). In 1989, the gene was localised to chromosome 7 and found to consist of approximately 250,000 base pairs that encode an mRNA of 6.5 kb.[1–3] Over 1000 mutations have been identified, resulting in aberrant transcription, translation, cellular trafficking, and/or ion channel function.[4] At the cellular level, CFTR is an apical membrane protein, found in several types of lung epithelial cells, that serves as a regulated chloride channel.[5] Through interactions between CFTR and the amiloride-sensitive epithelial sodium channel (ENaC), absence of functional CFTR in CF epithelial cells results in both sodium hyper-absorption and lack of cyclic adenosine monophosphate (cAMP)-mediated chloride secretion.[6] Moreover, it is now clear that CFTR is important for other cellular functions, including post-translational processing of high molecular weight glycoconjugates and cell surface receptors, pH regulation of intracellular organelles and airway surface liquid, regulation of membrane trafficking, secretion of mucus, and regulation of glutathione transport.[7,8] Further, there is an increasing amount of data demonstrating that constitutive and stimulated release of soluble inflammatory mediators is increased from CF airway epithelial cells.[9,10] This may reflect, in part, increased basal and stimulated cell signalling resulting from increased NFκB activity, AMP-dependent kinase activity, or altered antioxidant homeostasis in cells with defective CFTR.[11] Although the mechanisms for many of these effects remain incompletely understood, they evidence the complexity and multiple cellular effects that need to be considered with approaches to correct abnormal CFTR.

At the organ level, CFTR in lung is localised primarily in the ciliated cells of the proximal airways and of the submucosal glands. Whether CFTR is substantially expressed in other epithelial cell types remains unclear. Earlier studies suggesting expression in non-ciliated epithelial cells of the airways and glands, including basal epithelial cells, have been contradicted by more recent studies demonstrating that, although CFTR mRNA may be found in these cells, levels of CFTR protein expression are low and of uncertain significance.[12,13] Moreover, although CFTR can be detected in type 2 alveolar epithelial cells, expression is generally low compared to either proximal or distal airways.[13] This suggests that targeting the ciliated airway epithelial cells and the submucosal glands will have the most effect on regulation of mucus and airway surface liquid, and presumably ameliorate the most relevant pathophysiologic respiratory effects of defective CFTR. However, there is turnover of the differentiated ciliated epithelial cells, and targeting the underlying basal progenitor epithelial cells may be a more viable approach to provide for longer lasting or even indefinite CFTR correction. As further discussed below, progress has been made towards identifying endogenous progenitor cells resident in both proximal and distal airways and also identifying potential defects in endogenous progenitor cells in CF lungs. Most recently, it has been demonstrated that recombinant adeno-associated virus vectors may preferentially target endogenous distal progenitor cells in mice.[14] This suggests a potential route for intervention but as yet no viable strategy for specifically targeting these cells in humans has been identified.

It has also become evident that correcting defective CFTR alone may not fully aid airways disease. This is exemplified by the development of the ENaC overexpressing mouse.[15,16] Unlike the CFTR knockout or delta F transgenic mice, which do not develop substantial airways disease, increased airway sodium absorption in the airways of the ENaC mice resulted in airway surface liquid volume depletion, increased mucus concentration, delayed mucus transport and mucus adhesion to airway surfaces.[15] The mice also developed severe spontaneous lung disease, comparable to that in patients with CF, including mucus obstruction, goblet cell metaplasia, neutrophilic inflammation and poor bacterial clearance. This evidences the important role of ENaC in the development of many of the clinical manifestations of CF lung disease. Whether replacing defective CFTR will subsequently result in appropriate regulation of ENaC remains unclear.[17,18]

2. Gene Therapy Approaches for Cystic Fibrosis

Isolation of the CF gene led to speculation that the abnormal CFTR gene could be replaced *in situ* with gene transfer techniques utilizing either viral or non-viral gene transfer vectors to deliver a normal copy of the CF gene and replace the defective gene.[19–22] As the pulmonary disease in CF, characterised by abnormal mucus secretion, chronic bacterial infection, and airway inflammation, is the major cause of morbidity and mortality, the lung was the first target organ for gene replacement.[23] Although initial demonstrations of successful gene transfer in cultured lung epithelial cells and in small animal models generated enthusiasm, further studies in primate models and in patients have been discouraging. In particular, despite a number of clinical trials, a viable therapeutic strategy has yet to emerge. This reflects a number of obstacles to successful, sustained, and repeatable gene transfer in the lung, some of which may only be overcome with great difficulty[19,22] (Figs. 1 and 2). As such, research efforts have largely shifted focus from clinical to more fundamental studies that seek to delineate the cell and molecular biology of gene transfer to airway epithelium. Nonetheless, the experience with gene therapy has set the stage for recent attempts to utilise cell therapy approaches to replace defective airway epithelial cells *in situ* with cells containing normal CFTR.[22,24]

3. Endogenous Lung Progenitor Cells in Cystic Fibrosis

Endogenous tissue stem cells are undifferentiated cells that have been identified in nearly all tissues and are thought to contribute to tissue maintenance and repair. These are rare, highly specialised cells that are often localised to specialised niches within each tissue and that usually cycle infrequently. These cells exhibit self-renewal capacity — they can produce more unspecialised cells — and can also give rise to daughter cells known as progenitor cells or transit amplifying cells. Progenitor cells have a finite life span, more restricted differentiation potential, and higher rates of proliferation compared to stem cells. Both stem and progenitor cells may give rise to the more specialised, or differentiated, cells of the organ[25–27] (see also Chapter 4).

The focus in lung has been predominantly on epithelial progenitor cells but increasing evidence suggests potential vascular and mesenchymal progenitor cell populations as well. Moreover, as the lung is a complex organ, several airway epithelial stem and progenitor cell hierarchies have been identified along the tracheobronchial tree in mouse models.[25–27] In trachea

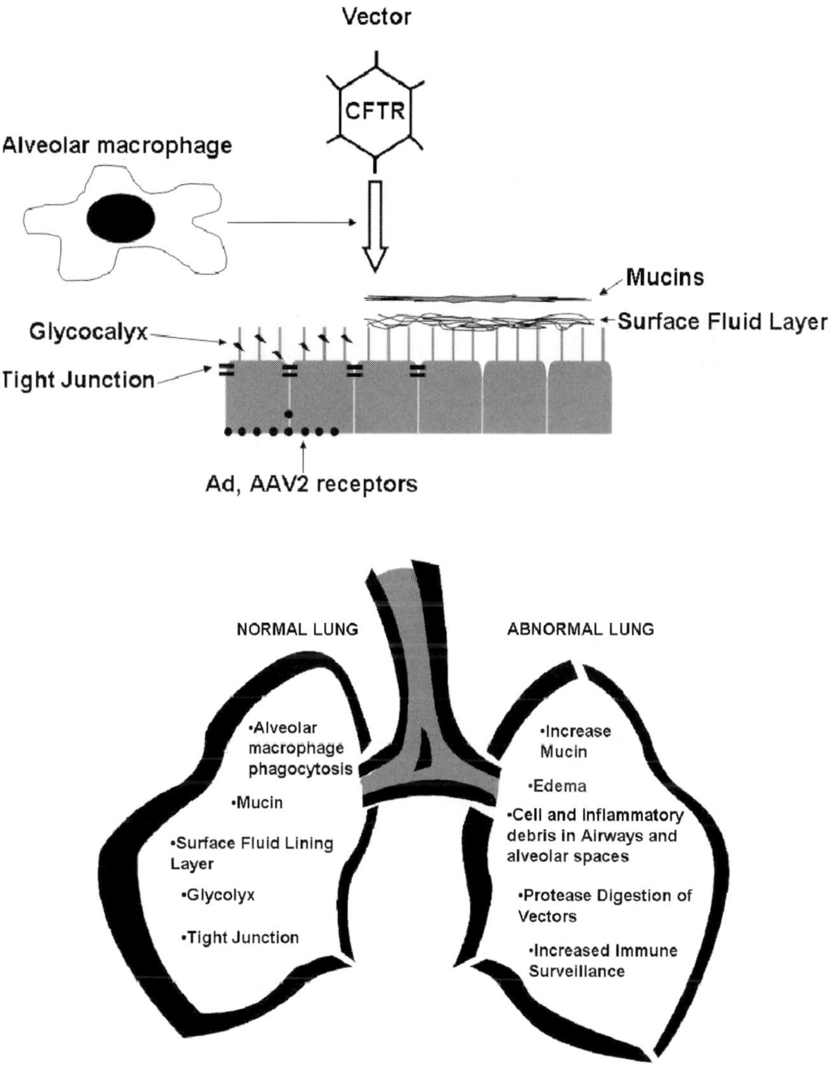

Figure 1: Schematics of barriers to gene and potentially also to cell delivery in normal and injured lungs. Reprinted with permission from Sueblinvong *et al.* (2007).[22]

and large airways, a sub-population of basal epithelial cells that express cytokeratins 5 and 14 has been implicated.[28–30] In lower airways in mice, Clara cells exhibit characteristics of transit-amplifying cells following injury to differentiated ciliated airway epithelial cells. However, unlike

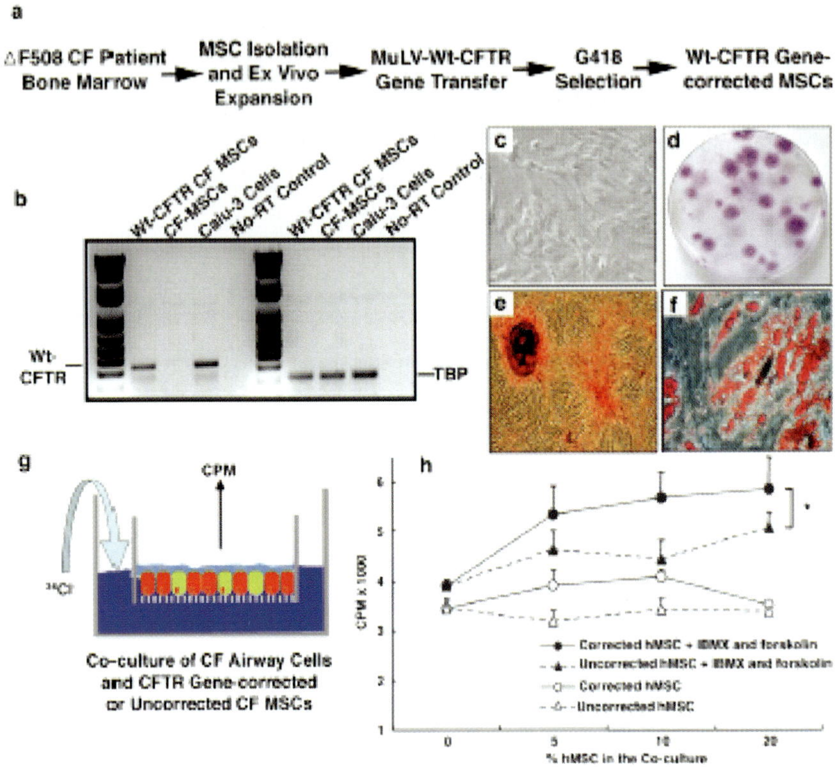

Figure 2: CFTR-corrected CF-patient MSCs retained their multipotency and responded to cAMP stimulation by secreting chloride to the apical side. (a) Schematic for CF patient MSC isolation, expansion, gene-correction and positive drug selection. (b) RT-PCR to verify the successful CFTR gene transfer. RT-PCR was performed to amplify wild-type CFTR transcripts but not ΔF508 mutant transcripts. The gene corrected CF MSCs and positive control Calu 3 cells have wild-type CFTR transcription, while non-gene-corrected CF MSCs and the no-RT control show negative amplification. In the RT-PCR control for the TATA-box binding protein (TBP), all the samples except the no-RT control show positive PCR products. (c) Phase-contrast microscopic view of the CFTR gene-corrected CF-patient MSCs. (d) Photomicrograph of a representative CFU-F assay. Purple-stained foci are the MSC colonies. (e) Osteogenesis of the CFTR gene-corrected CF-patient MSCs. After differentiation in an osteogenic medium, cells had mineral deposits visualised in red by Alizarin Red staining. (f) Adipogenesis of the CFTR gene-corrected CF-patient MSCs. After differentiation in an adipogenic medium, cells had lipid droplet accumulation stained in red with Oil Red O. (g and h) CFTR-gene corrected MSCs from CF patients contributed to the apical cAMP-stimulated Cl^- secretion. CFTR-gene corrected CF-patient MSCs or non-gene corrected CF-patient MSCs were mixed with ΔF508 CF airway epithelial cells at different ratios as indicated. After a month in culture at the air–liquid interface, chloride efflux assays were performed. Two-way ANOVA test revealed that co-cultures with the CFTR gene-corrected CF-patient MSCs[62] had a greater chloride secretion in response

transit-amplifying cells in tissues with higher rates of epithelial turnover, such as intestine, Clara cells exhibit a low proliferative frequency in the steady-state, are broadly distributed throughout the bronchiolar epithelium, and contribute to the specialised tissue function. In more distal airways, toxin (i.e. naphthalene)-resistant variant Clara cells have been identified as having stem cell functions and have been termed as bronchiolar stem cells.[26,31] Naphthalene-resistant cells are also located within discrete microenvironments within bronchioles that include the neuroepithelial body (NEB) and bronchioalveolar duct junction (BADJ).[26,32] Another population of naphthalene-resistant cells that stain for both Clara cell secretory protein (CCSP) and pro-surfactant protein C (SPC), termed bronchioalveolar stem cells (BASCs), has also been described at the BADJ in mice.[33] It is possible that toxin-resistant Clara cells, BASCs, and other cells may represent different interpretations of the same cell population(s) and highlight the need for both rigorous methods of lineage tracing and further underscoring the importance of the *in vivo* microenvironment on cell behaviour. Most recently, another population of putative progenitor cells expressing CCSP, stem cell antigen (SCA-1), stage-specific embryonic antigen 1 (SSEA-1), and the embryonic stem cell marker Oct-4 have been identified in neonatal mice.[34] These cells were able to form epithelial colonies and differentiate into both type 1 and type 2 alveolar epithelial cells. Interestingly, these cells were susceptible to infection with the SARS (severe acute respiratory syndrome) virus, raising the possibility that endogenous lung progenitor cells may be specific disease targets. The possibility remains that other endogenous stem or progenitor populations exist, and there is much to be learned about regulatory mechanisms and pathways, as have been elucidated in other epithelial progenitor cell populations, notably skin and intestine.[25–27] Importantly, the human correlates of the endogenous airway progenitor populations described in mice are less well understood. Defining human airway progenitor populations is a critical step and the focus of intense research activity.

Figure 2: (*Continued*) to the IBMX and forskolin stimulation than the co-cultures with non-gene corrected CF-patient MSCs (N = 4, $p < 0.05$). Figure reprinted with permission from Wang *et al.* (2005).[62] Copyright (2005) National Academy of Sciences, USA.

Recent investigations have begun clarifying cell signalling and other mechanisms regulating putative lung progenitor populations. For example, tumorigenic insults, including deletion of MAPK, p18 deletion, and p27 oncogenic mutation, have been shown to induce an expansion of CCSP/pro-SPC dual-labelled BADJ cells number and enhance lung tumorigenesis.[35–37] Most recently, it has been demonstrated that conditional potentiation of beta-catenin signalling in the embryonic lung results in the amplification of airway stem cells through attenuated differentiation rather than augmented proliferation.[38] However, the precise role of these and other pathways in endogenous lung progenitor cell homeostasis and response to injury remains to be determined.

Less information is available on the differences in endogenous stem and progenitor cells in different clinical lung diseases, including CF. The airway epithelium in CF patients contains primitive cuboidal cells that express primitive cell markers including thyroid transcription factor and cytokeratin 7.[39] Neuroepithelial cells also express CFTR, which appears to play a role in neuropeptide secretion.[40,41] CFTR knockout mice also contain fewer pulmonary neuroendocrine cells during embryonic development but increased numbers of these cells are seen postnatally.[42] This suggests that endogenous progenitor cell pathways in CF lungs may be altered, but this has not been extensively investigated.

Endogenous progenitor cells may also be attractive candidates for targeting with gene transfer vectors that provide sustained expression. Using adult transgenic Rosa26-Flox/LacZ reporter mice, Liu and colleagues have recently demonstrated that airway-based administration of Cre-expressing recombinant adeno-associated virus vectors (rAAV1Cre and rAAV5Cre) preferentially transduced type 2 alveolar epithelial cells and cells in the conducting but not larger airways.[14] Notably, the number of β-gal expressing conducting airway cells, predominantly Clara cells, and the overall amount of β-gal activity in lung homogenates, steadily increased over a six-month period reaching, respectively for rAAV1Cre and rAAV5Cre, 3% and 5% of total activity measured in positive control Rosa26-LacZ reporter mice, despite the absence of detectable Cre in Clara cells. Speculating that this might in part result from rAAV-mediated transduction of airway progenitor cells, naphthalene was administered to the Rosa26-Flox/LacZ mice to selectively deplete Clara cells but leave in place toxin-resistant variant Clara cells that might serve as precursors. Administration of either rAAV1Cre or rAAV5Cre along with BRDU identified both Lac-Z positive and negative label-retaining cells in bronchiolar airways, at bronchioalveolar duct junctions, and also in some type 2 alveolar epithelial cells. Importantly, Lac-Z positive label-retaining cells in bronchiolar airways were found whether the vector was given before or after naphthalene and BRDU administration. Immunostaining demonstrated that a subset of Lac-Z positive label-retaining cells in the bronchiolar airways and BADJs stained for CCSP and that these infrequent cells were associated with large transgene-expressing patches of cells, consistent with clonal expansion of the rAAV-transduced stem/progenitor cells in the regenerating airway epithelium. Furthermore, primary cultures of epithelial cells obtained from extra-lobar bronchi from rAAV1Cre and rAAV5Cre transduced Rosa26-Flox/LacZ mice demonstrated a 6–7 fold increase in colony-forming efficiency when cultured at the air–liquid interface. These results suggest that rAAV1 and rAAV5 selectively transduce airway epithelial cells *in vivo* with higher proliferative capacity and characteristics suggestive of distal airway progenitor cells. These exciting findings support the possibility of selectively or preferentially transducing stem/progenitor cell populations in lung. While much further investigation needs to be done, this provides a new potential therapeutic approach for diseases affecting airway and alveolar epithelium. However, overall, there remains much to be learned about endogenous stem and lung progenitor cells including clarification of human counterparts to the cells identified in mouse models. Furthermore, relatively little is known about the behaviour of endogenous stem or progenitor cells in clinical lung diseases or clinical disease models.

4. Cell-Based Therapies for CF Lung Disease

A potential therapeutic approach for CF and other lung diseases has been stimulated by reports demonstrating in mouse models that several cell populations derived from adult bone marrow or from umbilical cord blood, including stromal-derived mesenchymal stem cells (MSCs), endothelial progenitor cells, and circulating fibrocytes, can localise to

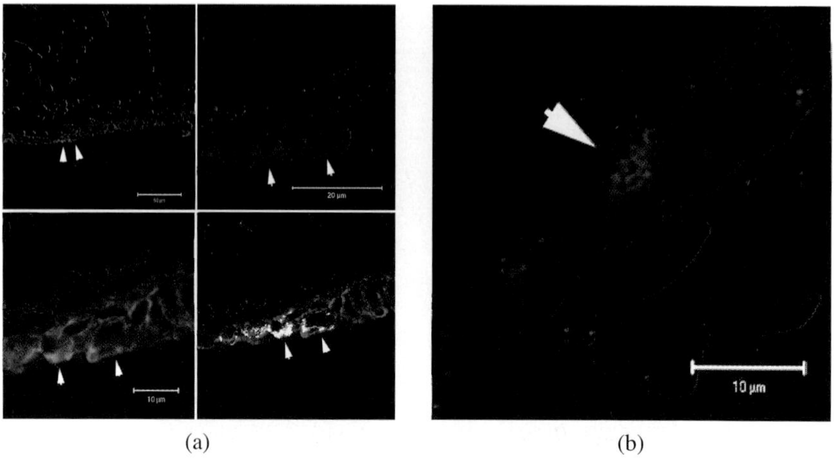

Figure 4: Human β2-microglobulin and CFTR-positive cells can be detected in NOD-SCID mice airways following systemic administration of human cord blood-derived MSCs. (a) NOD-SCID lung section three months following CB-MSC administration. Blue = DAPI nuclear stain, green = pancytokeratin, red = β2-microglobulin, white = co-localisation of pancytokeratin and β2-microglobulin. (b) NOD-SCID lung sections one month following CB-MSC administration. Blue = DAPI nuclear stain, green = CFTR, red = β2-microglobulin. Original magnifications 40×; 200×. Figure adapted with permission from Sueblinvong *et al.* (2008).[52] © American Thoracic Society.

CFTR-expressing lentiviral vector. This suggests that cord blood stem cells obtained from the births of infants with homozygous CFTR defects can be corrected *ex vivo* and subsequently utilised for autologous administration.[52] In comparable studies, Bruscia and colleagues found only rare engraftment of airway epithelium following transplantation of total marrow into lethally irradiated CFTR KO mice.[63] Interestingly, despite only rare engraftment also being found in intestinal epithelium, CFTR activity was detected by potential difference measurements in the rectal mucosa of recipient mice. This observation remains unexplained, as it is difficult to reconcile the rare engrafted cell observed with a physiologic change in potential difference measurements. Bruscia and colleagues also found that transplant of whole marrow into one-day-old CFTR KO mice did not increase the number of donor origin cells engrafted as either respiratory or intestinal epithelium.[64]

These studies demonstrate that the systemic administration of several different types of adult bone marrow or cord blood-derived cells did not result in significant structural engraftment as CFTR-expressing cells as either airway or intestinal epithelium in mouse models. Although not yet investigated in CF mouse models, several recent studies have also found only low levels of airway epithelial engraftment following direct airway administration of donor marrow-derived cells in mice.[55,65] Overall these observations suggest that, unless significant advances are made in understanding the mechanisms by which cells are recruited to airway epithelium and induced to undergo phenotypic conversion to functional airway epithelial cells, correction of CF lung disease by structural engraftment of adult marrow or cord blood-derived stem cells is not likely at present. Alternatively, it has been established that both mouse and human embryonic stem cells can be induced in culture to develop markers of lung epithelium. Further, human embryonic stem cell lines containing the deltaF508 CFTR mutation have been established, although these have not yet been comprehensively studied.[66–74] However, there has not yet been convincing evidence of significant structural engraftment of lung epithelium by either systemically or intratracheally administered embryonic stem cells. Further, investigations utilizing human embryonic stem cells are currently limited by scientific, ethical and political considerations in the United States and other countries. Recent studies with induced pluripotent stem cells (iPS) suggest that these may provide a more feasible alternative to embryonic stem cells, particularly as autologous iPS can conceivably be developed from patients with CF or other diseases.[75,76] However, to date no studies have demonstrated development of lung cell phenotypes from iPS cells. Tissue engineering approaches to grow functional lung tissue *ex vivo* utilizing either embryonic or adult stem cells may provide another alternative but this has not yet been investigated for CF.[77–85]

Nonetheless, despite rare engraftment of airway or alveolar epithelium, there are an increasing number of studies demonstrating a functional role of adult marrow-derived cells in mitigation of lung injury. This has been described in models of acute lung inflammation, emphysema, and fibrosis and has been observed mostly with MSCs.[65,86–92] Notably, systemic administration of MSCs immediately after intratracheal bleomycin administration decreased subsequent lung collagen accumulation, fibrosis, and levels of matrix metalloproteinases.[88,90] Secretion of IL-1 receptor antagonist by the MSCs is hypothesised to account for at least some of these effects.[90] Comparably, intratracheal administration of MSCs four hours after intratracheal endotoxin administration to mice decreased mortality, tissue inflammation, and concentration of pro-inflammatory mediators, such as TNFα and MIP-1β, in bronchoalveolar lavage fluid compared to endotoxin-only treated mice.[65] *Ex vivo* transduction of MSCs to express angiopoietin-1 further decreased endotoxin-mediated lung injury, presumably through abrogation of endotoxin-mediated endothelial injury.[92] These results suggest that MSCs can have significant immunomodulatory effects in the lung in the absence of engraftment. However, the mechanisms by which this occurs are largely unknown and there is no available information concerning immunomodulatory MSC effects in CF lungs. This would seem a promising area for study.

5. Conclusion and the Way Forward

Cell therapy approaches for CF lung disease remain promising. However, there is much to be learned both about the role of endogenous progenitor cells in CF lung disease and whether therapeutic engraftment of exogenously administered stem cells can become a viable therapeutic approach. Immunomodulation of CF lung inflammation and infection by adult MSCs or other stem cell type may also prove to be a useful strategy but is as yet of unproven benefit.

References

1. Riordan, J.R., Rommens, J.M., Kerem, B.S., *et al.* Identification of the cystic fibrosis gene: cloning and characterization of complementary DNA. *Science* **245**, 1066–1073 (1989).
2. Rommens, J.M., Iannuzzxi, M.C., Kerem, B.S., Drumm, M.L., Melmer, G., Dean, M., Rozmahel, R., Cole, J.L., Kennedy, D., Hidaka, N., Zsiga, M., Buchwald, M.J., Riordan, R.L., Tsui, C., and Collins, F.S. Identification of the cystic fibrosis gene: chromosome walking and jumping. *Science* **245**, 1059–1065 (1989).

3. Collins, F.S. Cystic fibrosis: molecular biology and therapeutic implications. *Science* **256**, 774–779 (1992).
4. Welsh, M.J. and Smith, A.E. Molecular mechanisms of CFTR chloride channel dysfunction in cystic fibrosis. *Cell* **73**, 1251–1254 (1993).
5. Rich, D.P., Anderson, M.P., Gregory, R.J., Cheng, S.H., Paul, S., Jefferson, D.M., McCann, J.D., Klinger, K.W., Smith, A.E., and Welsh, M.J. Expression of cystic fibrosis transmembrane conductance regulator corrects defective chloride channel regulation in cystic fibrosis airway epithelial cells. *Nature* **347**, 358–363 (1990).
6. Boucher, R.C. Airway epithelial fluid transport. *Am Rev Respir Dis* **150**, 581–593 (1994).
7. Mehta, A. CFTR: More than just a chloride channel. *Pediatr Pulmonol* **39**(4), 292–298 (2005).
8. Riordan, J.R. CFTR function and prospects for therapy. *Annu Rev Biochem* **77**, 701–726 (2008).
9. Machen, T.E. Innate immune response in CF airway epithelia: hyperinflammatory? *Am J Physiol Cell Physiol* **291**(2), C218–C230 (2006).
10. Jacquot, J., Tabary, O., Le Rouzic, P., and Clement, A. Airway epithelial cell inflammatory signalling in cystic fibrosis. *Int J Biochem Cell Biol* **40**(9), 1703–1715, (2008).
11. Nichols, D., Chmiel, J., and Berger, M. Chronic inflammation in the cystic fibrosis lung: alterations in inter- and intracellular signaling. *Clin Rev Allergy Immunol* **34**(2), 146–62, (2008).
12. Anson, D.S., Smith, G.J., and Parsons, D.W. Gene therapy for cystic fibrosis airway disease — is clinical success imminent? *Curr Gene Ther* **6**(2), 161–179 (2006).
13. Kreda, S.M., Mall, M., Mengos, A., *et al.* Characterization of wild-type and deltaF508 cystic fibrosis transmembrane regulator in human respiratory epithelia. *Mol Biol Cell* **16**(5), 2154–2167 (2005).
14. Liu, X., Luo, M., Guo, C., Yan, Z., Wang, Y., Lei-Butters, D.C.M., and Engelhardt, J.F. Analysis of adeno-associated virus progenitor cell transduction in mouse lung. *Mol Ther* **17**(2), 285–293 (2009).
15. Mall, M., Grubb, B.R., Harkema, J.R., *et al.* Increased airway epithelial Na+ absorption produces cystic fibrosis-like lung disease in mice. *Nat Med* **10**(5), 487–493 (2004).

16. Donaldson, S.H. and Boucher, R.C. Sodium channels and cystic fibrosis. *Chest* **132**(5), 1631–1636 (2007).
17. Amaral, M.D. and Kunzelmann, K. Molecular targeting of CFTR as a therapeutic approach to cystic fibrosis. *Trends Pharmacol Sci* **28**(7), 334–341 (2007).
18. Rubenstein, R.C. Targeted therapy for cystic fibrosis: cystic fibrosis transmembrane conductance regulator mutation-specific pharmacologic strategies. *Mol Diagn Ther* **10**(5), 293–301 (2006).
19. Weiss, D.J. and Pilewski, J. Current status of gene therapy for cystic fibrosis. *Semin Pulm Crit Care Med* **24**, 749–770 (2003).
20. Griesenbach, U., Geddes, D.M., and Alton, E.W. Gene therapy progress and prospects: cystic fibrosis. *Gene Ther* **13**(14), 1061–1067.
21. Rosenecker, J., Huth, S., and Rudolph, C. Gene therapy for cystic fibrosis lung disease: current status and future perspectives. *Curr Opin Mol Ther* **8**(5), 439–445 (2006).
22. Sueblinvong, V., Suratt, B.T., and Weiss, D.J. Novel therapies for the treatment of cystic fibrosis: new developments in gene and stem cell therapy. *Clin Chest Med* **28**(2), 361–379 (2007).
23. Vender, R.L. Cystic fibrosis lung disease in adult patients. *Postgrad Med* **120**(1), 64–74 (2008).
24. Thelin, W.R. and Boucher, R.C. The epithelium as a target for therapy in cystic fibrosis. *Curr Opin Pharmacol* **7**(3), 290–295 (2007).
25. Rawlins, E.L. and Hogan, B.L. Epithelial stem cells of the lung: privileged few or opportunities for many? *Development* **133**(13), 2455–2465 (2006).
26. Stripp, B.R. Hierarchical organization of lung progenitor cells: is there an adult lung tissue stem cell? *Proc Am Thorac Soc* **5**(6), 695–698 (2008).
27. Kim, C.F. Paving the road for lung stem cell biology: bronchioalveolar stem cells and other putative distal lung stem cells. *Am J Physiol Lung Cell Mol Physiol* **293**(5), L1092–L1098 (2007).
28. Borthwick, D.W., Shahbazian, M., Krantz, Q.T., Dorin, J.R., and Randell, S.H. Evidence for stem-cell niches in the tracheal epithelium. *Am J Respir Cell Mol Biol* **24**(6), 662–670 (2001).
29. Schoch, K.G., Lori, A., Burns, K.A., Eldred, T., Olsen, J.C., and Randell, S.H. A subset of mouse tracheal epithelial basal cells generates large colonies *in vitro*. *Am J Physiol Lung Cell Mol Physiol* **286**(4), L631–L642 (2004).

30. Engelhardt, J.F., Schlossberg, H., Yankaskas, J.R., and Dudus, L. Progenitor cells of the adult human airway involved in submucosal gland development. *Development* **121**(7), 2031–2046 (1995).
31. Giangreco, A., Reynolds, S.D., and Stripp, B.R. Terminal bronchioles harbor a unique airway stem cell population that localizes to the bronchoalveolar duct junction. *Am J Pathol* **161**(1), 173–182 (2002).
32. Hong, K.U., Reynolds, S.D., Giangreco, A., Hurley, C.M., and Stripp, B.R. Clara cell secretory protein-expressing cells of the airway neuroepithelial body microenvironment include a label-retaining subset and are critical for epithelial renewal after progenitor cell depletion. *Am J Respir Cell Mol Biol* **24**(6), 671–681 (2001).
33. Kim, C.F., Jackson, E.L., Woolfenden, A.E., Lawrence, S., Babar, I., Vogel, S., Crowley, D., Bronson, R.T., and Jacks, T. Identification of bronchioalveolar stem cells in normal lung and lung cancer. *Cell* **121**(6), 823–835 (2005).
34. Ling, T.Y., Kuo, M.D., Li, C.L., Yu, A.L., Huang, Y.H., Wu, T.J., Lin, Y.C., Chen, S.H., and Yu, J. Identification of pulmonary Oct-4+ stem/progenitor cells and demonstration of their susceptibility to SARS coronavirus (SARS-CoV) infection *in vitro*. *Proc Natl Acad Sci USA* **103**(25), 9530–9535 (2006).
35. Besson, A., Hwang, H.C., Cicero, S., Donovan, S.L., Gurian-West, M., Johnson, D., Clurman, B.E., Dyer, M.A., and Roberts, J.M. Discovery of an oncogenic activity in p27Kip1 that causes stem cell expansion and a multiple tumor phenotype. *Genes Dev* **21**(14), 1731–1746 (2007).
36. Ventura, J.J, Tenbaum, S., Perdiguero, E., Huth, M., Guerra, C., Barbacid, M., Pasparakis, M., and Nebreda, A.R. p38alpha MAP kinase is essential in lung stem and progenitor cell proliferation and differentiation. *Nat Genet* **39**(6), 750–758 (2007).
37. Yanagi, S., Kishimoto, H., Kawahara, K., Sasaki, T., Sasaki, M., Nishio, M., Yajima, N., Hamada, K., Horie, Y., Kubo, H., Whitsett, J.A., Mak, T.W., Nakano, T., Nakazato, M., and Suzuki, A. Pten controls lung morphogenesis, bronchioalveolar stem cells, and onset of lung adenocarcinomas in mice. *J Clin Invest* **117**(10), 2929–2940 (2007).
38. Reynolds, S.D., Zemke, A.C., Giangreco, A., Brockway, B.L., Teisanu, R.M., Drake, J.A., Mariani, T., Di, P.Y., Taketo, M.M., and Stripp, B.R. Conditional stabilization of beta-catenin expands the pool of lung stem cells. *Stem Cells* **26**(5), 1337–1346 (2008).

39. Hollande, E., Cantet, S., Ratovo, G., Daste, G., Bremont, F., and Fanjul, M. Growth of putative progenitors of type II pneumocytes in culture of human cystic fibrosis alveoli. *Biol Cell* **96**(6), 429–441 (2004).
40. Pan, J., Bear, C., Farragher, S., Cutz, E., and Yeger, H. Cystic fibrosis transmembrane conductance regulator modulates neurosecretory function in pulmonary neuroendocrine cell-related tumor cell line models. *Am J Respir Cell Mol Biol* **27**(5), 553–560 (2002).
41. Yeger, H., Pan, J., Fu, X.W., Bear, C., and Cutz, E. Expression of CFTR and Cl(−) conductances in cells of pulmonary neuroepithelial bodies. *Am J Physiol Lung Cell Mol Physiol* **281**(3), L713–L721 (2001).
42. Pan, J., Luk, C., Kent, G., Cutz, E., and Yeger, H. Pulmonary neuroendocrine cells, airway innervation, and smooth muscle are altered in Cftr null mice. *Am J Respir Cell Mol Biol* **35**(3), 320–326 (2006).
43. Weiss, D.J., Berberich, M.A., Borok, Z., Gail, D.B., Kolls, J.K., Penland, C., and Prockop, D.J. Adult stem cells, lung biology, and lung disease. NHLBI/Cystic Fibrosis Foundation Workshop. *Proc Am Thorac Soc* **3**(3), 193–207 (2006).
44. Weiss, D.J., Kolls, J.K., Ortiz, L.A., Panoskaltis-Mortari, A., and Prockop, D.J. Stem cells and cell therapy approaches for lung diseases. Conference report. *Proc Am Thorac Soc* **5**, 637–667 (2008).
45. Suratt, B.T., Cool, C.D., Serls, A.E., Chen, L., Varella-Garcia, M., Shpall, E.J., Brown, K.K., and Worthen, G.S. Human pulmonary chimerism after hematopoietic stem cell transplantation. *Am J Respir Crit Care Med* **168**(3), 318–322 (2003).
46. Spencer, H., Rampling, D., Aurora, P., Bonnet, D., Hart, S.L., and Jaffe, A. Transbronchial biopsies provide longitudinal evidence for epithelial chimerism in children following sex mismatched lung transplantation. *Thorax* **60**(1), 60–62 (2005).
47. Albera, C., Polak, J.M., Janes, S., Griffiths, M.J., Alison, M.R., Wright, N.A., Navaratnarasah, S., Poulsom, R., Jeffery, R., Fisher, C., Burke, M., and Bishop, A.E. Repopulation of human pulmonary epithelium by bone marrow cells: a potential means to promote repair. *Tissue Eng* **11**(7–8), 1115–1121 (2005).
48. Kleeberger, W., Versmold, A., Rothamel, T., Glockner, S., Bredt, M., Haverich, A., Lehmann, U., and Kreipe, H. Increased chimerism of bronchial and alveolar epithelium in human lung allografts undergoing chronic injury. *Am J Pathol* **162**(5), 1487–1494 (2003).

49. Kotton, D.N., Fabian, A.J., and Mulligan, R.C. Failure of bone marrow to reconstitute lung epithelium. *Am J Respir Cell Mol Biol* **33**(4), 328–334 (2005).
50. Chang, J.C., Summer, R., Sun, X., Fitzsimmons, K., and Fine, A. Evidence that bone marrow cells do not contribute to the alveolar epithelium. *Am J Respir Cell Mol Biol* **33**(4), 335–342 (2005).
51. Loi, R., Beckett, T., Goncz, K.K., Suratt, B.T., and Weiss, D.J. Limited restoration of cystic fibrosis lung epithelium *in vivo* with adult bone marrow-derived cells. *Am J Respir Crit Care Med* **173**(2), 171–179 (2006).
52. Sueblinvong, V., Loi, R., Eisenhauer, P.L., Bernstein, I.M., Suratt, B.T., Spees, J.L., and Weiss, D.J. Derivation of lung epithelium from human cord blood-derived mesenchymal stem cells. *Am J Resp Crit Care Med* **177**, 701–711 (2008).
53. Yan, C., Lian, X., Dai, Y., Wang, X., Qu, P., White, A., Qin, Y., and Du, H. Gene delivery by the hSP-B promoter to lung alveolar type II epithelial cells in LAL-knockout mice through bone marrow mesenchymal stem cells. *Gene Ther* **14**(20), 1461–1470 (2007).
54. Spees, J.L., Whitney, M.J., Sullivan, D.E., Lasky, J.A., Laboy, M., Ylostalo, J., and Prockop, D.J. Bone marrow progenitor cells contribute to repair and remodeling of the lung and heart in a rat model of progressive pulmonary hypertension. *FASEB J* **22**, 1226–1236 (2007).
55. Wong, A.P., Dutly, A.E., Sacher, A., Lee, H., Hwang, D.M., Liu, M., Keshavjee, S., Hu, J., and Waddell, T.K. Targeted cell replacement with bone marrow cells for airway epithelial regeneration. *Am J Physiol Lung Cell Mol Physiol* **293**(3), L740–L752 (2007).
56. Serikov, V.B., Popov, B., Mikhailov, V.M., Gupta, N., and Matthay, M.A. Evidence of temporary airway epithelial repopulation and rare clonal formation by BM-derived cells following naphthalene injury in mice. *Anat Rec (Hoboken)* **290**(9), 1033–1045 (2007).
57. Wang, X.X., Zhang, F.R., Shang, Y.P., Zhu, J.H., Xie, X.D., Tao, Q.M., Zhu, J.H., and Chen, J.Z. Transplantation of autologous endothelial progenitor cells may be beneficial in patients with idiopathic pulmonary arterial hypertension: a pilot randomized controlled trial. *J Am Coll Cardiol* **49**(14), 1566–1571 (2007).
58. Zhao, Y.D., Courtman, D.W., Deng, Y., Kugathasan, L., Zhang, Q., and Stewart, D.J. Rescue of monocrotaline-induced pulmonary arterial

hypertension using bone marrow-derived endothelial-like progenitor cells: efficacy of combined cell and eNOS gene therapy in established disease. *Circ Res* **96**(4), 442–450 (2005).

59. Spees, J.L., Olson, S.D., Ylostalo, J., Lynch, P.J., Smith, J., Perry, A., Peister, A., Wang, M.Y., and Prockop, D.J. Differentiation, cell fusion, and nuclear fusion during *ex vivo* repair of epithelium by human adult stem cells from bone marrow stroma. *Proc Natl Acad Sci USA* **100**(5), 2397–2402 (2003).
60. Harris, R.G., Herzog, E.L., Bruscia, E.M., Grove, J.E., Van Arnam, J.S., and Krause, D.S. Lack of a fusion requirement for development of bone marrow-derived epithelia. *Science* **305**(5680), 90–93 (2004).
61. Herzog, E.L., Van Arnam, J., Hu, B., Zhang, J., Chen, Q., Haberman, A.M., and Krause, D.S. Lung-specific nuclear reprogramming is accompanied by heterokaryon formation and Y chromosome loss following bone marrow transplantation and secondary inflammation. *FASEB J* **21**(10), 2592–2601 (2007).
62. Wang, G., Bunnell, B.A., Painter, R.G., Tom, S., Lanson, N.A., Spees, J.L., Bertucci, D., Peister, A., Weiss, D.J., Valentine, V.G., Prockop, D.J., and Kolls, J.K. Adult stem cells from bone marrow stroma differentiate into airway epithelial cells: potential therapy for cystic fibrosis. *Proc Natl Acad Sci* **102**(1), 186–191 (2005).
63. Bruscia, E.M., Ziegler, E.C., Price, J.E., Weiner, S., Egan, M.E., and Krause, D.S. Engraftment of donor-derived epithelial cells in multiple organs following bone marrow transplantation into newborn mice. *Stem Cells* **24**(10), 2299–2308 (2006).
64. Bruscia, E.M., Price, J.E., Cheng, E.C., Weiner, S., Caputo, C., Ferreira, E.C., Egan, M.E., and Krause, D.S. Assessment of cystic fibrosis transmembrane conductance regulator (CFTR) activity in CFTR-null mice after bone marrow transplantation. *Proc Natl Acad Sci USA* **103**(8), 2965–2970 (2006).
65. Gupta, N., Su, X., Popov, B., Lee, J.W., Serikov, V., and Matthay, M.A. Intrapulmonary delivery of bone marrow-derived mesenchymal stem cells improves survival and attenuates endotoxin-induced acute lung injury in mice. *J Immunol* **179**(3), 1855–1863 (2007).
66. Rippon, H.J., Polak, J.M., Qin, M., and Bishop, A.E. Derivation of distal lung epithelial progenitors from murine embryonic stem cells using

a novel three-step differentiation protocol. *Stem Cells* **24**(5), 1389–1398 (2006).

67. Samadikuchaksaraei, A., Cohen, S., Isaac, K., Rippon, H.J., Polak, J.M., Bielby, R.C., and Bishop, A.E. Derivation of distal airway epithelium from human embryonic stem cells. *Tissue Eng* **12**(4), 867–875 (2006).
68. Van Vranken, B.E., Romanska, H.M., Polak, J.M., Rippon, H.J., Shannon, J.M., and Bishop, A.E. Coculture of embryonic stem cells with pulmonary mesenchyme: a microenvironment that promotes differentiation of pulmonary epithelium. *Tissue Eng* **11**(7–8), 1177–1187 (2005).
69. Ali, N.N., Edgar, A.J., Samadikuchaksaraei, A., Timson, C.M., Romanska, H.M., Polak, J.M., and Bishop, A.E. Derivation of type II alveolar epithelial cells from murine embryonic stem cells. *Tissue Eng* **8**(4), 541–550 (2002).
70. Denham, M., Cole, T.J., and Mollard, R. Embryonic stem cells form glandular structures and express surfactant protein C following culture with dissociated fetal respiratory tissue. *Am J Physiol Lung Cell Mol Physiol* **290**(6), L1210–L1215 (2006).
71. Coraux, C., Nawrocki-Raby, B., Hinnrasky, J., Kileztky, C., Gaillard, D., Dani, C., and Puchelle, E. Embryonic stem cells generate airway epithelial tissue. *Am J Respir Cell Mol Biol* **32**(2), 87–92 (2005).
72. Wang, D., Haviland, D.L., Burns, A.R., Zsigmond, E., and Wetsel, R.A. A pure population of lung alveolar epithelial type II cells derived from human embryonic stem cells. *Proc Natl Acad Sci USA* **104**(11), 4449–4454 (2007).
73. Pickering, S.J., Minger, S.L., Patel, M., Taylor, H., Black, C., Burns, C.J., Ekonomou, A., and Braude, P.R. Generation of a human embryonic stem cell line encoding the cystic fibrosis mutation deltaF508, using preimplantation genetic diagnosis. *Reprod Biomed Online* **10**(3), 390–397 (2005).
74. Mateizel, I., De Temmerman, N., Ullmann, U., Cauffman, G., Sermon, K., Van de Velde, H., De Rycke, M., Degreef, E., Devroey, P., Liebaers, I., and Van Steirteghem, A. Derivation of human embryonic stem cell lines from embryos obtained after IVF and after PGD for monogenic disorders. *Hum Reprod* **21**(2), 503–511 (2006).
75. Takahashi, K. and Yamanaka, S. Induction of pluripotent stem cells from mouse embryonic and adult fibroblast cultures by defined factors. *Cell* **126**(4), 663–676 (2006).

76. Takahashi, K., Tanabe, K., Ohnuki, M., Narita, M., Ichisaka, T., Tomoda, K., and Yamanaka, S. Induction of pluripotent stem cells from adult human fibroblasts by defined factors. *Cell* **131**(5), 861–872 (2007).
77. Nichols, J.E. and Cortiella, J. Engineering of a complex organ: progress toward development of a tissue-engineered lung. *Proc Am Thorac Soc* **5**(6), 723–730 (2008).
78. Macchianni, P., Jungebluth, P., Go, T., *et al.* Clinical transplantation of a tissue-engineered airway. *Lancet Online Publ* **673**(08), 61599–61608 (2008).
79. Mondrinos, M.J., Koutzaki, S.H., Poblete, H.M., Crisanti, M.C., Lelkes, P.I., and Finck, C.M. *In vivo* pulmonary tissue engineering: contribution of donor-derived endothelial cells to construct vascularization. *Tissue Eng Part A* **14**(3), 361–368 (2008).
80. Hoganson, D.M., Pryor, H.I., 2nd, and Vacanti, J.P. Tissue engineering and organ structure: a vascularized approach to liver and lung. *Pediatr Res* **63**(5), 520–526 (2008).
81. Mondrinos, M.J., Koutzaki, S., Lelkes, P.I., and Finck, C.M. A tissue-engineered model of fetal distal lung tissue. *Am J Physiol Lung Cell Mol Physiol* **293**(3), L639–L650 (2007).
82. Mondrinos, M.J., Koutzaki, S., Jiwanmall, E., Li, M., Dechadarevian, J.P., Lelkes, P.I., and Finck, C.M. Engineering three-dimensional pulmonary tissue constructs. *Tissue Eng* **12**(4), 717–728 (2006).
83. Lin, Y.M., Boccaccini, A.R., Polak, J.M., Bishop, A.E., and Maquet, V. Biocompatibility of poly-DL-lactic acid (PDLLA) for lung tissue engineering. *J Biomater Appl* **21**(2), 109–118 (2006).
84. Cortiella, J., Nichols, J.E., Kojima, K., Bonassar, L.J., Dargon, P., Roy, A.K., Vacant, M.P., Niles, J.A., and Vacanti, C.A. Tissue-engineered lung: an *in vivo* and *in vitro* comparison of polyglycolic acid and pluronic F-127 hydrogel/somatic lung progenitor cell constructs to support tissue growth. *Tissue Eng* **12**(5), 1213–1225 (2006).
85. Andrade, C.F., Wong, A.P., Waddell, T.K., Keshavjee, S., and Liu, M. Cell-based tissue engineering for lung regeneration. *Am J Physiol Lung Cell Mol Physiol* **292**(2), L510–518 (2007).
86. Ishizawa, K., Kubo, H., Yamada, M., Kobayashi, S., Numasaki, M., Ueda, S., Suzuki, T., and Sasaki, H. Bone marrow-derived cells contribute to lung regeneration after elastase-induced pulmonary emphysema. *FEBS Lett* **556**(1–3), 249–252 (2004).

87. Yamada, M., Kubo, H., Kobayashi, S., Ishizawa, K., Numasaki, M., Ueda, S., Suzuki, T., and Sasaki, H. Bone marrow-derived progenitor cells are important for lung repair after lipopolysaccharide-induced lung injury. *J Immunol* **172**(2), 1266–1272 (2004).
88. Ortiz, L.A., Gambelli, F., McBride, C., Gaupp, D., Baddoo, M., Kaminski, N., and Phinney, D.G. Mesenchymal stem cell engraftment in lung is enhanced in response to bleomycin exposure and ameliorates its fibrotic effects. *Proc Natl Acad Sci USA* **100**(14), 8407–8411 (2003).
89. Rojas, M., Xu, J., Woods, C.R., Mora, A.L., Spears, W., Roman, J., and Brigham, K.L. Bone marrow-derived mesenchymal stem cells in repair of the injured lung. *Am J Respir Cell Mol Biol* **33**(2), 145–152 (2005).
90. Ortiz, L.A., Dutreil, M., Fattman, C., Pandey, A.C., Torres, G., Go, K., and Phinney, D.G. Interleukin 1 receptor antagonist mediates the antiinflammatory and antifibrotic effect of mesenchymal stem cells during lung injury. *Proc Natl Acad Sci USA* **104**(26), 11002–11007 (2007).
91. Xu, J., Woods, C.R., Mora, A.L., Joodi, R., Brigham, K.L., Iyer, S., and Rojas, M. Prevention of endotoxin-induced systemic response by bone marrow-derived mesenchymal stem cells in mice. *Am J Physiol Lung Cell Mol Physiol* **293**(1), L131–L141 (2007).
92. McCarter, S.D., Mei, S.H., Lai, P.F., Zhang, Q.W., Parker, C.H., Suen, R.S., Hood, R.D., Zhao, Y.D., Deng, Y., Han, R.N., Dumont, D.J., and Stewart, D.J. Cell-based angiopoietin-1 gene therapy for acute lung injury. *Am J Respir Crit Care Med* **175**(10), 1014–1026 (2007).

Chapter 12
Pulmonary Hypertension and Stem Cell Therapy

Manoj M. Lalu, Rohit Moudgil and Duncan J. Stewart

Pulmonary arterial hypertension (PAH) is a progressive and lethal condition characterised by elevated pulmonary pressures and right ventricular hypertrophy. Significant advances have been made in the treatment of this condition; however, the prognosis remains extremely poor for the majority of patients. Over the last decade a number of preclinical studies have explored the potential for cell therapy in PAH. These have included novel investigations using somatic and stem/progenitor cells, either alone or transfected with potentially therapeutic genes. The success of early preclinical work has recently prompted the first clinical trials of cell therapy for PAH. This chapter will review the basic pathophysiology of PAH, the current pharmacotherapy, and then examine the preclinical and clinical investigations of cell therapy for this disease.

Keywords: Pulmonary arterial hypertension; endothelial progenitor cells; endothelial nitric oxide synthase.

Outline

1. Introduction

The transplantation of cells to repair or replace damaged tissue is an old concept that has only lately reached fruition. Dating back to the 16th century, Paracelsus stated that "the heart heals the heart, the lung heals the lung…like cures like", thus suggesting that living tissue may be the best way to treat injured organs. Despite these early ideas, our understanding of cellular therapy has only recently allowed for its successful application. Currently there is a great deal of interest in treating severe systemic diseases by this approach. To date a number of trials have investigated cellular therapy for a wide range of clinical conditions, from acute myocardial infarction[1] to diabetic foot ulcers.[2] This type of treatment offers a new therapeutic avenue for diseases that have been refractory to current modalities of treatment.

Despite recent improvements in therapies, pulmonary hypertension continues to have poor outcomes and current treatments do not address the fundamental structural and functional abnormalities of this disease. This chapter will provide an update on the basic mechanisms of pulmonary hypertension, review its current pharmacologic treatments, and then summarise preclinical and clinical investigations of cellular therapy for this disease.

2. Pulmonary Arterial Hypertension — A Brief Overview

Pulmonary arterial hypertension (PAH) is defined by elevated pulmonary arterial pressure (>25 mmHg at rest and >30 mmHg with exertion) caused by abnormalities in the distal pulmonary arteries. Unfortunately, pulmonary hypertension is a progressive disease that leads to right ventricular failure and is uniformly fatal if left untreated. Clinically, the initial symptoms are quite nonspecific, with more than 60% of patients reporting dyspnoea on exertion as their main complaint.[3] Other symptoms include fatigue, peripheral edema, chest pain, dizziness and palpitations.

Table 1: Etiological classification of pulmonary hypertension.

Category 1	Pulmonary arterial hypertension • Idiopathic • Familial • Pulmonary hypertension associated with collagen vascular disease, congenital shunt, portal hypertension, HIV infection, drugs (e.g. anorexigens), "other" (e.g. hereditary haemmorhagic telangiectasia) • Pulmonary hypertension associated with venous or capillary involvement: pulmonary venoocclusive disease, pulmonary capillary haemangiomatosis
Category 2	Pulmonary venous hypertension
Category 3	Pulmonary hypertension associated with respiratory system disorder and/or hypoxaemia
Category 4	Pulmonary hypertension due to chronic thrombotic and/or embolic disease
Category 5	Pulmonary arterial hypertension due to diseases directly affecting the pulmonary vasculature (e.g. sarcoidosis, lymphangiomatosis)

Table 2: World Health Organisation – Functional classification of pulmonary hypertension.

Class I:	No limitation of physical activity. Ordinary physical activity does not cause undue dyspnoea or fatigue, chest pain, or near syncope.
Class II:	Slight limitation of physical activity but comfortable at rest. Ordinary physical activity causes undue dyspnoea or fatigue, chest pain, or near syncope.
Class III:	Marked limitation of physical activity but comfortable at rest. Less than ordinary activity causes undue dyspnoea or fatigue, chest pain, or near syncope.
Class IV:	Unable to carry out any physical activity without symptoms. Manifest signs of right-heart failure. Dyspnoea and/or fatigue may even be present at rest.

Currently, pulmonary hypertension can be classified clinically by its etiology and also by its functional effects (Tables 1 and 2).

Although the pathological changes underlying pulmonary hypertension were first described in the late 19th century,[4,5] it was not until the 1950s that the true histopathological characteristics of pulmonary hypertension were described in its present form (Fig. 1).[6] Initial insights into the underlying molecular causes of this disease came from studies in the 1970s–1980s.[7–9] During this period tools such as cell culture were introduced and the critical

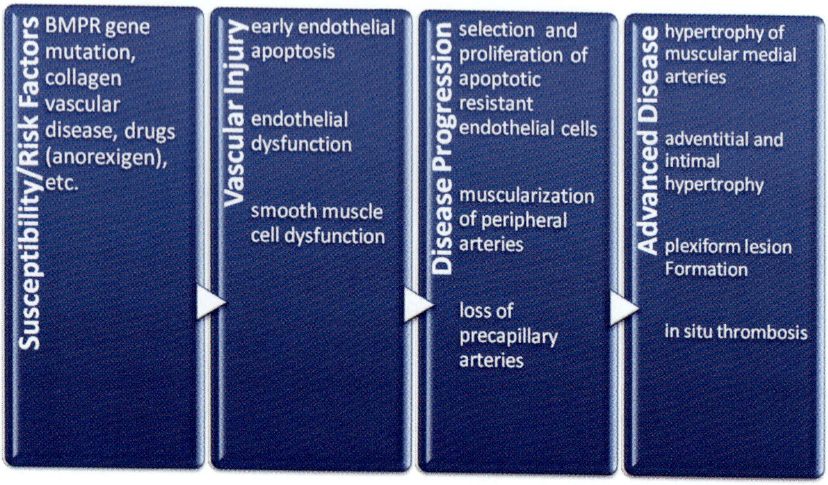

Figure 1: Pathobiology of pulmonary hypertension.

role of endothelial cells (ECs) in the regulation of vascular structure and function was first appreciated.

Early anatomical studies of PAH identified SMCs as a primary target in the pathogenesis of this disease. Beyond the terminal bronchioles the arterioles are normally only partially muscularised, and smaller intra-acinar arteries are devoid of smooth muscle cells. In contrast, advanced PAH is characterised by a marked increase in muscularisation, particularly at the level of the distal arteriole. This precapillary segment of the pulmonary vascular bed is normally the site of the greatest pressure decrease along the pulmonary circulation,[10] and small changes in arteriolar diameter due to arteriolar remodelling can lead to large elevations of pulmonary vascular resistance (as per Poiseuille's law). However, more recent molecular and functional studies have increasingly pointed to the endothelium as the site of the initial lesions in the disease. Thus, the loss of endothelial vasodilator mechanisms (or endothelial dysfunction) in this critical region is thought to be a key mechanism of the initiation of PAH.[11] Molecular mechanisms underlying this endothelial dysfunction have not been fully delineated; however, a number of endothelial factors such as nitric oxide,[12] prostaglandins[13,14] and endothelin[15] have been implicated.

Perhaps the most important new insights into the molecular mechanisms of PAH have come from studies linking the bone morphogenetic protein (BMP) receptor with familial PAH. Two independent groups identified heterozygous germline mutations in the BMP type II receptor (BMPR2), a member of the transforming growth factor beta superfamily of receptors, in patients with familial pulmonary hypertension.[16,17] Mutations in the BMPR2 gene have been found in approximately 70% of families with PAH.[18] In addition, up to 25% of patients with apparently sporadic idiopathic pulmonary hypertension have been found to harbour similar mutations.[19] The majority (~70%) of BMPR2 coding mutations are frame-shift and nonsense mutations, many of which would be expected to produce a transcript susceptible to nonsense-mediated mRNA decay.[20,21] Thus, haplo-insufficiency for BMPR2 represents the predominant molecular mechanism underlying an inherited predisposition to familial PAH. However, how loss of signalling via this pathway leads to the development of PAH is still a matter of heated debate.

Recent studies have shown that while BMPR2 mutations are associated with increased SMC growth and survival, these also predispose to EC apoptosis and dysfunction,[22] thereby linking this mutation to the critical role that the endothelium plays in PAH. These effects fit well with the emerging paradigm of PAH pathobiology, which suggests that EC apoptosis is the initial trigger of this disease.[23] However, over time, widespread and progressive EC loss creates a selection pressure which favours the emergence of apoptotic-resistant ECs, ultimately contributing to intimal hyperplasia and formation of complex plexiform lesions in more advanced disease.[24,25] In some cases these are characterised by clonal expansion of hyperproliferative and apoptosis-resistant ECs.[26] In a small study (three controls and five idiopathic pulmonary hypertension subjects), cultured ECs isolated from the pulmonary arteries of patients with idiopathic pulmonary hypertension showed increased proliferation based on BrdU incorporation and decreased apoptosis based on caspase-3 activity compared to ECs isolated from controls.[27] Furthermore, direct evidence of microsatellite instability and concomitant perturbation of growth and apoptosis gene expression has been obtained with analysis of plexiform lesions from patients with PAH.[28]

The initial evidence supporting this "angioproliferative" paradigm came from the seminal experiments of the Voelkel group, in which the vascular endothelial growth factor (VEGF) receptor was blocked in rats subjected to chronic hypoxia to determine the role of this pathway in PAH.[29] Rather than inhibiting the increase in pulmonary pressures in this model, VEGF receptor blockade potentiated PAH and resulted in a paradoxical increase in EC proliferation. This was associated with evidence of early EC apoptosis, and a severe PAH phenotype that could be rescued inhibiting caspases, key mediators of apoptosis. Moreover, in other studies overexpression of VEGF (or indeed other angiogenic factors)[30,31] by adenovirus mediated gene transfer[32] or cell-based transfer,[33] reduced early apoptosis and prevented the disease development in an experimental model of PAH.

Cumulatively, the evidence suggests that ECs play an integral role in pathogenesis of PAH. The current concept suggests that apoptosis of ECs represents the initial response to injury, followed later by the emergence of apoptotic-resistant and hyperproliferative ECs. These hyperproliferative ECs contribute to intimal hyperplasia, and the formation of plexiform lesions which are characteristic (although not universal) lesions in late stage disease.[26] EC apoptosis may be particularly important at the level of intra-acinar arterioles, which, by virtue of the lack of muscularisation, are likely to be particularly vulnerable to injury and degeneration. The loss of BMP signalling may predispose to exaggerated EC apoptosis at this level, which is likely aggravated by environmental stress or other genetic influences (i.e. second hit). To the extent that pulmonary EC injury and loss underlies the development of PAH, an ideal treatment would enhance the repair and regeneration of microvascular endothelium. Thus, treatments such as progenitor cell therapy might offer unique advantages over current pharmacological therapies which do not address the fundamental vascular pathology of PAH.

3. Pulmonary Arterial Hypertension — Evolution of Treatments

Current pharmacological treatments of PAH include anticoagulants, calcium channel blockers, prostanoids, endothelin receptor antagonists, and

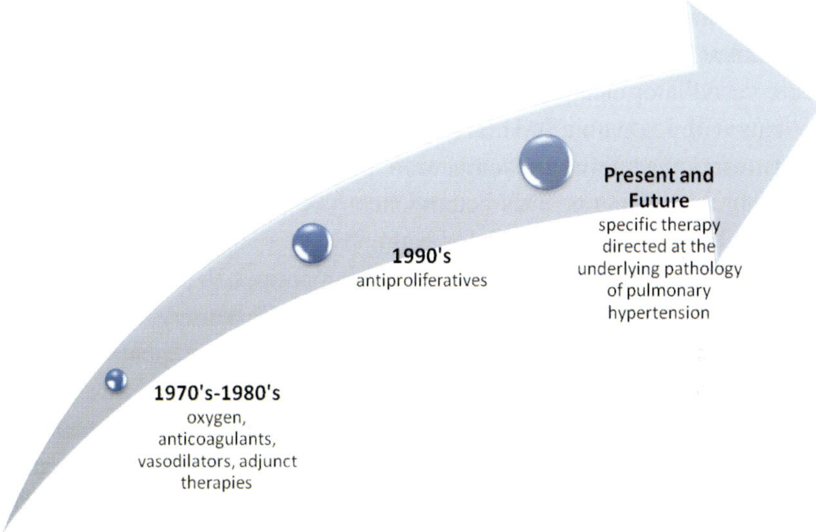

Figure 2: The evolution of treatment for pulmonary hypertension.

selective phosphodiesterase type V inhibitors. These therapeutic approaches have been developed over the last three decades (Fig. 2) and will be briefly reviewed below. Most of the agents were developed as vasodilators at a time when the vasoconstriction was thought to be a dominant mechanism of this disease; however, many of the agents also have effects on vascular cell growth and thus may influence arterial remodelling to some extent over the long-term.

Anticoagulation has long been an adjunct therapy for PAH and likely addresses *in situ* thrombosis which is inherent in advanced PAH, and relates in part to the prothrombotic properties of dysfunctional endothelium.[34] The most commonly used anticoagulant, warfarin, has been suggested to improve survival in idiopathic PAH[35] although this has never been proven in a rigorously controlled study. Other adjunct therapies include digoxin for heart failure and diuretics to reduce fluid overload in the lungs. Although these treatments likely improve quality of life for PAH patients, they have not been shown to prolong survival.

Calcium channel blockers (nifedipine, diltiazem) are classical vasodilators and have been used in the treatment of PAH for many years. However, these are indicated only in patients with evidence of a strong vasomotor component to their disease, reflected by a significant positive response to acute vasodilator challenge (NO or prostacyclin). Unfortunately <10% of patients with idiopathic PAH (and even fewer patients with other forms of PAH) respond acutely to vasodilators,[36] and of those less than 50% maintain long term benefits. This heterogeneous response to calcium channel blockers underscores the complex pathophysiology of PAH.

Analogues of prostacyclin (a vasodilator, antiplatelet and antiproliferative molecule) are often used as a first line of therapy for Class IV patients and remain the gold standard for the pharmacological therapy of PAH. Epoprostenol was the first approved agent for this disease, but it needs to be administered as a continuous infusion due to its short half-life, which not only makes it cumbersome, but also results in significant morbidity related to the chronically implanted catheter. Nonetheless, this drug improves functional capacity and survival.[37,38] However, after initiation of parenteral therapy, many patients develop dependence, and even brief discontinuation due to pump failure or catheter blockage can result in a "rebound" increase in pulmonary pressures that can prove fatal. As well, nuisance side effects are frequent, including jaw pain, diarrhoea, flushing and headache, and often limit the tolerated dose. Longer acting analogues (treprostinil, iloprost) have come into use more recently, and the efficacy of a number of these agents as inhaled or even oral therapy is under intense investigation.

Antagonists of endothelin receptors were the first effective oral agents to be introduced for the therapy of PAH. These agents include bosentan (which blocks both A and B type receptors), ambrisentan and sitaxsentan (selective A type blockers), and they block the potent vasoconstrictor and smooth muscle mitogenic actions of endothelin-1. Due to their convenience, these agents are often used as first line treatment for Class III patients.[39] Bosentan has been the most widely studied and was shown to improve exercise capacity and possibly survival as well, compared to historical controls.[40,41] Although usually well tolerated, this class of medication can cause severe hepatotoxicity requiring discontinuation of therapy in approximately 10% of patients.

A more recent development has been the introduction of phosphodiesterase V inhibitors such as sildenafil. Phosphodiesterase V is responsible for degrading cyclic guanosine monophsophate (the second messenger for NO) in pulmonary vasculature (and, more famously, in the penile vasculature). Thus, by inhibiting this enzyme, sildenafil potentiates the vasodilatory and antiproliferative effects of nitric oxide. This oral medication is being used increasingly for Class II and III disease and has been demonstrated to improve exercise capacity and haemodynamics in a one-year study.[42]

Despite improvements in the treatment of PAH, particularly over the last decade, the response to these pharmacological therapies is usually incomplete and this disease remains progressive and ultimately lethal. Thus there is a great deal of interest in combined therapies to improve the magnitude and durability of response.[43,44] However, such combination therapies can be associated with increased side effects and idiosyncratic reactions. Ultimately, many patients fail all available therapies and require more drastic therapies, such as septostomy and lung transplantation. Unfortunately, given the current shortage of organ donors, the majority of patients that are listed for lung transplantation never receive an organ. Moreover, the prognosis of patients post lung transplantation remains relatively guarded, with a 50% post-transplant survival of 4.3 years.[45]

In light of these limitations there has been renewed interest in the development of novel treatment strategies that specifically address the fundamental pathological abnormalities underlying pulmonary hypertension.[46] The remainder of this chapter will focus on cell therapy, which is one avenue of treatment that could address the underlying structural and functional vascular abnormalities of the disease.

4. Pulmonary Arterial Hypertension — Endothelial Progenitor Cell Therapy

To date, the majority of preclinical work and all of the clinical studies exploring the utility of cell therapy for PAH have focused on endothelial progenitor cells (EPCs). These cells were first identified in 1997 by Asahara *et al.*[47] as a subset of CD34+ bone marrow-derived mononuclear progenitor cells. When peripheral circulating cells were cultured with VEGF on fibronectin-coated petri dishes, they differentiated into cells expressing a variety of EC characteristics. This included expression of EC

number of complications that might be expected following bone marrow transplantation (including pulmonary embolism), the patient did show improvement in exercise tolerance and angina. However, there was little haemodynamic benefit, and pulmonary artery pressures were not changed six months after therapy (mean pressure pre-therapy: 45–50 mmHg; post-therapy: 44 mmHg). Although the cell preparation was intended to provide CD34+ haematopoietic stem cells to repopulate the bone marrow and "reprogram" the immune system, it is tempting to speculate that some of the benefits noted may have derived from EPCs.

Indeed, given the importance of endothelial injury in the pathogenesis PAH, there is a strong rationale for treatments that may potentially repair or regenerate the endothelial lining. The first clinical experience with EPC therapy was published by Chen's group. Adult patients with PAH were randomised in a non-blinded manner to receive either conventional pharmacotherapy alone or together with EPCs derived from peripheral blood mononuclear cells (cultured *ex vivo* for five days under conditions similar to those described above for early growth EPCs). The resulting cell suspension ($4–22 \times 10^7$ cells) was infused over several minutes intravenously. In this non-blinded study, patients receiving cell therapy demonstrated modest (5–15%) but significant improvements in all the endpoints measured, including six-minute walk distances and pulmonary haemodynamics. In contrast, the group treated with conventional pharmacotherapy alone group showed no significant improvement. Perhaps most importantly, this therapy was well tolerated despite the lack of screening for intra-cardiac shunting (which could increase the risk of systemic embolisation) or haemodynamic monitoring during the delivery procedure (which by necessity involves some degree of microembolisation of the pulmonary bed).[89]

These promising results in adult patients were reinforced in an open-label pilot study of paediatric patients (aged 7–17) with PAH. A similar protocol was used to isolate and characterise early-growth EPCs, with a range of $2–13 \times 10^7$ cells ultimately being infused intravenously. Again, patients were subjected to six-minute walk tests and invasive haemodynamic measurements before and twelve weeks after cell therapy. Following treatment, walk distance increased by approximately 40 m with similar improvements in haemodynamic endpoints as well.[90]

Together, these two trials suggest that progenitor cell therapy may be a safe and effective clinical treatment for PAH. However, overall observed benefits were rather modest and the lack of rigorous blinding or even a control group in the latter study precludes drawing any definitive conclusions about efficacy.

Moreover, given the strong preclinical evidence suggesting substantial additional benefit with the use of transfected EPCs, it is possible that appropriately engineered cells could produce an added benefit. Indeed, this may be even more important in the context of clinical therapy since, as discussed above, the pathophysiology underlying PAH may interfere with both the number and efficacy of patient-derived EPCs. Thus "enhancing" autologous EPC regenerative function may be necessary to benefit from the full potential of cell therapy.

The PHACeT Trial (Pulmonary Hypertension: Assessment of Cell Therapy) is designed to address this issue and has been ongoing at two sites in Canada (Toronto and Montreal) since 2006. This phase 1 trial builds on the preclinical studies reviewed above, and is examining the safety of eNOS-transfected EPCs in patients with idiopathic pulmonary hypertension, familial pulmonary hypertension, or anorexigen-induced pulmonary hypertension with refractory Class III or IV symptoms with maximal pharmacologic treatment (eligibility criteria were recently amended to include PAH associated with systemic sclerosis). A number of exclusion criteria were included in the protocol to address potential safety concerns, notably the presence of a definite intra-cardiac communication by echo-contrast studies. The primary endpoints of this phase 1 trial are safety and tolerability; however, the data collected will also preliminarily address clinical efficacy (i.e. by evaluation of exercise capacity, pulmonary haemodynamics and quality of life).

For the PHACeT trial, we are harvesting autologous mononuclear cells from the peripheral blood by apheresis, and EPCs selected by differential culture in a manner similar to what has been performed in the supporting preclinical studies. After five days of culture, cells are transfected with 650 μg/mL of plasmid DNA containing the human eNOS gene by electropration (MaxCyte Incorporated). After transfection, cells are delivered directly into the right ventricular cavity via a Swan-Ganz catheter in aliquouts over three days. Based on the extensive preclinical data already described, it is believed that these cells are efficiently filtered and trapped within the pulmonary circulation at the level of the precapillary arteriole, and about a third of these cells will persist for an extended period of time in the lung tissue. However, for safety reasons the transfection is by design only transient, and eNOS transgene expression is not expected to persist beyond one week.

An open-label, dose escalating protocol is being used to address the primary endpoints of safety and tolerability. The design of this protocol allows a wide range of doses to be studied, ranging from 7×10^6 cells up to 150×10^6 cells (Fig. 3); however, based on previous experience, we do not anticipate the need to progress beyond the 50 million dose panel. The

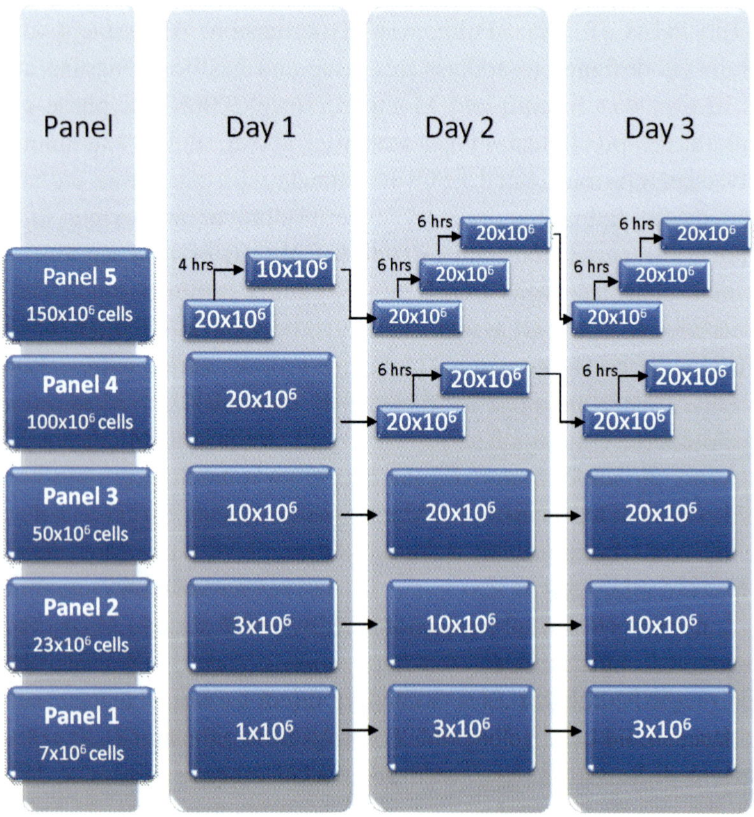

Figure 3: Escalating cell dosing protocol for the PHACeT Trial.

study design allows for careful monitoring of the acute haemodynamic effects of cell delivery during the entire procedure. Follow-up visits are planned regularly over the first year post-cell delivery, with echocardiograms, ventilation perfusion scanning, pulmonary function tests and routine blood work. Repeat invasive haemodynamic measurements are scheduled for three months post cell delivery. Also, dyspnoea self-assessment scores, quality of life scoring, and a six-minute walk test will be analysed. Following this first year of intensive follow-up patients will be asked to return biannually and then annually for a total of fifteen years to monitor for side effects.

7. Conclusions and Future Directions

Preclinical and clinical studies exploring cell therapy in PAH have thus far yielded encouraging results. However, efficacy and safety have not yet been firmly established in rigorous clinical trials. Moreover, we do not fully understand the mechanisms behind the potential beneficial effects of cell therapy. In the same way that we understand pharmacotherapy via classic pharmacodynamic and pharmacokinetic parameters (e.g. mechanism of action, half-life, metabolism, etc.) we should aim to define the effects of cell therapy with a similar degree of scientific rigour. Nonetheless, cell therapy has already shown promise in early clinical trials, and in future larger and more highly powered trials are warranted to address further the efficacy of cell therapy in PAH.

References

1. Hristov, M., Heussen, N., Schober, A., and Weber, C. Intracoronary infusion of autologous bone marrow cells and left ventricular function after acute myocardial infarction: a meta-analysis. *J Cell Mol Med* **10**, 727–733 (2006).
2. Vojtassak, J., Danisovic, L., Kubes, M., Bakos, D., Jarabek, L., Ulicna, M., and Blasko, M. Autologous biograft and mesenchymal stem cells in treatment of the diabetic foot. *Neuro Endocrinol Lett* **27**(Suppl 2), 134–137 (2006).
3. Rich, S., Dantzker, D.R., Ayres, S.M., Bergofsky, E.H., Brundage, B.H., Detre, K.M., Fishman, A.P., Goldring, R.M., Groves, B.M., Koerner, S.K.,

et al. Primary pulmonary hypertension. A national prospective study. *Ann Intern Med* **107**, 216–223 (1987).

4. Romberg. Uber Sklerose der Lungen Arterie [German]. *Dtsch Arch Klin Med* **48**, 197–206 (1891).
5. Fishman, A.P. A century of pulmonary hemodynamics. *Am J Respir Crit Care Med* **170**, 109–113 (2004).
6. Dresdale, D.T., Schultz, M., and Michtom, R.J. Primary pulmonary hypertension. I. Clinical and hemodynamic study. *Am J Med* **11**, 686–705 (1951).
7. Kombe, A.H., Smith, P., Heath, D., and Biggar, R. Endothelial cell pavement pattern in the pulmonary trunk in rats in chronic hypoxia. *Br J Dis Chest* **74**, 362–368 (1980).
8. Reid, L. and Meyrick, B. Hypoxia and pulmonary vascular endothelium. *Ciba Found Symp* **78**, 37–60 (1980).
9. Jaenke, R.S. and Alexander, A.F. Fine structural alterations of bovine peripheral pulmonary arteries in hypoxia-induced hypertension. *Am J Pathol* **73**, 377–398 (1973).
10. Wheeler, E.C. and Brenner, Z.R. Peripheral vascular anatomy, physiology, and pathophysiology. *AACN Clin Issues* **6**, 505–514 (1995).
11. Reneman, R.S., Arts, T., and Hoeks, A.P. Wall shear stress — an important determinant of endothelial cell function and structure — in the arterial system *in vivo*. Discrepancies with theory. *J Vasc Res* **43**, 251–269 (2006).
12. Giaid, A. and Saleh, D. Reduced expression of endothelial nitric oxide synthase in the lungs of patients with pulmonary hypertension. *N Engl J Med* **333**, 214–221 (1995).
13. Christman, B.W., McPherson, C.D., Newman, J.H., King, G.A., Bernard, G.R., Groves, B.M., and Loyd, J.E. An imbalance between the excretion of thromboxane and prostacyclin metabolites in pulmonary hypertension. *N Engl J Med* **327**, 70–75 (1992).
14. Szczeklik, J., Dubiel, J.S., Mysik, M., Pyzik, Z., Krol, R., and Horzela, T. Effects of prostaglandin E1 on pulmonary circulation in patients with pulmonary hypertension. *Br Heart J* **40**, 1397–1401 (1978).
15. Stewart, D.J., Levy, R.D., Cernacek, P., and Langleben, D. Increased plasma endothelin-1 in pulmonary hypertension: marker or mediator of disease? *Ann Intern Med* **114**, 464–469 (1991).
16. Deng, Z., Haghighi, F., Helleby, L., Vanterpool, K., Horn, E.M., Barst, R.J., Hodge, S.E., Morse J.H., and Knowles J.A. Fine mapping of PPH1, a gene

for familial primary pulmonary hypertension, to a 3-cM region on chromosome 2q33. *Am J Respir Crit Care Med* **161**, 1055–1059 (2000).

17. Lane, K.B., Machado, R.D., Pauciulo, M.W., Thomson, J.R., Phillips, J.A., 3rd, Loyd, J.E., Nichols, W.C., and Trembath, R.C. Heterozygous germline mutations in BMPR2, encoding a TGF-beta receptor, cause familial primary pulmonary hypertension. The International PPH Consortium. *Nat Genet* **26**, 81–84 (2000).
18. Machado, R.D., Aldred, M.A., James, V., Harrison, R.E., Patel, B., *et al.* Mutations of the TGF-beta type II receptor BMPR2 in pulmonary arterial hypertension. *Hum Mutat* **27**, 121–132 (2006).
19. Thomson, J.R., Machado, R.D., Pauciulo, M.W., Morgan, N.V., Humbert, M., *et al.* Sporadic primary pulmonary hypertension is associated with germline mutations of the gene encoding BMPR-II, a receptor member of the TGF-beta family. *J Med Genet* **37**, 741–745 (2000).
20. Aldred, M.A., Vijayakrishnan, J., James, V., Soubrier, F., Gomez-Sanchez, M.A., Martensson, G., Galie, N., Manes, A., Corris, P., Simonneau, G., Humbert, M., Morrell, N.W., and Trembath, R.C. BMPR2 gene rearrangements account for a significant proportion of mutations in familial and idiopathic pulmonary arterial hypertension. *Hum Mutat* **27**, 212–213 (2006).
21. Cogan, J.D., Vnencak-Jones, C.L., Phillips, J.A., 3rd, Lane, K.B., Wheeler, L.A., Robbins, I.M., Garrison, G., Hedges, L.K., and Loyd, J.E. Gross BMPR2 gene rearrangements constitute a new cause for primary pulmonary hypertension. *Genet Med* **7**, 169–174 (2005).
22. Teichert-Kuliszewska, K., Kutryk, M.J., Kuliszewski, M.A., Karoubi, G., Courtman, D.W., Zucco, L., Granton, J., and Stewart, D.J. Bone morphogenetic protein receptor-2 signaling promotes pulmonary arterial endothelial cell survival: implications for loss-of-function mutations in the pathogenesis of pulmonary hypertension. *Circ Res* **98**, 209–217 (2006).
23. Jeffery, T.K. and Morrell, N.W. Molecular and cellular basis of pulmonary vascular remodeling in pulmonary hypertension. *Prog Cardiovasc Dis* **45**, 173–202 (2002).
24. Taraseviciene-Stewart, L., Kasahara, Y., Alger, L., Hirth, P., McMahon, G., Waltenberger, J., Voelkel, N.F., and Tuder, R.M. Inhibition of the VEGF receptor 2 combined with chronic hypoxia causes cell death-dependent pulmonary endothelial cell proliferation and severe pulmonary hypertension. *FASEB J* **15**, 427–438 (2001).

25. Sakao, S., Taraseviciene-Stewart, L., Lee, J.D., Wood, K., Cool, C.D., and Voelkel, N.F. Initial apoptosis is followed by increased proliferation of apoptosis-resistant endothelial cells. *FASEB J* **19**, 1178–1180 (2005).
26. Tuder, R.M., Cool, C.D., Yeager, M., Taraseviciene-Stewart, L., Bull, T.M., and Voelkel, N.F. The pathobiology of pulmonary hypertension. Endothelium. *Clin Chest Med* **22**, 405–418 (2001).
27. Masri, F.A., Xu, W., Comhair, S.A., Asosingh, K., Koo, M., Vasanji, A., Drazba, J., Anand-Apte, B., and Erzurum, S.C. Hyperproliferative apoptosis-resistant endothelial cells in idiopathic pulmonary arterial hypertension. *Am J Physiol Lung Cell Mol Physiol* **293**, L548–554 (2007).
28. Yeager, M.E., Halley, G.R., Golpon, H.A., Voelkel, N.F., and Tuder, R.M. Microsatellite instability of endothelial cell growth and apoptosis genes within plexiform lesions in primary pulmonary hypertension. *Circ Res* **88**, E2–E11 (2001).
29. Sakao, S., Taraseviciene-Stewart, L., Lee, J.D., Wood, K., Cool, C.D., and Voelkel, N.F. Vascular endothelial growth factor receptor blockade by SU5416 combined with pulsatile shear stress causes apoptosis and subsequent proliferation of apoptosis-resistant endothelial cells. *Chest* **128**, 610S–611S (2005).
30. Zhao, Y.D., Campbell, A.I., Robb, M., Ng, D., and Stewart, D.J. Protective role of angiopoietin-1 in experimental pulmonary hypertension. *Circ Res* **92**, 984–991 (2003).
31. Kugathasan, L., Dutly, A.E., Zhao, Y.D., Deng, Y., Robb, M.J., Keshavjee, S., and Stewart, D.J. Role of angiopoietin-1 in experimental and human pulmonary arterial hypertension. *Chest* **128**, 633S–642S (2005).
32. Partovian, C., Adnot, S., Raffestin, B., Louzier, V., Levame, M., Mavier, I.M., Lemarchand, P., and Eddahibi, S. Adenovirus-mediated lung vascular endothelial growth factor overexpression protects against hypoxic pulmonary hypertension in rats. *Am J Respir Cell Mol Biol* **23**, 762–771 (2000).
33. Campbell, A.I., Zhao, Y., Sandhu, R., and Stewart, D.J. Cell-based gene transfer of vascular endothelial growth factor attenuates monocrotaline-induced pulmonary hypertension. *Circulation* **104**, 2242–2248 (2001).
34. Herve, P., Humbert, M., Sitbon, O., Parent, F., Nunes, H., Legal, C., Garcia, G., and Simonneau, G. Pathobiology of pulmonary hypertension. The role of platelets and thrombosis. *Clin Chest Med* **22**, 451–458 (2001).

35. Johnson, S.R., Mehta, S., and Granton, J.T. Anticoagulation in pulmonary arterial hypertension: a qualitative systematic review. *Eur Respir J* **28**, 999–1004 (2006).
36. Sitbon, O., Humbert, M., Jais, X., Ioos, V., Hamid, A.M., Provencher, S., Garcia, G., Parent, F., Herve, P., and Simonneau, G. Long-term response to calcium channel blockers in idiopathic pulmonary arterial hypertension. *Circulation* **111**, 3105–3111 (2005).
37. Barst, R.J., Rubin, L.J., Long, W.A., McGoon, M.D., Rich, S., Badesch, D.B., Groves, B.M., Tapson, V.F., Bourge, R.C., Brundage, B.H., *et al.* A comparison of continuous intravenous epoprostenol (prostacyclin) with conventional therapy for primary pulmonary hypertension. The Primary Pulmonary Hypertension Study Group. *N Engl J Med* **334**, 296–302 (1996).
38. McLaughlin, V.V., Shillington, A., and Rich, S. Survival in primary pulmonary hypertension: the impact of epoprostenol therapy. *Circulation* **106**, 1477–1482 (2002).
39. Puri, A., McGoon, M.D., and Kushwaha, S.S. Pulmonary arterial hypertension: current therapeutic strategies. *Nat Clin Pract Cardiovasc Med* **4**, 319–329 (2007).
40. Channick, R.N., Simonneau, G., Sitbon, O., Robbins, I.M., Frost, A., Tapson, V.F., Badesch, D.B., Roux, S., Rainisio, M., Bodin, F., and Rubin, L.J. Effects of the dual endothelin-receptor antagonist bosentan in patients with pulmonary hypertension: a randomised placebo-controlled study. *Lancet* **358**, 1119–1123 (2001).
41. Sitbon, O., McLaughlin, V.V., Badesch, D.B., Barst, R.J., Black, C., Galie, N., Humbert, M., Rainisio, M., Rubin, L.J., and Simonneau, G. Survival in patients with class III idiopathic pulmonary arterial hypertension treated with first line oral bosentan compared with an historical cohort of patients started on intravenous epoprostenol. *Thorax* **60**, 1025–1030 (2005).
42. Galie, N., Ghofrani, H.A., Torbicki, A., Barst, R.J., Rubin, L.J., Badesch, D., Fleming, T., Parpia, T., Burgess, G., Branzi, A., Grimminger, F., Kurzyna, M., and Simonneau, G. Sildenafil citrate therapy for pulmonary arterial hypertension. *N Engl J Med* **353**, 2148–2157 (2005).
43. Ghofrani, H.A., Wiedemann, R., Rose, F., Olschewski, H., Schermuly, R.T., Weissmann, N., Seeger, W., and Grimminger, F. Combination therapy with oral sildenafil and inhaled iloprost for severe pulmonary hypertension. *Ann Intern Med* **136**, 515–522 (2002).

44. Wilkens, H., Guth, A., Konig, J., Forestier, N., Cremers, B., Hennen, B., Bohm, M., and Sybrecht, G.W. Effect of inhaled iloprost plus oral sildenafil in patients with primary pulmonary hypertension. *Circulation* **104**, 1218–1222 (2001).
45. Yung, G. Lung transplantation and pulmonary hypertension. *Semin Cardiothorac Vasc Anesth* **11**, 149–156 (2007).
46. Michelakis, E.D., Wilkins, M.R., and Rabinovitch, M. Emerging concepts and translational priorities in pulmonary arterial hypertension. *Circulation* **118**, 1486–1495 (2008).
47. Asahara, T., Murohara, T., Sullivan, A., Silver, M., van der Zee, R., Li, T., Witzenbichler, B., Schatteman, G., and Isner, J.M. Isolation of putative progenitor endothelial cells for angiogenesis. *Science* **275**, 964–967 (1997).
48. Bhattacharya, V., McSweeney, P.A., Shi, Q., Bruno, B., Ishida, A., Nash, R., Storb, R.F., Sauvage, L.R., Hammond, W.P., and Wu, M.H. Enhanced endothelialization and microvessel formation in polyester grafts seeded with CD34(+) bone marrow cells. *Blood* **95**, 581–585 (2000).
49. Dimmeler, S., Aicher, A., Vasa, M., Mildner-Rihm, C., Adler, K., Tiemann, M., Rutten, H., Fichtlscherer, S., Martin, H., and Zeiher, A.M. HMG-CoA reductase inhibitors (statins) increase endothelial progenitor cells via the PI 3-kinase/Akt pathway. *J Clin Invest* **108**, 391–397 (2001).
50. Kalka, C., Masuda, H., Takahashi, T., Kalka-Moll, W.M., Silver, M., Kearney, M., Li, T., Isner, J.M., and Asahara, T. Transplantation of *ex vivo* expanded endothelial progenitor cells for therapeutic neovascularization. *Proc Natl Acad Sci USA* **97**, 3422–3427 (2000).
51. Murohara, T., Ikeda, H., Duan, J., Shintani, S., Sasaki, K., Eguchi, H., Onitsuka, I., Matsui, K., and Imaizumi, T. Transplanted cord blood-derived endothelial precursor cells augment postnatal neovascularization. *J Clin Invest* **105**, 1527–1536 (2000).
52. Crosby, J.R., Kaminski, W.E., Schatteman, G., Martin, P.J., Raines, E.W., Seifert, R.A., and Bowen-Pope, D.F. Endothelial cells of hematopoietic origin make a significant contribution to adult blood vessel formation. *Circ Res* **87**, 728–730 (2000).
53. Murayama, T., Tepper, O.M., Silver, M., Ma, H., Losordo, D.W., Isner, J.M., Asahara, T., and Kalka, C. Determination of bone marrow-derived endothelial progenitor cell significance in angiogenic growth factor-induced neovascularization *in vivo*. *Exp Hematol* **30**, 967–972 (2002).

54. Hur, J., Yoon, C.H., Kim, H.S., Choi, J.H., Kang, H.J., Hwang, K.K., Oh, B.H., Lee, M.M., and Park, Y.B. Characterization of two types of endothelial progenitor cells and their different contributions to neovasculogenesis. *Arterioscler Thromb Vasc Biol* **24**, 288–293 (2004).
55. Kocher, A.A., Schuster, M.D., Szabolcs, M.J., Takuma, S., Burkhoff, D., Wang, J., Homma, S., Edwards, N.M., and Itescu, S. Neovascularization of ischemic myocardium by human bone-marrow-derived angioblasts prevents cardiomyocyte apoptosis, reduces remodeling and improves cardiac function. *Nat Med* **7**, 430–436 (2001).
56. Yoon, C.H., Hur, J., Park, K.W., Kim, J.H., Lee, C.S., Oh, I.Y., Kim, T.Y., Cho, H.J., Kang, H.J., Chae, I.H., Yang, H.K., Oh, B.H., Park, Y.B., and Kim, H.S. Synergistic neovascularization by mixed transplantation of early endothelial progenitor cells and late outgrowth endothelial cells: the role of angiogenic cytokines and matrix metalloproteinases. *Circulation* **112**, 1618–1627 (2005).
57. Loomans, C.J., Wan, H., de Crom, R., van Haperen, R., de Boer, H.C., Leenen, P.J., Drexhage, H.A., Rabelink, T.J., van Zonneveld, A.J., and Staal, F.J. Angiogenic murine endothelial progenitor cells are derived from a myeloid bone marrow fraction and can be identified by endothelial NO synthase expression. *Arterioscler Thromb Vasc Biol* **26**, 1760–1767 (2006).
58. Schatteman, G.C., Dunnwald, M., and Jiao, C. Biology of bone marrow-derived endothelial cell precursors. *Am J Physiol Heart Circ Physiol* **292**, H1–18 (2007).
59. Fujiyama, S., Amano, K., Uehira, K., Yoshida, M., Nishiwaki, Y., Nozawa, Y., Jin, D., Takai, S., Miyazaki, M., Egashira, K., Imada, T., Iwasaka, T., and Matsubara, H. Bone marrow monocyte lineage cells adhere on injured endothelium in a monocyte chemoattractant protein-1-dependent manner and accelerate reendothelialization as endothelial progenitor cells. *Circ Res* **93**, 980–989 (2003).
60. He, T., Smith, L.A., Harrington, S., Nath, K.A., Caplice, N.M., and Katusic, Z.S. Transplantation of circulating endothelial progenitor cells restores endothelial function of denuded rabbit carotid arteries. *Stroke* **35**, 2378–2384 (2004).
61. Walter, D.H., Rittig, K., Bahlmann, F.H., Kirchmair, R., Silver, M., Murayama, T., Nishimura, H., Losordo, D.W., Asahara, T., and Isner, J.M. Statin therapy accelerates reendothelialization: a novel effect involving

mobilization and incorporation of bone marrow-derived endothelial progenitor cells. *Circulation* **105**, 3017–3024 (2002).

62. Gill, M., Dias, S., Hattori, K., Rivera, M.L., Hicklin, D., Witte, L., Girardi, L., Yurt, R., Himel, H., and Rafii, S. Vascular trauma induces rapid but transient mobilization of VEGFR2(+)AC133(+) endothelial precursor cells. *Circ Res* **88**, 167–174 (2001).
63. Iba, O., Matsubara, H., Nozawa, Y., Fujiyama, S., Amano, K., Mori, Y., Kojima, H., and Iwasaka, T. Angiogenesis by implantation of peripheral blood mononuclear cells and platelets into ischemic limbs. *Circulation* **106**, 2019–2025 (2002).
64. Kawamoto, A., Murayama, T., Kusano, K., Ii, M., Tkebuchava, T., Shintani, S., Iwakura, A., Johnson, I., von Samson, P., Hanley, A., Gavin, M., Curry, C., Silver, M., Ma, H., Kearney, M., and Losordo, D.W. Synergistic effect of bone marrow mobilization and vascular endothelial growth factor-2 gene therapy in myocardial ischemia. *Circulation* **110**, 1398–1405 (2004).
65. Yamaguchi, J., Kusano, K.F., Masuo, O., Kawamoto, A., Silver, M., Murasawa, S., Bosch-Marce, M., Masuda, H., Losordo, D.W., Isner, J.M., and Asahara, T. Stromal cell-derived factor-1 effects on *ex vivo* expanded endothelial progenitor cell recruitment for ischemic neovascularization. *Circulation* **107**, 1322–1328 (2003).
66. Hur, J., Yoon, C.H., Lee, C.S., Kim, T.Y., Oh, I.Y., Park, K.W., Kim, J.H., Lee, H.S., Kang, H.J., Chae, I.H., Oh, B.H., Park, Y.B., and Kim, H.S. Akt is a key modulator of endothelial progenitor cell trafficking in ischemic muscle. *Stem Cells* **25**, 1769–1778 (2007).
67. Gulati, R., Jevremovic, D., Peterson, T.E., Witt, T.A., Kleppe, L.S., Mueske, C.S., Lerman, A., Vile, R.G., and Simari, R.D. Autologous culture-modified mononuclear cells confer vascular protection after arterial injury. *Circulation* **108**, 1520–1526 (2003).
68. Kinnaird, T., Stabile, E., Burnett, M.S., Lee, C.W., Barr, S., Fuchs, S., and Epstein, S.E. Marrow-derived stromal cells express genes encoding a broad spectrum of arteriogenic cytokines and promote *in vitro* and *in vivo* arteriogenesis through paracrine mechanisms. *Circ Res* **94**, 678–685 (2004).
69. Heil, M., Ziegelhoeffer, T., Mees, B., and Schaper, W. A different outlook on the role of bone marrow stem cells in vascular growth: bone marrow delivers software not hardware. *Circ Res* **94**, 573–574 (2004).

70. Urbich, C., Aicher, A., Heeschen, C., Dernbach, E., Hofmann, W.K., Zeiher, A.M., and Dimmeler, S. Soluble factors released by endothelial progenitor cells promote migration of endothelial cells and cardiac resident progenitor cells. *J Mol Cell Cardiol* **39**, 733–742 (2005).
71. Karra, R., Vemullapalli, S., Dong, C., Herderick, E.E., Song, X., Slosek, K., Nevins, J.R., West, M., Goldschmidt-Clermont, P.J., and Seo, D. Molecular evidence for arterial repair in atherosclerosis. *Proc Natl Acad Sci USA* **102**, 16789–16794 (2005).
72. Rauscher, F.M., Goldschmidt-Clermont, P.J., Davis, B.H., Wang, T., Gregg, D., Ramaswami, P., Pippen, A.M., Annex, B.H., Dong, C., and Taylor, D.A. Aging, progenitor cell exhaustion, and atherosclerosis. *Circulation* **108**, 457–463 (2003).
73. Tepper, O.M., Galiano, R.D., Capla, J.M., Kalka, C., Gagne, P.J., Jacobowitz, G.R., Levine, J.P., and Gurtner, G.C. Human endothelial progenitor cells from type II diabetics exhibit impaired proliferation, adhesion, and incorporation into vascular structures. *Circulation* **106**, 2781–2786 (2002).
74. Junhui, Z., Xingxiang, W., Guosheng, F., Yunpeng, S., Furong, Z., and Junzhu, C. Reduced number and activity of circulating endothelial progenitor cells in patients with idiopathic pulmonary arterial hypertension. *Respir Med* **102**, 1073–1079 (2008).
75. Seeger, F.H., Haendeler, J., Walter, D.H., Rochwalsky, U., Reinhold, J., Urbich, C., Rossig, L., Corbaz, A., Chvatchko, Y., Zeiher, A.M., and Dimmeler, S. p38 mitogen-activated protein kinase downregulates endothelial progenitor cells. *Circulation* **111**, 1184–1191 (2005).
76. Seeger, F.H., Zeiher, A.M., and Dimmeler, S. Cell-enhancement strategies for the treatment of ischemic heart disease. *Nat Clin Pract Cardiovasc Med* **4**(Suppl 1), S110–113 (2007).
77. Wilson, D.W., Segall, H.J., Pan, L.C., Lame, M.W., Estep, J.E., and Morin, D. Mechanisms and pathology of monocrotaline pulmonary toxicity. *Crit Rev Toxicol* **22**, 307–325 (1992).
78. Lame, M.W., Jones, A.D., Wilson, D.W., Dunston, S.K., and Segall, H.J. Protein targets of monocrotaline pyrrole in pulmonary artery endothelial cells. *J Biol Chem* **275**, 29091–29099 (2000).
79. Campbell, A.I., Kuliszewski, M.A., and Stewart, D.J. Cell-based gene transfer to the pulmonary vasculature: Endothelial nitric oxide synthase

overexpression inhibits monocrotaline-induced pulmonary hypertension. *Am J Respir Cell Mol Biol* **21**, 567–575 (1999).

80. Nagaya, N., Kangawa, K., Kanda, M., Uematsu, M., Horio, T., Fukuyama, N., Hino, J., Harada-Shiba, M., Okumura, H., Tabata, Y., Mochizuki, N., Chiba, Y., Nishioka, K., Miyatake, K., Asahara, T., Hara, H., and Mori, H. Hybrid cell-gene therapy for pulmonary hypertension based on phagocytosing action of endothelial progenitor cells. *Circulation* **108**, 889–895 (2003).
81. Zhao, Y.D., Courtman, D.W., Deng, Y., Kugathasan, L., Zhang, Q., and Stewart, D.J. Rescue of monocrotaline-induced pulmonary arterial hypertension using bone marrow-derived endothelial-like progenitor cells: efficacy of combined cell and eNOS gene therapy in established disease. *Circ Res* **96**, 442–450 (2005).
82. Yip, H.K., Chang, L.T., Sun, C.K., Sheu, J.J., Chiang, C.H., Youssef, A.A., Lee, F.Y., Wu, C.J., and Fu, M. Autologous transplantation of bone marrow-derived endothelial progenitor cells attenuates monocrotaline-induced pulmonary arterial hypertension in rats. *Crit Care Med* **36**, 873–880 (2008).
83. Takahashi, M., Nakamura, T., Toba, T., Kajiwara, N., Kato, H., and Shimizu, Y. Transplantation of endothelial progenitor cells into the lung to alleviate pulmonary hypertension in dogs. *Tissue Eng* **10**, 771–779 (2004).
84. Zhao, Q., Liu, Z., Wang, Z., Yang, C., Liu, J., and Lu, J. Effect of preprocalcitonin gene-related peptide-expressing endothelial progenitor cells on pulmonary hypertension. *Ann Thorac Surg* **84**, 544–552 (2007).
85. Deng, W., St Hilaire, R.C., Chattergoon, N.N., Jeter, J.R., Jr., and Kadowitz, P.J. Inhibition of vascular smooth muscle cell proliferation *in vitro* by genetically engineered marrow stromal cells secreting calcitonin gene-related peptide. *Life Sci* **78**, 1830–1838 (2006).
86. Keating, A. Mesenchymal stromal cells. *Curr Opin Hematol* **13**, 419–425 (2006).
87. Kanki-Horimoto, S., Horimoto, H., Mieno, S., Kishida, K., Watanabe, F., Furuya, E., and Katsumata, T. Implantation of mesenchymal stem cells overexpressing endothelial nitric oxide synthase improves right ventricular impairments caused by pulmonary hypertension. *Circulation* **114**, I181–185 (2006).
88. Tamm, M., Gratwohl, A., Tichelli, A., Perruchoud, A.P., and Tyndall, A. Autologous haemopoietic stem cell transplantation in a patient with severe

pulmonary hypertension complicating connective tissue disease. *Ann Rheum Dis* **55**, 779–780 (1996).
89. Wang, X.X., Zhang, F.R., Shang, Y.P., Zhu, J.H., Xie, X.D., Tao, Q.M., and Chen, J.Z. Transplantation of autologous endothelial progenitor cells may be beneficial in patients with idiopathic pulmonary arterial hypertension: a pilot randomized controlled trial. *J Am Coll Cardiol* **49**, 1566–1571 (2007).
90. Zhu, J.H., Wang, X.X., Zhang, F.R., Shang, Y.P., Tao, Q.M., and Chen, J.Z. Safety and efficacy of autologous endothelial progenitor cells transplantation in children with idiopathic pulmonary arterial hypertension: open-label pilot study. *Pediatr Transplant* **12**, 650–655 (2008).

Chapter 13

Molecular Induction of Alveolar Regeneration

Matthew Hind and Siân Stinchcombe

Alveolar regeneration using developmentally active small molecules is an attractive therapeutic option for the treatment of diseases where lung development is disrupted, such as bronchopulmonary dysplasia or where alveolar surface area is lost, such as pulmonary emphysema. Understanding programmes of alveolar development, maintenance and regeneration will be critical in exploiting these novel therapeutic opportunities.

Keywords: Retinoic acid; oestrogens; granulocyte macrophage colony stimulating factor; adrenomodulin; alveolar plasticity; lung regeneration.

Outline

1. Introduction
2. Lung Development
3. Retinoic Acid (RA) has a Master Role in Patterning the Developing Lung
4. RA has a Role in Alveolar Development
5. The Adult Lung Structure Maintenance Programme
6. RA in the Adult Structure Maintenance Programme
7. Evidence of Molecular Plasticity: Endogenous Programmed Alveolar Regression and Regeneration
8. Calorie Restriction
9. Oestrogen
10. Experimental Evidence of Alveolar Regeneration

1. Introduction

In this chapter, we highlight regeneration-inducing properties of a number of different molecules in mammalian models of lung disease. We discuss regeneration in a developmental context and highlight salient findings from lower vertebrates which suggest that evolutionary conserved mechanisms of regeneration may exist across species. In many biological systems, regeneration to a greater or lesser degree recapitulates development; therefore to discuss molecules that are required for lung regeneration we must first consider lung development.

2. Lung Development

(See also Chapters 2 and 3)

The molecular cues that guide the development of the respiratory system are conserved through evolution from the branching structure of the respiratory organ in flies to mammalian lung morphogenesis.[1] There is now great interest in basic lung developmental biology with a view to understanding lung cell fate not only during development but also in the adult state. Recent evidence suggests that there may be a greater degree of plasticity in the mature organ than was previously thought. In adult pancreas, the use of three defined transcription factors has been demonstrated to reprogramme differentiated exocrine pancreas cells into cells that share structure and function with islet beta cells,[2] illustrating cellular dedifferentiation and reprogramming as a potential non-stem cell mechanism of organ regeneration.

A comprehensive review of lung development is beyond the scope of this chapter, and therefore we will highlight some important principles emerging from early developmental studies. Lung development requires a series of molecular and genetic cues released from endoderm and mesoderm that pattern the presumptive developing lung. Initial development of future conducting airways (trachea, bronchi and bronchioles) is followed by later development of gas-exchanging tissues (respiratory bronchioles, alveolar ducts and alveoli). A number of morphologically distinct, though continuous, developmental stages have been described, including branching morphogenesis, pseudoglandular, glandular, saccular and alveolar stages. To date, perhaps 50 genes and molecules have been described that influence different stages of lung development. Many more candidate factors have been identified in genomic screening experiments with, as yet, undefined roles. Interestingly, many of these genes are not restricted to the lung but have roles in other developing and regenerating biological systems. Organ specificity and patterning is therefore probably generated by multiple temporo-spatial gene and protein interactions rather than expression of lung-specific genes. Clearly, there is a shared genetic heritage with other foregut-derived structures such as stomach, dorsal and ventral pancreas and liver. Understanding regional specification of foregut, why the presumptive lung becomes lung and not liver, requires further understanding of gene expression and associated fate mapping studies. Factors with early lung developmental roles such as retinoic acid (RA), TGF-β, Hox genes, fibroblast growth factors (FGF) including FGF10 and Wnt signalling are used repeatedly at different stages of lung development, thus each may have multiple roles.[3]

3. Retinoic Acid (RA) has a Master Role in Patterning the Developing Lung

(See also Chapter 7)

The importance of RA signalling in lung development was first established by the examination of foetuses from vitamin A-deficient mothers and the observation of frequent characteristic lung malformations, including bilateral lung hypoplasia, left lung agenesis and agenesis of the oesophago-tracheal septum.[4] These dramatic effects on lung patterning reveal a role for endogenous RA in early lung development. RA synthesis and activation are reported during foregut specification around embryonic day 8.5–9.5 in the mouse.[5] In selective knockouts of foregut RA signalling, lung buds fail to form, placing RA upstream of Hox, Tbx and FGF10.[6] If foregut explants are cultured in the presence of an RA receptor antagonist then lung bud formation is blocked. The use of selective retinoic acid receptor (RAR) agonists suggests that this occurs through RARβ[7] resulting in selective regulation of fibroblast growth factor 10 (FGF10)[8] through a transforming growth factor β (TGF-β) mediated pathway.[9] If RA levels are maintained in embryonic lung culture as in early lung bud then secondary branching fails.[5] This suggests that in common with other developing biological systems that utilise RA signalling, RA acts as a molecular master switch: RA signal is required to allow initial lung bud development, and is then switched off to allow correct secondary branching.

4. RA has a Role in Alveolar Development (See also Chapter 7)

Several strands of *in vitro* and *in vivo* investigation have provided compelling evidence for a further role of endogenous RA signalling during a later period of lung development in the regulation of post-natal alveologenesis. Alveoli are formed by a number of mechanisms that appear to be differentially regulated. Alveolar septation, the subdivision of alveolar saccules, occurs during a developmentally restricted period that occurs in the first two post-natal weeks in rats and mice.[10,11] In humans, this process extends until at least 18 months of age.[12]

Studies indicate that the molecular components required for RA signalling are temporally associated with the period of alveolar septation in rats, mice and humans. Studies in the mouse have demonstrated that RARα1, RARβ2, RARβ4 and RARγ2 isoforms are present from post-natal day 1 (P1) to P15, peak from P4 to P9, and can be localised by *in situ* hybridisation to cells within the alveolar walls.[13] RARβ and RARγ mRNA have also been shown to increase in rat lung tissue at P2, preceding the onset of secondary septation.[14] Cellular retinol binding protein 1 (CRBP-1), CRBP-2 and cellular RA binding protein 1 (CRABP-1) are significantly upregulated in the early post-natal mouse lung and CRBP-1 and CRABP-1 can be identified by immuno-histochemistry in erupting and elongating septa.[13] The temporal pattern of expression is similar in whole rat lung tissue[15] and isolated lipid-laden interstitial fibroblasts (LIFs) during this period.[14] These findings imply that local RA production is increased at this time, perhaps in a cell-specific manner. Dexamethasone (Dex) dosing of rat pups from P4–14 inhibits alveolar septation[16] and downregulates CRBP-1 and CRABP-1 expression.[17] The RA-synthesising enzyme retinaldehyde dehydrogenase 1 (RALDH-1) is also upregulated within the alveolar septa of the post-natal mouse.[18] LIFs are known to store retinol in the form of retinyl esters,[19] and to synthesise and secrete RA in the form of all-trans RA (atRA) and 4-*oxo*-RA.[20] Dex treatment of cultured LIFs has been shown to halve the amount of RA released into the medium.[20] Retinoid stores in LIFs change markedly in the early post-natal period: retinyl ester concentrations decline from their late gestational peak; retinol levels peak between P2–8; and atRA and 9-*cis*-RA levels peak on P2, decline between P3 and P7, and then rise again on P8.[14,21] This dynamic profile is consistent with a significant increase in endogenous RA requirement, synthesis and utilisation during the rapid period of alveolar secondary septation.

Since LIFs which provide the endogenous source of RA are frequently found at the base of elongating septa, are often in close proximity to alveolar epithelial type II (AETII) cells, and have cyotoplasmic protrusions extending to the endothelium and epithelium, it is plausible that they establish a distal lung signalling centre which may act by autocrine and paracrine mechanisms to influence gene expression and coordinate the processes required for alveolar septation.

Pulmonary microvasculature endothelial cells upregulate CRBP-1 in response to exogenous RA *in vitro*. The same response is generated by a serum-free medium conditioned by cultured LIFs, and is inhibited by addition of RAR pan-antagonist or RXR pan-antagonist,[20] indicating that RA secretion by LIFs can regulate gene expression in endothelial cells. RA increases proliferation in human foetal lung capillary endothelial precursor cells *in vitro*.[22] Neonatal AETII cells have also been shown to proliferate *in vitro* in the presence of RA,[23] but little is known about the *in vivo* response of alveolar epithelial cells to endogenous RA. In addition, RA significantly increases ^{3}H-thymidine incorporation in cultured post-natal rat fibroblasts in a platelet-derived growth factor (PDGF)-dependent manner, suggesting that endogenous RA may induce autocrine LIF proliferation or paracrine myofibroblast proliferation through a PDGF-mediated mechanism during developmental alveologenesis.[24]

Further investigations have clearly demonstrated the effect of RA on elastin synthesis. When LIFs are cultured with exogenous RA, there is a significant rise in the rate of elastin transcription, leading to a 2.5-fold increase in the steady state level of tropoelastin mRNA.[25] Conversely, if cultured P8 rat lung fibroblasts are treated with alcohol dehydrogenase (ADH) inhibitors and aldehyde dehydrogenase (ALDH) inhibitors to prevent endogenous RA synthesis, elastin gene expression is suppressed.[26]

Strong *in vivo* evidence for a central role for RA in septation is provided by studies assessing the effects of manipulating RA levels during the critical period for alveolarisation. Administration of systemic exogenous RA to rat pups from P3 to P13 resulted in upregulation of CRBP-1 expression,[17] a 50% increase in alveolar number and a 47% reduction in mean alveolar volume in lungs examined at P14.[27] In contrast, mice receiving disulfiram, an inhibitor of RA synthesis, from P2 to P14 showed a 26% increase in the mean alveolar diameter (also known as the mean chord length (Lm)) when examined on P23,[28] consistent with impaired secondary septation. Conditional post-natal overexpression of a dominant negative pan-RAR also leads to fewer, larger alveoli and diminished alveolar SA.[29] The conservation of retinoid-regulated mechanisms for alveologenesis between rodent and human lung has been illustrated by the discovery of homozygous stimulated by retinoic acid 6 (STRA6) mutations in a pleiotropic malformation syndrome which includes lung hypoplasia and failed alveolar septation.[30] This syndrome represents the first human lung phenotype associated with mutations in the retinoic acid signalling pathway.

5. The Adult Lung Structure Maintenance Programme

Alveolar tissue in the healthy adult lung has classically been considered to be relatively static, with little turnover of cellular or extracellular components, other than in response to acute injury. However, recent findings have highlighted a previously unforeseen degree of plasticity suggesting that dynamic programmes are required to maintain adult lung structure. Alveolar structural homeostasis requires integration of molecular mechanisms to protect against damage by oxidative stress, protease activity, inflammation and pathological lung cell apoptosis, with active mechanisms to maintain structural cells, blood vessels and extracellular matrix (ECM) composition. A number of studies now demonstrate that the factors required for normal alveolar development have significant roles in adult alveolar maintenance, though the specific molecular cues that are involved in this programme are not yet known. This concept has led to the emergence of a new paradigm of impaired or dysregulated alveolar repair and regeneration in the generation and progression of seemingly diverse pathologic lung diseases that include emphysema and pulmonary fibrosis.

6. RA in the Adult Structure Maintenance Programme

Retinoids are derivatives of Vitamin A and include the biologically active metabolite RA. Mammals are unable to synthesise retinoids *de novo*, so they must be obtained from the diet, either from animal sources (retinyl esters and retinol) or plant sources (β-carotene). Early Vitamin A deprivation studies in rats demonstrated the requirement for retinoids in the lungs of adult animals.[31] The most striking defect observed was the metaplastic transformation of pseudostratified squamous tracheobronchial epithelium into keratinising squamous epithelium. This effect is entirely reversed with dosing with retinol.[32] These early studies were performed without inflation-fixation, making it difficult to analyse lung parenchymal structure. Recent studies examining the distal lung of weanling rats maintained on a Vitamin A deficient (VAD) diet for eight weeks report patchy emphysema with associated peribronchial lymphocytic infiltrates analogous to human COPD.[33] In a study of more prolonged Vitamin A deficiency, a group of rats were fed with a VAD diet from 12 weeks for 12 months, at which point severe retinoid deficiency was confirmed by HPLC analysis. Marked histological changes in alveolar structure were evident throughout the lung, with significantly enlarged airspaces with an increased alveolar mean chord length (Lm), reduced alveolar surface area (Sa) and marked thinning of the alveolar walls compared to Vitamin A sufficient controls.[34] Does this apparent structural abnormality alter lung function? The answer appears to be yes, as studies in mild VAD demonstrate diminished elastin and collagen content of the lung and impaired tissue mechanics. Importantly, RA dosing in these adult animals reversed both the structural and functional abnormalities.[35]

cell replication are upregulated, and within three hours there is significant expression supportive of angiogenesis, ECM remodelling and guided cell motion, i.e. four processes required in the production of new septal tissue.[49] These findings demonstrate an innate ability of the mammalian lung to rapidly regenerate alveolar tissue in response to increased O_2 demand.

Is there any evidence of calorie-manipulated alveolar plasticity in man? Surprisingly, the answer is yes. Careful studies performed by Jewish physicians in the Warsaw ghetto during the Second World War, where typical food intake was only 800 kCal/day, recorded an average 30–40% diminution of oxygen uptake, and diagnosed emphysema in 13.5% of those examined, based on either radiological or post mortem appearances of the lung.[52] A comparative study of 21 female anorexia nervosa (AN) patients and 16 age-matched female controls showed significantly higher CT measurements of emphysema in the AN group, which correlated closely with lower body mass index (BMI) and reduced diffusing capacity of the lung for carbon monoxide (DL_{CO}).[53] Other case reports have also demonstrated emphysematous change and impaired gas exchange capacity in chronically malnourished patients.[54] There are several caveats to consider in these observations. Firstly, neither of these scenarios involved pure calorie restriction. Since total food intake was reduced in both cases and detailed nutritional data was not recorded, deficiency of essential vitamins (including retinoid content) and minerals cannot be excluded, unlike animal CR studies. Secondly, many female patients in both cases may have been amenorrhoeic. Oestrogen is known to influence alveolar morphology (see below) and the disturbance of menstrual cycle may be a confounding factor.

So it would seem that the pathway for calorie restriction-associated alveolar loss is conserved from mice to man. Whether adult humans have the capacity to regenerate alveoli is not yet known but it appears to be easily testable by evaluating the effects of refeeding on lung structure and function.

9. Oestrogen

Although oxygen uptake almost doubles in female rats during pregnancy and lactation, body mass-specific alveolar Sa is identical in virgin, pregnant and lactating female rats.[55] Preparation for the future metabolic demands of pregnancy takes place at the onset of sexual maturity, when female rats and mice develop a higher body mass-specific number of alveoli and alveolar Sa than age-matched males despite their similar body mass-specific oxygen uptake. This sexual dimorphism in alveolar morphology is mediated by oestrogen, acting via oestrogen receptor α and β.[56–60] Therefore, ovariectomy at P21 in sexually immature female rats[56] and mice[58] diminishes the formation of alveoli, but oestrogen replacement therapy from the time of ovariectomy allows alveolar sexual dimorphism to develop.[56] Within three weeks of ovariectomy of sexually mature adult female mice there is significant loss of alveoli and alveolar Sa without loss of lung volume. Subsequent oestrogen replacement reverses the architectural effects of ovariectomy and normal alveolar morphology is restored.[58] These findings indicate that oestrogen is required for maintenance of alveolar structure in the adult female, and is capable of regenerating lost alveolar tissue. This represents a further example of adult alveolar plasticity based on an endogenous signalling pathway.

Clinical studies of ageing indicate that oestrogen plays a significant role in the maintenance of lung function in women. Diffusing capacity, a functional correlate of alveolar Sa, naturally falls with age in non-smokers. Men lose diffusing capacity at a rate of 6% per decade throughout life, whilst the rate of loss in women is only 2% per decade before menopause, but accelerates to 6% per decade post-menopause.[61] These rates of loss parallel the age-related loss of maximum oxygen consumption.[62] Measurements of FEV_1, an index of small airways obstruction which is influenced by lung tissue elastic recoil and tethering by alveolar attachments, has demonstrated that elderly women receiving hormone replacement therapy (HRT) have a higher FEV_1 than similar-aged women not receiving HRT.[63] HRT has also been shown to increase FEV_1 and forced vital capacity in post-menopausal women after three months of treatment.[64] Furthermore, the use of an oestrogen-containing oral contraceptive pill in 24 to 35-year-old women has been shown to measurably increase forced expiratory flow rates, particularly at low lung volumes.[65] In contrast, low concentrations of ovarian hormones may play a role in the development of COPD. Women constitute approximately 75% of never-smokers over 55 years of age with clinical and functional evidence of COPD.[66,67] Although these studies provide indirect functional evidence rather than visual proof, they strongly suggest that the ability of oestrogen to preserve and regenerate alveolar tissue is conserved between species.

The mechanisms underlying the structural adaptation of adult alveolar tissue in response to these metabolic and hormonal manipulations are not yet clear. However, these models demonstrate the existence of intrinsic programmes for alveolar regeneration, and raise the possibility that such programmes may be therapeutically induced to replace absent alveolar tissue in diseases such as emphysema and bronchopulmonary dysplasia.

10. Experimental Evidence of Alveolar Regeneration

The ability to regenerate individual tissue types and whole organs varies widely between animal species. In adult mammals, apart from the unique phenomenon of antler regeneration in deer, organ regeneration is thought to be limited to the liver. This is in striking contrast to urodele amphibians which are able to regenerate whole limbs, tails, jaws, spinal cords, forebrain, lens, retina and gills.[68] However, mammals are able to continually replace many tissues, including the epidermis, gut endoderm and blood cells. It is commonly believed that regeneration did not evolve as a separate phenomenon involving novel gene pathways, but as a repair response that re-activates pathways originally used in the development of that organ or tissue. This concept and the role of RA as a regeneration-inducing molecule have been elegantly demonstrated in studies of limb development and regeneration: limb bud development is dependent upon RA synthesis;[69] RA is detected in the regenerating amphibian limb by HPLC[70] and a reporter gene system;[71] and when RA synthesis is inhibited by disulfiram, regeneration stops.[72] This role for RA in regeneration also occurs in adult mammals, as RA signalling has also been detected during deer antler regeneration,[73] reviewed in Ref. 74.

Evidence demonstrating a pivotal role for RA in post-natal alveologenesis, an ongoing requirement for Vitamin A sufficiency in the adult lung, and the existence of endogenous programmes for alveolar regeneration, has lead to the hypothesis that exogenous RA may be able to induce alveolar regeneration in the adult lung by recapitulating developmental programmes of alveologenesis. Such a process would require coordination of the cellular and extracellular components of existing alveolar walls to produce new alveolar septa.

This remarkable possibility was first observed by Massaro and colleagues, who demonstrated that RA is able to reverse the effects of elastase-induced damage in the rat lung.[75] In this study adult male rats were dosed with intratracheal elastase to destroy alveolar tissue, resulting in characteristic changes of human and experimental emphysema: 18% increase in lung volume, reflecting a loss of elastic recoil; larger but less numerous alveoli, quantified by an increase in alveolar Lm and 45% reduction in alveolar number; and diminished volume-corrected alveolar Sa. The elastase model of emphysema in rats has been shown to progressively worsen over 1–2 months, with no spontaneous recovery thereafter.[76] When these rats were treated 25 days later with 0.5 mg/kg intraperitoneal RA for 12 days and examined at the end of this dosing period, their lung volume, Lm, alveolar number and gas-exchanging Sa had returned to the normal range. Histological examination revealed more numerous, smaller alveoli in elastase-RA treated rats but not elastase-vehicle treated controls consistent with the formation of new septa from the walls of enlarged alveoli.

This study has subsequently been repeated by another group, who confirmed that RA abrogates experimental emphysema in rats.[77] Their analysis of the cellular response to RA showed areas of proliferation within the alveoli of elastase-RA treated rats but not elastase-vehicle treated rats, identified by: proliferating cell nuclear antibody (PCNA); immunohistochemistry; and a threefold increase in apoptosis of alveolar wall cells within 24 hours of initial RA treatment, which then rapidly subsided, detected by *in situ* hybridisation 3UTP nick end labelling (TUNEL). These preliminary findings suggest that significant alveolar remodelling takes place during RA treatment. This group also reported change in the ratio of alveolar-to-arterial partial pressure of O_2 (the A–a gradient for pO_2) following RA treatment, suggesting a functional improvement in diffusion capacity of the lungs with alveolar regeneration.

In a further repeat study of RA-induced alveolar regeneration in the elastase rat model of emphysema, several parameters of lung function were observed for three months following RA treatment.[78] Pause (a non-invasive indicator of airflow obstruction) increased, indicating progression of airways obstruction in elastase-treated rats that was then halted by RA treatment. Residual volume and total lung capacity were increased by elastase instillation, but significantly restored back towards normal range values by RA. Lung tissue density measurement by CT scan of inflation-fixed lungs was noticeably reduced following elastase, and the density measurements of rats treated with RA were between elastase and normal values. Although respiratory system compliance was increased, and DL_{CO} and FEV_1 were decreased, in elastase-induced emphysema, no significant changes were seen in these parameters following RA treatment.

Elastase has also been used to generate experimental emphysema in mice, and RA treatment has successfully induced alveolar regeneration in this model.[79] Elastase application to a cultured human airway epithelial cell line and a human alveolar epithelial cell line causes reduced cell viability and increased apoptosis. However, when RA is added to these cultures it preserves viability, inhibits apoptosis and inhibits caspase 3 induction.[80]

It has been clearly established that RA is also able to induce alveolar regeneration in other animal models of emphysema. The tight skin (Tsk) mouse carries an autosomal dominant tandem repeat mutation in the fibrillin-1 gene, which disrupts elastin polymerisation, and is associated with multiple connective tissue defects, such as hyperplasia of tendon sheaths, cartilage and bone and skin fibrosis.[81,82] Heterozygotes for the Tsk mutation have mildly enlarged primary pulmonary saccules at P4, and by one month of age display marked emphysematous changes, with enlarged alveoli, thinned or broken alveolar walls, disorganised and broken elastic fibres and increased lung compliance.[83,84] The Tsk mouse therefore represents a genetically determined model of emphysema. When these mice were treated with 0.5 mg/kg RA daily from P40–51 and examined the following day, alveoli were 2.7-fold smaller and 3.5-fold more numerous than in untreated controls, indicating that novel alveolar septation had been induced.[85]

RA is also able to rescue pharmacologically impaired failure of septation. Dex is known to inhibit alveolar septation and accelerate alveolar wall thinning in the critical period for alveolarisation. A study of male rats treated with Dex from P4–13 showed complete inhibition of septation which was not reversed after Dex was stopped, resulting in a permanent threefold increase in mean alveolar volume. When these rats later received 0.5 mg/kg RA daily from P24–36 and were examined on P37, their mean alveolar volume reduced by 1.6-fold compared to Dex vehicle treated litter mates, and other parameters of alveolar morphometry were also partially recovered.[85] RA-induced alveolar septation in the Dex rat model has similarly been demonstrated by measurement of radial alveolar counts.[86]

A Dex mouse model of alveolar insufficiency has also been shown to undergo alveolar regeneration when treated with RA.[28] In this study, mouse pups were treated with Dex daily from P2–14 (with a two-day break on P8 and P9), resulting in fewer, larger airspaces, grossly simplified alveolar structure and significantly reduced body mass-specific alveolar SA. Mice that then received 2 mg/kg RA daily from P42–54 (with a two-day break on P48 and P49) and were assessed four weeks later showed clear histological evidence of alveolar regeneration, reflected by a significant reduction in Lm and an increase in body mass-specific Sa which completely restored this parameter to normal.

Subsequent use of this model and dosing protocol has demonstrated that 4-*oxo*-RA is as effective as atRA (the isoform used in all previous regeneration studies) at inducing alveolar regeneration, but other natural retinoids including 9-*cis*-RA (the ligand for RXR) and all-*trans*-retinol (the parent Vitamin A molecule) have no significant effect on alveolar morphology.[87] In addition, selective agonists for RARα, β, and γ have differential degrees of effect depending upon the morphological parameter observed. Thus, RARα selective agonist is as effective as RA at restoring Lm, the ratio of alveolar Sa-to-lung volume (Sa/LV) and body mass-specific alveolar Sa; RARβ selective agonist restores Lm and Sa/LV to a lesser (but significant) degree but has no significant impact on body mass-specific Sa; and RARγ selective agonist induces the most complete restoration of body mass-specific Sa of any compound in this study but has a lesser effect on Lm and Sa/LV. Pan-RXR agonist is essentially inactive. These results imply that the mechanism of induction of alveolar regeneration by retinoids involves the activation of a RAR, and cannot occur with isolated RXR activation. In addition, RARβ null mutant mice treated with post-natal Dex have been shown to undergo alveolar regeneration when treated with RA as adults, but not to a greater degree than

Regenerating organs and tissues adopt different strategies to replenish populations of specialised cell types. Some organs employ local dedifferentiation and proliferation of committed cell lines (e.g. amphibian limb regeneration,[99] mammalian liver regeneration,[100] whilst others make use of scattered stem or progenitor cells throughout the tissue, e.g. gastrointestinal epithelium).[101] Tissue-specific stem cells remain quiescent until called upon to proliferate and produce daughter cells which then differentiate into restricted cell phenotypes. Reports of repopulation of many diverse tissues by bone marrow (BM)-derived multipotent stem cells have also introduced a new non-organ based paradigm for regeneration.[102,103]

If RA is able to re-induce developmental mechanisms of alveologenesis, this would suggest that several intrinsic alveolar cell types respond in a coordinated manner to generate new alveolar septa in alveolar regeneration. However, the cellular composition of alveolar tissue in the adult lung differs from that of the post-natal lung, in both healthy tissue and models of emphysema. Evidence of BM-derived stem cell involvement and of intrinsic progenitor cell activity in alveolar injury models suggests that alternative strategies may be available for induction by RA.

12. Bone Marrow-Derived Stem Cell Engraftment

Studies using genetically labelled BM transplantation in mice introduced the possibility that BM-derived pluripotent stem cells may have the potential to engraft within the lung and transdifferentiate into lung epithelial, endothelial and interstitial cell phenotypes. In an initial study, a single BM-derived haematopoietic stem cell from an adult male mouse was transplanted into a female recipient mouse which had previously been irradiated to ablate all resident bone marrow.[103] One month later fluorescent *in situ* hybridisation (FISH) for the Y chromosome indicated that the single donor cell had not only repopulated all bone marrow cell lineages, but that donor-derived cells had also engrafted into several organs, including the lung. Up to 20% of all AETII cells in the recipient mouse (defined by surfactant protein B (SPB) expression) were reported to contain a Y chromosome. A similar study infused LacZ-labelled donor BM cells into wild type recipient mice after bleomycin-induced lung injury, and detected rare engraftment of flat cells into the alveolar walls which co-expressed the LacZ BM tracking marker and T1α, an AETI cell marker,[102] therefore challenging the theory that adult AETI cells invariably arise from local precursor cells. Subsequent reports presented further evidence that circulating BM-derived cells are able to contribute to lung tissue repair after injury by bleomycin,[104,105] elastase,[106] endotoxins,[107,108] naphthalene[109] and radiation.[110,111] Models in humans have also been provided by sex-mismatched transplant patients: female recipients of male BM were shown to have pulmonary chimerism in epithelial and endothelial tissue;[112] and male recipients of female lung allografts were found to have greater numbers of Y chromosome-positive cells in areas of bronchial and alveolar epithelium injured by rejection or infection.[113]

Further work in this field has questioned the reproducibility of lung epithelial engraftment and the methods used for its detection. Immunophenotypic analysis of recipient lungs in single haematopoietic stem cell transplantation and parabiotic mouse models has demonstrated that all BM or blood-derived cells in lung tissue express CD45, a pan-haematopoietic marker which is never present in normal airway or alveolar epithelium.[114] Deconvolution fluorescence microscopy has also been used to resolve three-dimensional reconstructions of cells in tissue section which appear to express both the donor bone marrow reporter gene and an AETII cell-specific marker. In all cases apparent co-expression was shown to be an artefact of overlapping fluorescent signal from adjacent cells.[115] Furthermore, repeat studies of engraftment in the bleomycin-injured lung have revealed that autofluorescence and rare non-specific binding of fluorescently tagged antibodies can also introduce false-positive artefact.[116] When these erroneous elements were excluded, and donor BM-derived cells from a mouse line expressing green fluorescent protein (GFP) under the regulatory control of the human SPC promoter were used to exclude donor-derived blood cells in alveolar capillaries from cell counts, no reconstitution of AETII cells by BM-derived cells was detected. It is also possible that donor BM cells fuse with recipient cells to form heterokaryons, rather than engrafting and transdifferentiating to acquire lung epithelial phenotypes. A Cre/lox system was used with GFP expression in transgenic mice to identify fused plasticity of cell fate amongst differentiated cell types. Whilst a number of putative stem and progenitor cells have been identified in the distal lung, lineages and mechanisms for alveolar epithelial maintenance and repair remain unclear.

New alveolar epithelium, capillaries and interstitial tissue are required to construct new alveolar septa in RA-induced regeneration. Whilst endothelial and interstitial cells are thought to be capable of self-renewal, AETI cells, which make up 93% of the alveolar epithelium, are considered to be terminally differentiated and unable to proliferate. The AETII cell is thought to act as the progenitor of AETI cells in alveolar development and repair, and it is therefore reasonable to suggest that these cells respond to RA to increase alveolar Sa in regeneration. This induced expansion of epithelial cells could conceivably provide the stimulus for accompanying angiogenesis and ECM deposition.

Studies of alveolar cell ^{3}H-thymidine labelling after injury by exposure to nitrogen dioxide or hyperoxia in mature rodent lungs have shown initial proliferation of AETII cells followed by a large increase in AETI cell labelling several days later.[125,126] In contrast to AETII cells, AETI cells showed minimal ^{3}H-thymidine labelling and an absence of mitotic figures. These results are interpreted as evidence that AETI cells are non-proliferative and terminally differentiated, and that sister cells from AETII cell division can transform into AETI cells. Similarly alveolar epithelial injury by ozone exposure causes AETI cell necrosis, followed by a marked increase in mitotic activity within alveolar septal tissue, which is mostly accounted for by AETII cell proliferation.[127] Subsequent studies of AETII cells isolated from hyperoxia-exposed adult rat lungs have identified a subpopulation of AETII cells which are proliferative, resistant to hyperoxic injury, and have high levels of telomerase activity,[128] and therefore possess the properties of progenitor cells capable of repopulating and repairing damaged alveolar epithelium.

Analysis of proliferation in the prenatal developing rat lung has demonstrated maximal DNA synthesis and cell division when alveoli are lined exclusively with AETII cells, and lower levels later in the AETI cell epithelium,[129] consistent with the theory that foetal AETII cells act as progenitors for AETI cells. During developmental alveologenesis AETII cells are often seen at the base of outgrowing septa in close proximity to

LIFs,[130–133] which provide an endogenous source of RA.[19] This is thought to stimulate proliferation and downstream gene expression in local cells by autocrine and paracrine signalling, and *in vitro* evidence suggests that exogenous RA could upregulate these paracrine mechanisms again in AETII cells during alveolar regeneration. *In vitro* cultured AETII cells proliferate in response to RA in a dose-dependent manner.[23] This effect is associated with reductions in insulin-like growth factor binding protein 2 (IGFBP-2) and CKI $p21^{CIP1}$ expression. These proteins exert anti-proliferative actions and have been shown to accumulate in various situations that block *in vitro* AETII cell proliferation, such as oxidant exposure,[134] TGFβ-1 treatment,[23] TNFα treatment[135] and glucocorticoid treatment.[136] RA abrogates AETII cell growth arrest by hyperoxia,[134] TGF-β1[23] and TNFα[135] by modulating production of IGFBP-2 or CKI $p21^{CIP1}$. Similar actions during alveolar regeneration *in vivo* may permit AETII cell proliferation to proceed.

14. Alveolar Regeneration by Adrenomedullin

Adrenomedullin (AM) is a multifunctional regulatory peptide that was originally isolated from human phaeochromocytoma[137] and has subsequently been detected in many human and rodent tissues, including endothelium[138] and the adult and foetal lung.[139–141] In particular, AM expression has been demonstrated in AETII cells of the normal human lung[142] and rat lung.[143]

AM belongs to the calcitonin family and signals via a specific receptor, which is a member of the G-protein linked receptor superfamily. To date its known actions include induction of angiogenesis,[144] vasodilatation,[145] inhibition of endothelial cell apoptosis[146,147] and autocrine modulation of surfactant secretion by AETII cells.[143] Its expression in the lung is upregulated by pro-inflammatory cytokines, lipopolysaccharide and hypoxia.[148] *In vitro* studies of intracellular pathways have demonstrated that AM expression is under the control of hypoxia-inducible factor and triiodothyronine in several cell lines.[149] RA has also been shown to influence AM expression, increasing AM production by cultured vascular smooth muscle cells,[150] non-human primate cells[151] and murine macrophages.[152] These findings suggest that AM may play a role in alveolar maintenance and repair.

The reparative abilities of AM have been studied in elastase-treated mice following GFP+ BM transplantation.[153] Using a protocol similar to Ishizawa *et al.*,[79] mice received intratracheal elastase followed immediately by continuous infusion of AM (or vehicle) for 14 days. After a further 14 days, functional and histological analyses revealed that AM had significantly attenuated the increases in lung volume, static lung compliance and Lm typically caused by elastase. There was also clear evidence of increased BM cell mobilisation in peripheral blood samples on flow cytometry and lung tissue sections by immunohistochemistry. In addition, serum vascular endothelial growth factor (VEGF) concentration was unaffected by elastase instillation but significantly increased by AM treatment. *In vitro* AM was shown to protect against the effects of elastase exposure, decreasing the frequency of apoptosis and attenuating caspase-3 activation in a human AETII cell line (A549) and a human umbilical vein endothelial cell line (HUVEC). This comprehensive assessment indicates that AM is able to partially prevent and/or repair elastase-induced acute alveolar tissue damage, and suggests that this may be achieved by direct effects on epithelial and endothelial cells and indirect effects brought about by BM-derived cells.

To investigate the effects of AM on established emphysema, an additional group of mice received a 12-day continuous infusion of AM (or vehicle) starting 25 days after elastase instillation. Morphometric analysis after completion of treatment revealed a small but statistically significant improvement in the Lm of AM-treated mice. Lung volume and static lung compliance were significantly restored towards normal values by AM. GFP+ cell numbers were noticeably increased in lung tissue sections from AM-treated mice. Therefore, remarkably, AM appears to induce morphological and functional alveolar regeneration in the elastase mouse model following ablative irradiation and BM transplantation. However, alveolar Sa values were not reported in this paper, and so no objective assessment was made of the ability of AM to increase alveolar Sa. It is possible that the apparent reduction in Lm was a result of reduced tissue compliance and lung volume, rather than *de novo* alveolar septation.

The regenerative abilities of AM have not yet been tested in other animal models of emphysema to determine whether these results are model-specific or more widely applicable. Whilst underlying mechanisms of AM action in acute lung injury have been investigated, these findings may not apply to the pathological backdrop in established emphysema.

If we accept the conclusion that AM induces alveolar regeneration, this study raises several interesting questions. Does AM activate the same regenerative programmes as RA? Is AM an essential downstream effector in RA-induced alveolar regeneration? Does AM act in concert on epithelial, endothelial and interstitial elements of alveolar tissue to promote new alveolar septation? Or is septal growth driven by angiogenesis? Alternatively, are the beneficial effects of AM partly or wholly enacted by paracrine signalling or the immunomodulatory actions of mobilised BM-derived cells? If so, do such mechanisms play a role in RA-induced alveolar regeneration as well? These possibilities are intriguing and clearly warrant further study.

15. Conclusion and the Way Forward

In this chapter, we have reviewed evidence describing a previously unrecognised degree of structural plasticity in the distal lung. Specific small molecule regulators of alveolar development, maintenance, senescence and regeneration have now been reported. Further understanding of the molecular regulation of these complex programmes in both lung health and disease may enable the use of molecular therapeutics to direct the intrinsic cellular regenerative response. This approach requires not just identification of specific regenerative signals, but the presence or absence of effector cell populations. The first human lung regeneration study in patients with COPD funded by the NIH demonstrated that RA has biological activity and is well tolerated in this group of patients.[154] Further, much larger Pharma sponsored studies using RAR specific agonists are underway and outcomes are awaited. Despite the current driver of "translation" the basic cell and molecular biology of lung regeneration remains poorly understood.

References

1. Metzger, R.J., Klein, O.D., *et al.* The branching programme of mouse lung development. *Nature* **453**(7196), 745–750 (2008).

2. Zhou, Q., Brown, J., *et al. In vivo* reprogramming of adult pancreatic exocrine cells to beta-cells. *Nature* **455**(7213), 627–632 (2008).
3. Maeda, Y., Dave, V., *et al.* Transcriptional control of lung morphogenesis. *Physiol Rev* **87**(1), 219–244 (2007).
4. Wilson, J.G., Roth, C.B., and Warkany, J. An analysis of the syndrome of malformations induced by maternal Vitamin A deficiency. Effects of restoration of Vitamin A at various times during gestation. *Am J Anat* **92**, 28 (1953).
5. Malpel, S., Mendelsohn, C., and Cardoso, W.V. Regulation of retinoic acid signaling during lung morphogenesis. *Development* **127**, 3057–3067 (2000).
6. Wang, Z., Dolle, P., *et al.* Retinoic acid regulates morphogenesis and patterning of posterior foregut derivatives. *Dev Biol* **297**(2), 433–445 (2006).
7. Desai, T.J., Chen, F., *et al.* Distinct roles for retinoic acid receptors alpha and beta in early lung morphogenesis. *Dev Biol* **291**(1), 12–24 (2006).
8. Desai, T.J., Malpel, S., *et al.* Retinoic acid selectively regulates Fgf10 expression and maintains cell identity in the prospective lung field of the developing foregut. *Dev Biol* **273**(2), 402–415 (2004).
9. Chen, F., Desai, T.J. *et al.* Inhibition of Tgf beta signaling by endogenous retinoic acid is essential for primary lung bud induction. *Development* **134**(16), 2969–2979 (2007).
10. Burri, P.H., Dbaly, J., *et al.* The postnatal growth of the rat lung. I. Morphometry. *Anat Rec* **178**(4), 711–730 (1974).
11. Amy, R.W., Bowes, D., *et al.* Postnatal growth of the mouse lung. *J Anat* **124**(Pt 1), 131–151 (1977).
12. Zeltner, T.B. and Burri, P.H. The postnatal development and growth of the human lung. II. Morphology. *Respir Physiol* **67**(3), 269–282 (1987).
13. Hind, M., Corcoran, J., *et al.* Temporal/spatial expression of retinoid binding proteins and RAR isoforms in the postnatal lung. *Am J Physiol Lung Cell Mol Physiol* **282**(3), L468–L476 (2002).
14. McGowan, S.E., Harvey, C.S., *et al.* Retinoids, retinoic acid receptors, and cytoplasmic retinoid binding proteins in perinatal rat lung fibroblasts. *Am J Physiol* **269**(4 Pt 1), L463–L472 (1995).
15. Ong, D.E. and Chytil, F. Changes in levels of cellular retinol and retinoic acid binding proteins of the liver and lung during perinatal development. *Proc Natl Acad Sci USA* **73**, 2 (1976).

16. Massaro, D., Teich, N., Maxwell, S., Massaro, G.D., and Whitney, D. Postnatal development of pulmonary alveoli: regulation and evidence for a "critical period" in rats. *J Clin Invest* **76**, 1297–1305 (1985).
17. Whitney, D., Massaro, G.D., Massaro, D., and Clerch, L. Gene expression of retinoid binding proteins; modulation by retinoic acid and dexamethasone in neonatal rat lung. *Pediatr Res* **45**(1), 2–7 (1999).
18. Hind, M., Corcoran, J., *et al.* Alveolar proliferation, retinoid synthesizing enzymes, and endogenous retinoids in the postnatal mouse lung. Different roles for Aldh-1 and Raldh-2. *Am J Respir Cell Mol Biol* **26**(1), 67–73 (2002).
19. Okabe, T., Yorifuji, H., Yamada, E., and Takaku, F. Isolation and characterization of Vitamin-A-storing lung cells. *Exp Cell Res* **154**, 125–135 (1984).
20. Dirami, G., Massaro, G.D., *et al.* Lung retinol storing cells synthesize and secrete retinoic acid, an inducer of alveolus formation. *Am J Physiol Lung Cell Mol Physiol* **286**(2), L249–L256 (2004).
21. Chytil, F. Retinoids in lung development. *FASEB J* **10**(9), 986–992 (1996).
22. Cho, S.J., George, C.L., Snyder, J.M., and Acarregui, M.J. Retinoic acid and erythropoietin maintain alveolar development in mice treated with an angiogenesis inhibitor. *Am J Respir Cell Mol Biol* **33**, 622–628 (2005).
23. Nabeyrat, E., Besnard, V., *et al.* Retinoic acid induced proliferation of lung alveolar epithelial cells: relation with the IGF system. *Am J Physiol Lung Cell Mol Physiol* **275**, L71–L79 (1998).
24. Liebeskind, A., Srinivasan, S., *et al.* Retinoic acid stimulates immature lung fibroblast growth via a PDGF-mediated autocrine mechanism. *Am J Physiol Lung Cell Mol Physiol* **279**(1), L81–L90 (2000).
25. Liu, B., Harvey, C.S., and McGowan, S.E. Retinoic acid increases elastin in neonatal rat lung fibroblast cultures. *Am J Physiol* **265**, 430–437 (1993).
26. McGowan, S.E., Doro, M.M., *et al.* Endogenous retinoids increase perinatal elastin gene expression in rat lung fibroblasts and fetal explants. *Am J Physiol* **273**(2 Pt 1), L410–L416 (1997).
27. Massaro, G.D. and Massaro, D. Postnatal treatment with retinoic acid increases the number of pulmonary alveoli in rats. *Am J Physiol* **270**(2 Pt 1), L305–L310 (1996).
28. Hind, M. and Maden, M. Retinoic acid induces alveolar regeneration in the adult mouse lung. *Eur Respir J* **23**(1), 20–27 (2004).

29. Yang, L., Naltner, A., *et al.* Overexpression of a dominant negative retinoic acid receptor alpha acuses alveolar abnormality in transgenic neonatal lungs. *Endocrinology* **144**, 3004–3011 (2003).
30. Pasutto, F., Sticht, H., *et al.* Mutations in STRA6 cause a broad spectrum of malformations including anophthalmia, congenital heart defects, diaphragmatic hernia, alveolar capillary dysplasia, lung hypoplasia, and mental retardation. *Am J Hum Genet* **80**(3), 550–560 (2007).
31. Wolbach, S.B. and Howe, P.R. Tissue changes following deprevation of fat-soluble A vitamin. *J Exp Med* **42**, 753–757 (1925).
32. Wolbach, S.B. and Howe, P.R. Epithelial repair and recovery from Vitamin A deficiency. *J Exp Med* **57**, 25 (1932).
33. Baybutt, R.C., Hu, L., *et al.* Vitamin A deficiency injures lung and liver parenchyma and impairs function of rat type II pneumocytes. *J Nutr* **130**(5), 1159–1165 (2000).
34. Maden, M. and Hind, M. Retinoic acid in alveolar development, maintenance and regeneration. *Philos Trans R Soc Lond B Biol Sci* **359**(1445), 799–808 (2004).
35. McGowan, S.E. and Holmes, A.J. Vitamin A deficiency alters pulmonary parenchymal collagen and tissue mechanics. *Respir Physiol Neurobiol* **156**(3), 312–319 (2007).
36. Morabia, A., Menkes, M.J., Comstock, G.W., and Tockman, M.S. Serum retinol and airway obstruction. *Am J Epidemiol* **132**, 77–82 (1990).
37. Guenegou, A., Leynaert, B., *et al.* Serum carotenoids, Vitamins A and E, and 8 year lung function decline in a general population. *Thorax* **61**(4), 320–326 (2006).
38. Morabia, A., Sorenson, A., Kumanyika, S.K., Abbey, H., Cohen, B.H., and Chee, E. Vitamin A, cigarette smoking, and airway obstruction. *Am Rev Respir Dis* **140**, 1312–1316 (1989).
39. Tenney, S.M. and Remmers, J.E. Comparative quantitative morphology of the mammalian lung diffusion area. *Nature* **197**, 2 (1963).
40. Weibel, E.R., Taylor, C.R., *et al.* The concept of symmorphosis: a testable hypothesis of structure-function relationship. *Proc Natl Acad Sci USA* **88**(22), 10357–10361 (1991).
41. Massaro, D. and Massaro, G.D. Invited review: pulmonary alveoli: formation, the "call for oxygen," and other regulators. *Am J Physiol Lung Cell Mol Physiol* **282**(3), L345–L358 (2002).

42. Gonzales-Pacheco, D.M., Buss, W.C., *et al.* Energy restriction reduces metabolic rate in adult male Fisher-344 rats. *J Nutr* **123**(1), 90–97 (1993).
43. Munch, I.C., Markussen, N.H., *et al.* Resting oxygen consumption in rats during food restriction, starvation and refeeding. *Acta Physiol Scand* **148**(3), 335–340 (1993).
44. Sahebjami, H. and Wirman, J.A. Emphysema-like changes of starved rats. *Am Rev Respir Dis* **124**, 619–624 (1981).
45. Harkema, J.R., Mauderly, J.L., Gregory, R.E., and Pickerell, J.A. A comparison of starvation and elastase models of emphysema in the rat. *Am Rev Respir Dis* **129**, 584–591 (1984).
46. Kerr, J.S., Riley, D.J., Lanza-Jacoby, S., Berg, R.A., Spilker, W.C., and Yu, S.Y. Nutritional emphysema in the rat. Influence of protein depletion and impaired lung growth. *Am Rev Respir Dis* **131**, 644–650 (1985).
47. Karlinsky, J.B., Goldstein, R.H., Ojserkis, B., and Snider, G.L. Lung mechanics and connective tissue changes in starvation-induced emphysema in hamsters. *Am J Physiol* **251**, R282–R288 (1986).
48. Massaro, D., Massaro, G.D., *et al.* Calorie-related rapid onset of alveolar loss, regeneration, and changes in mouse lung gene expression. *Am J Physiol Lung Cell Mol Physiol* **286**(5), L896–L906 (2004).
49. Massaro, D., Alexander, E., *et al.* Rapid onset of gene expression in lung, supportive of formation of alveolar septa, induced by refeeding mice after calorie restriction. *Am J Physiol Lung Cell Mol Physiol* **292**(5), L1313–L1326 (2007).
50. Lieberman, J. The ABCs of granule-mediated cytotoxicity: new weapons in the arsenal. *Nat Rev Immunol* **3**(5), 361–370 (2003).
51. Saetta, M. Airway inflammation in chronic obstructive pulmonary disease. *Am J Respir Crit Care Med* **160**(5 Pt 2), S17–S20 (1999).
52. Winick, M. *Hunger Disease. Studies by the Jewish Physicians in the Warsaw Ghetto* (New York, John Wiley and Sons, 1979).
53. Coxson, H.O., Chan, I.H., *et al.* Early emphysema in patients with anorexia nervosa. *Am J Respir Crit Care Med* **170**(7), 748–752 (2004).
54. Cook, V.J., Coxson, H.O., *et al.* Bullae, bronchiectasis and nutritional emphysema in severe anorexia nervosa. *Can Respir J* **8**(5), 361–365 (2001).
55. Massaro, G.D., Mortola, J.P., *et al.* Sexual dimorphism in the architecture of the lung's gas-exchange region. *Proc Natl Acad Sci USA* **92**(4), 1105–1107 (1995).

56. Massaro, G.D., Mortola, J.P., *et al.* Estrogen modulates the dimensions of the lung's gas-exchange surface area and alveoli in female rats. *Am J Physiol* **270**(1 Pt 1), L110–L114 (1996).
57. Patrone, C., Cassel, T.N., *et al.* Regulation of postnatal lung development and homeostasis by estrogen receptor beta. *Mol Cell Biol* **23**(23), 8542–8552 (2003).
58. Massaro, D. and Massaro, G.D. Estrogen regulates pulmonary alveolar formation, loss, and regeneration in mice. *Am J Physiol Lung Cell Mol Physiol* **287**(6), L1154–L1159 (2004).
59. Massaro, D. and Massaro, G.D. Estrogen receptor regulation of pulmonary alveolar dimensions: alveolar sexual dimorphism in mice. *Am J Physiol Lung Cell Mol Physiol* **290**(5), L866–L870 (2006).
60. Massaro, D., Clerch, L.B., *et al.* Estrogen receptor-alpha regulates pulmonary alveolar loss and regeneration in female mice: morphometric and gene expression studies. *Am J Physiol Lung Cell Mol Physiol* **293**, L222–L228 (2007).
61. Neas, L.M. and Schwartz, J. The determinants of pulmonary diffusing capacity in a national sample of U.S. adults. *Am J Respir Crit Care Med* **153**(2), 656–664 (1996).
62. Kasch, F.W., Boyer, J.L, *et al.* (1999). Ageing of the cardiovascular system during 33 years of aerobic exercise. *Age Ageing* **28**(6), 531–536.
63. Carlson, C.L., Cushman, M., *et al.* Hormone replacement therapy is associated with higher FEV1 in elderly women. *Am J Respir Crit Care Med* **163**(2), 423–428 (2001).
64. Pata, O., Atis, S., *et al.* The effects of hormone replacement therapy type on pulmonary functions in postmenopausal women. *Maturitas* **46**(3), 213–218 (2003).
65. Strinic, T. and Eterovic, D. Oral contraceptives improve lung mechanics. *Fertil Steril* **79**(5), 1070–1073 (2003).
65. von Hertzen, L., Reunanen, A., *et al.* Airway obstruction in relation to symptoms in chronic respiratory disease — a nationally representative population study. *Respir Med* **94**(4), 356–363 (2000).
67. Miravitlles, M., Ferrer, M., *et al.* Characteristics of a population of COPD patients identified from a population-based study. Focus on previous diagnosis and never smokers. *Respir Med* **99**(8), 985–995 (2005).

68. Goss, R.J. *Principles of Regeneration* (New York, Academic Press, 1969).
69. Maden, M., Sonneveld, E., *et al.* The distribution of endogenous retinoic acid in the chick embryo: implications for developmental mechanisms. *Development* **125**(21), 4133–4144 (1998).
70. Scadding, S.R. and Maden, M. Retinoic acid gradients during limb regeneration. *Dev Biol* **162**(2), 608–617 (1994).
71. Brockes, J.P. Introduction of a retinoid reporter gene into the urodele limb blastema. *Proc Natl Acad Sci USA* **89**, 1386–1390 (1992).
72. Maden, M. Retinoids as endogenous components of the regenerating limb and tail. *Wound Repair Regen* **6**(4), 358–365 (1998).
73. Allen, S.P., Maden, M., *et al.* A role for retinoic acid in regulating the regeneration of deer antlers. *Dev Biol* **251**(2), 409–423 (2002).
74. Maden, M. and Hind, M. Retinoic acid, a regeneration-inducing molecule. *Dev Dyn* **226**(2), 237–244 (2003).
75. Massaro, G.D. and Massaro, D. Retinoic acid treatment abrogates elastase-induced pulmonary emphysema in rats. *Nat Med* **3**(6), 675–677 (1997).
76. Karlinsky, J.B. and Snider, G.L. Animal models of emphysema. *Am Rev Respir Dis* **117**(6), 1109–1133 (1978).
77. Belloni, P.N., Garvin, L., *et al.* Effects of all-trans-retinoic acid in promoting alveolar repair. *Chest* **117**, 235–241 (2000).
78. Tepper, J., Pfeiffer, J., Aldrich, M., Tumas, D., Kern, J., Hoffman, E., McLennan, G., and Hyde, D. Can retinoic acid ameliorate the physiologic and morphologic effects of elastase instillation in the rat? *Chest* **117**(Suppl), 2 (2000).
79. Ishizawa, K., Kubo, H., Yamada, M., Kobayashi, S., Numasaki, M., Ueda, S., Suzuki, T., and Sasaki, H. Bone marrow derived cells contribute to lung regeneration after elastase-induced pulmonary emphysema. *FEBS Lett* **556**, 249–252 (2004).
80. Nakajoh, M., Fukushima, T., *et al.* Retinoic acid inhibits elastase-induced injury in human lung epithelial cell lines. *Am J Respir Cell Mol Biol* **28**(3), 296–304 (2003).
81. Green, M.C., Sweet, H.O., and Bunker, L.E. Tight-skin, a new mutation of the mouse causing excessive growth of connective tissue and skeleton. *Am J Pathol* **82**, 493–512 (1976).

82. Siracusa, L.D., McGrath, R., Ma, Q., Moskow, J.J., Manne, J., Christner, P.J., Buchberg, A.M., and Jimenez, S.A. A tandem duplication within the fibrillin 1 gene is associated with the mouse tight skin mutation. *Genome Res* **6**, 300–313 (1996).
83. Szapiel, S.V., Fulmer, J.D., *et al.* Hereditary emphysema in the tight-skin (Tsk/+) mouse. *Am Rev Respir Dis* **123**(6), 680–685 (1981).
84. Martorana, P.A., van Even, P., *et al.* A 16-month study of the development of genetic emphysema in tight-skin mice. *Am Rev Respir Dis* **139**(1), 226–232 (1989).
85. Massaro, G. and Massaro, D. Retinoic acid partially rescues failed septation in rats and in mice. *Am J Physiol Lung Cell Mol Physiol* **278**, L955–L960 (2000).
86. Garber, S.J., Zhang, H., *et al.* Hormonal regulation of alveolarization: structure-function correlation. *Respir Res* **7**, 47 (2006).
87. Maden, M. Retinoids have differing efficacies on alveolar regeneration in a dexamethasone-treated mouse. *Am J Respir Cell Mol Biol* **35**(2), 260–267 (2006).
88. Fujita, M., Shannon, J.M., *et al.* Overexpression of tumor necrosis factor-alpha produces an increase in lung volumes and pulmonary hypertension. *Am J Physiol Lung Cell Mol Physiol* **280**(1), L39–L49 (2001).
89. Fujita, M., Ye, Q., Ouchi, H., Nakashima, N., Hamada, N., Hagimoto, N., Kuwano, K., Mason, R.J., and Nakanishi, Y. Retinoic acid fails to reverse emphysema in adult mouse models. *Thorax* **59**, 224–230 (2004).
90. March, T.H., Bowen, L.E., *et al.* Effects of strain and treatment with inhaled all-trans-retinoic acid on cigarette smoke-induced pulmonary emphysema in mice. *COPD* **2**(3), 289–302 (2005).
91. Ofolue, A.F., Xiang, Y., Yang, N., and Belloni, P.N. Retinoids reverse cigarette smoke-induced emphysema in rats. *Am J Respir Crit Care Med* **165**, 137 (2002).
92. An, L., Zhang, H., *et al.* Prophylactic effect of retinoic acid on chronic obstructive pulmonary emphysema in rats. *Zhonghua Jie He He Hu Xi Za Zhi* **25**(6), 356–359 (2002).
93. Meshi, B., Vitalis, T.Z., Ionscu, D., Elliot, W.M., Liu, C., Wang, X.D., Hayashi, S., and Hogg, J.C. Emphysematous lung destruction by cigarette smoke: the effects of latent adenoviral on the inflammatory response. *Am J Respir Cell Mol Biol* **26**, 52–57 (2002).

94. Lucey, E.C., Goldstein, R.H., Breuer, R., Rexer, B.R., Ong, D.E., and Snider, G.L. Retinoic acid does not affect alveolar septation in adult FVB mice with elastase-induced emphysema. *Respiration* **70**, 200–205 (2003).
95. March, T.H., Cossey, P.Y., *et al.* Inhalation administration of all-transretinoic acid for treatment of elastase-induced pulmonary emphysema in Fischer 344 rats. *Exp Lung Res* **30**(5), 383–404 (2004).
96. Quintana, H.K., Cannet, C., *et al.* Proton MRI as a noninvasive tool to assess elastase-induced lung damage in spontaneously breathing rats. *Magn Reson Med* **56**(6), 1242–1250 (2006).
97. Nishia, Y., Boswella, V., Ansarib, T., Piprawalab, F., Satchi, S., and Page, C.P. Elastase-induced changes in lung function: relationship to morphometry and the effect of drugs. *Pulm Pharmacol Ther* **16**, 221–229 (2003).
98. Stinchcombe, S.V. and Maden, M. Retinoic acid induced alveolar regeneration: critical differences in strain sensitivity. *Am J Respir Cell Mol Biol* **38**(2), 185–191 (2008).
99. Brockes, J.P. and Kumar, A. Appendage regeneration in adult vertebrates and implications for regenerative medicine. *Science* **310**(5756), 1919–1923 (2005).
100. Fausto, N. Liver regeneration. *J Hepatol* **32**, 19–32 (2000).
101. Yen, T.H. and Wright, N.A. The gastrointestinal tract stem cell niche. *Stem Cell Rev* **2**(3), 203–212 (2006).
102. Kotton, D.N., Ma, B.Y., Cardoso, W.V., Sanderson, E.A., Summer, E.S., Williams, M.C., and Fine, A. Bone marrow derived cells as progenitors of lung alveolar epithelium. *Development* **128**, 5181–5188 (2001).
103. Krause, D.S., Theise, N.D., Collector, M.I., Henegariu, O., Hwang, S., Gardner, R., Neutzel, S., and Sharkis, S.J. Mulit-organ, multi-lineage engraftment by a single bone marrow derived stem cell. *Cell* **105**, 369–377 (2001).
104. Ortiz, L.A., Gambelli, F., *et al.* Mesenchymal stem cell engraftment in lung is enhanced in response to bleomycin exposure and ameliorates its fibrotic effects. *Proc Natl Acad Sci USA* **100**(14), 8407–8411 (2003).
105. Rojas, M., Xu, J., *et al.* Bone marrow-derived mesenchymal stem cells in repair of the injured lung. *Am J Respir Cell Mol Biol* **33**(2), 145–152 (2005).
106. Abe, S., Boyer, C., *et al.* Cells derived from the circulation contribute to the repair of lung injury. *Am J Respir Crit Care Med* **170**(11), 1158–1163 (2004).

107. Beckett, T., Loi, R., *et al.* Acute lung injury with endotoxin or NO_2 does not enhance development of airway epithelium from bone marrow. *Mol Ther* **12**(4), 680–686 (2005).
108. Xu, J., Woods, C.R., *et al.* Prevention of endotoxin-induced systemic response by bone marrow-derived mesenchymal stem cells in mice. *Am J Physiol Lung Cell Mol Physiol* **293**(1), L131–L141 (2007).
109. Loi, R., Beckett, T., *et al.* Limited restoration of cystic fibrosis lung epithelium *in vivo* with adult bone marrow-derived cells. *Am J Respir Crit Care Med* **173**(2), 171–179 (2006).
110. Theise, N.D., Henegariu, O., *et al.* Radiation pneumonitis in mice: a severe injury model for pneumocyte engraftment from bone marrow. *Exp Hematol* **30**(11), 1333–1338 (2002).
111. Herzog, E.L., Van Arnam, J., *et al.* Threshold of lung injury required for the appearance of marrow-derived lung epithelia. *Stem Cells* **24**(8), 1986–1992 (2006).
112. Suratt, B.T., Cool, C.D., *et al.* Human pulmonary chimerism after hematopoietic stem cell transplantation. *Am J Respir Crit Care Med* **168**(3), 318–322 (2003).
113. Kleeberger, W., Versmold, A., *et al.* Increased chimerism of bronchial and alveolar epithelium in human lung allografts undergoing chronic injury. *Am J Pathol* **162**(5), 1487–1494 (2003).
114. Wagers, A.J., Sherwood, R.I., *et al.* Little evidence for developmental plasticity of adult hematopoietic stem cells. *Science* **297**(5590), 2256–2259 (2002).
115. Chang, J.C., Summer, R., *et al.* Evidence that bone marrow cells do not contribute to the alveolar epithelium. *Am J Respir Cell Mol Biol* **33**(4), 335–342 (2005).
116. Kotton, D.N., Fabian, A.J., *et al.* Failure of bone marrow to reconstitute lung epithelium. *Am J Respir Cell Mol Biol* **33**(4), 328–334 (2005).
117. Harris, R.G., Herzog, E.L., *et al.* Lack of a fusion requirement for development of bone marrow-derived epithelia. *Science* **305**(5680), 90–93 (2004).
118. Herzog, E.L., Van Arnam, J., *et al.* Lung-specific nuclear reprogramming is accompanied by heterokaryon formation and Y chromosome loss following bone marrow transplantation and secondary inflammation. *FASEB J* **21**(10), 2592–2601 (2007).

119. Yuhgetsu, H., Ohno, Y., *et al.* Beneficial effects of autologous bone marrow mononuclear cell transplantation against elastase-induced emphysema in rabbits. *Exp Lung Res* **32**(9), 13–26 (2006).
120. Shigemura, N., Okumura, M., *et al.* Autologous transplantation of adipose tissue-derived stromal cells ameliorates pulmonary emphysema. *Am J Transplant* **6**(11), 2592–2600 (2006).
121. Direkze, N.C., Forbes, S.J., *et al.* Multiple organ engraftment by bone-marrow-derived myofibroblasts and fibroblasts in bone-marrow-transplanted mice. *Stem Cells* **21**(5), 514–520 (2003).
122. Epperly, M.W., Guo, H., *et al.* Bone marrow origin of myofibroblasts in irradiation pulmonary fibrosis. *Am J Respir Cell Mol Biol* **29**(2), 213–224 (2003).
123. Hashimoto, N., Jin, H., *et al.* Bone marrow-derived progenitor cells in pulmonary fibrosis. *J Clin Invest* **113**(2), 243–252 (2004).
124. Zhao, Y.D., Courtman, D.W., *et al.* Rescue of monocrotaline-induced pulmonary arterial hypertension using bone marrow-derived endothelial-like progenitor cells: efficacy of combined cell and eNOS gene therapy in established disease. *Circ Res* **96**(4), 442–450 (2005).
125. Evans, M.J., Cabral, L.J., *et al.* Renewal of alveolar epithelium in the rat following exposure to NO_2. *Am J Pathol* **70**(2), 175–198 (1973).
126. Adamson, I.Y. and Bowden, D.H. The type 2 cell as progenitor of alveolar epithelial regeneration. A cytodynamic study in mice after exposure to oxygen. *Lab Invest* **30**(1), 35–42 (1974).
127. Vincent, R. and Adamson, I.Y. Cellular kinetics in the lungs of aging Fischer 344 rats after acute exposure to ozone. *Am J Pathol* **146**(4), 1008–1016 (1995).
128. Reddy, R., Buckley, S., *et al.* Isolation of a putative progenitor subpopulation of alveolar epithelial type 2 cells. *Am J Physiol Lung Cell Mol Physiol* **286**(4), L658–L667 (2004).
129. Adamson, I.Y. and Bowden, D.H. Derivation of type 1 epithelium from type 2 cells in the developing rat lung. *Lab Invest* **32**(6), 736–745 (1975).
130. Burri, P.H. The postnatal growth of the rat lung. 3. Morphology. *Anat Rec* **180**(1), 77–98 (1974).
131. Vaccaro, C. and Brody, J.S. Ultrastructure of the developing alveoli. I. The role of the interstitial fibroblast. *Anat Rec* **192**, 13 (1978).
132. Maksvyitis, H.J., Vaccaro, C., and Brody, J.S. The isolation and characterisation of the lipid containing interstitial cell from the developing rat lung. *Lab Invest* **45**, 248–259 (1981).

133. Sannes, P.L. Structural and functional relationships between type II pneumocytes and components of extracellular matrices. *Exp Lung Res* **17**(4), 639–659 (1991).
134. Nabeyrat, E., Corroyer, S., *et al.* Retinoic acid protects against hyperoxia-mediated cell-cycle arrest of lung alveolar epithelial cells by preserving late G1 cyclin activities. *Am J Respir Cell Mol Biol* **25**(4), 507–514 (2001).
135. Besnard, V., Nabeyrat, E., *et al.* Protective role of retinoic acid from antiproliferative action of TNF-alpha on lung epithelial cells. *Am J Physiol Lung Cell Mol Physiol* **282**(4), L863–L871 (2002).
136. Mouhieddine, O.B., Cazals, V., *et al.* Insulin-like growth factor-II (IGF-II), type 2 IGF receptor, and IGF-binding protein-2 gene expression in rat lung alveolar epithelial cells: relation to proliferation. *Endocrinology* **135**(1), 83–91 (1994).
137. Kitamura, K., Kangawa, K., *et al.*. Adrenomedullin: a novel hypotensive peptide isolated from human pheochromocytoma. *Biochem Biophys Res Commun* **192**(2), 553–560 (1993).
138. Ishiyama, Y., Ohgi, T., *et al.* Pheochromocytoma associated with nocturnal hypertension. *Intern Med* **32**(10), 781–783 (1993).
139. Ichiki, Y., Kitamura, K., *et al.* Distribution and characterization of immunoreactive adrenomedullin in human tissue and plasma. *FEBS Lett* **338**(1), 6–10 (1994).
140. Sakata, J., Shimokubo, T., *et al.* Distribution and characterization of immunoreactive rat adrenomedullin in tissue and plasma. *FEBS Lett* **352**(2), 105–108 (1994).
141. Lopez, J. and Martinez, A. Cell and molecular biology of the multifunctional peptide, adrenomedullin. *Int Rev Cytol* **221**, 1–92 (2002).
142. Martinez, A., Miller, M.J., *et al.* Adrenomedullin receptor expression in human lung and in pulmonary tumors. *J Histochem Cytochem* **45**(2), 159–164 (1997).
143. Okumura, M., Kai, H., *et al.* Adrenomedullin increases phosphatidylcholine secretion in rat type II pneumocytes. *Eur J Pharmacol* **403**(3), 189–194 (2002).
143. Kim, W., Moon, S.O., *et al.* Angiogenic role of adrenomedullin through activation of Akt, mitogen-activated protein kinase, and focal adhesion kinase in endothelial cells. *FASEB J* **17**(13), 1937–1939 (2003).

145. Nishimatsu, H., Suzuki, E., *et al.* Adrenomedullin induces endothelium-dependent vasorelaxation via the phosphatidylinositol 3-kinase/Akt-dependent pathway in rat aorta. *Circ Res* **89**(1), 63–70 (2001).
146. Kato, H., Shichiri, M., *et al.* Adrenomedullin as an autocrine/paracrine apoptosis survival factor for rat endothelial cells. *Endocrinology* **138**(6), 2615–2620 (1997).
147. Sata, M., Kakoki, M., *et al.* Adrenomedullin and nitric oxide inhibit human endothelial cell apoptosis via a cyclic GMP-independent mechanism. *Hypertension* **36**(1), 83–88 (2000).
148. Hwang, I.S., Fung, M.L., *et al.* Age-related changes in adrenomedullin expression and hypoxia-inducible factor-1 activity in the rat lung and their responses to hypoxia. *J Gerontol A Biol Sci Med Sci* **62**(1), 41–49 (2007).
149. Otto, T. and Fandrey, J. Thyroid hormone induces hypoxia-inducible factor 1alpha gene expression through thyroid hormone receptor beta/retinoid × receptor alpha-dependent activation of hepatic leukemia factor. *Endocrinology* **149**(5), 2241–2250 (2008).
150. Minamino, N., Shoji, H., *et al.* Adrenocortical steroids, thyroid hormones and retinoic acid augment the production of adrenomedullin in vascular smooth muscle cells. *Biochem Biophys Res Commun* **211**(2), 686–693 (1995).
151. Zudaire, E., Martinez, A., *et al.* Characterization of adrenomedullin in non-human primates. *Biochem Biophys Res Commun* **321**(4), 859–869 (2004).
152. Kubo, A., Minamino, N., *et al.* Production of adrenomedullin in macrophage cell line and peritoneal macrophage. *J Biol Chem* **273**(27), 16730–16738 (1998).
153. Murakami, S., Nagaya, N., *et al.* Adrenomedullin regenerates alveoli and vasculature in elastase-induced pulmonary emphysema in mice. *Am J Respir Crit Care Med* **172**(5), 581–589 (2005).
154. Roth, M.D., Connett, J.E., *et al.* Feasibility of retinoids for the treatment of emphysema study. *Chest* **130**(5), 1334–1345 (2006).

Chapter 14

Assessment of Cell Engraftment

Robert Zweigerdt and William Rust

Demonstrating that a cell therapy can directly contribute to functional recovery of a diseased organ is made difficult by experimental methods that are liable to introduce false-positive data. In this chapter we analyze the methods used to prove that a transplanted cell has grafted into the target organ and highlight the potentials for producing misleading results. Furthermore, we discuss the advantages and pitfalls of supporting graft data by quantifying improvement of organ function. We conclude that a multi-tiered experimental approach is required to thoroughly dispel the notion that engraftment and functional improvement are artefactual. Because of the relative immaturity of the field of lung cell therapy, we draw many of our examples from tissues with related challenges, such as cardiomyocyte cell therapy.

Keywords: Stem cells; labeling; organ repair; monitoring; detection of engraftment; misleading results.

Outline

1. Introduction

Two general mechanisms for the use of cells to repair organ damage have been reported, both of which have been applied to the lung. The first strategy aims at tissue replacement by providing donor cells to an affected organ. While cells might be injected directly into the tissue, the least invasive technique is to introduce suitable stem or progenitor cells into the blood stream, which harbours the capacity to home to the lung and differentiate into, for example, alveolar epithelial cells.[1] Donor cells might originate from several sources such as the bone marrow, embryonic stem cells (ESC), or from the lung itself. They can further be genetically modified to repair a defective gene or to introduce a reporter gene to enable efficient cell tracking in an experimental setup.

The second method involves the engineering of replacement tissue *in vitro*, which is then transplanted to the lung.[2] A consequent long-term goal in the field is the engineering of a total bioartificial lung *in vitro* which can be used for surgical whole organ replacement.[3] However, in the current studies of cell or tissue transplantation, one major requirement is the functional integration of the replacement tissue into the complex organisation of the existing lung and to ensure efficient connection to the host vasculature. Furthermore, a significant challenge to researchers in the field is to demonstrate that the replacement tissue can not only integrate into the recipient's lung but can do so to a level sufficient to improve its function.

Demonstrating that a cell therapy can directly contribute to functional recovery of a diseased lung is not a trivial issue. Polymerase chain reaction (PCR)-based detection methods are insufficient to demonstrate successful organ engraftment. Methods of visual quantification of cellular differentiation and engraftment in biopsied samples are the most widely used. However, these methods are prone to introduce false positives and bias, thereby demanding demonstration of functional improvement in parallel (see Chapter 5). Assessment of functional recovery carries its own pitfalls as it can result from secondary effects of the transplant not related to long-term functional integration. It is therefore important to pay close attention to experimental design and to use multiple methods of evaluation.

In this chapter we will discuss the advantages and shortcomings of various methods of assessing donor cell engraftment in several organs. Despite the medical need, tissue engineering and cell therapy for lung repair is in its infancy. This might be due to the complexity of the organ, which consists of over 40 distinct cell types in the respiratory tract. The alveolar wall itself contains fibroblasts, myofibroblasts, and capillary endothelial cells embedded in an extracellular matrix. The alveolar epithelium is formed from two types of pneumocytes, type I which comprises 90% of the surface epithelial layer, and type II, the putative stem cells. Limited availability of replacement cells for lung repair is another reason for the early stage of the field. For example, the capacity of murine ESC to differentiate into airway epithelia containing both ciliated pneumocytes and Clara cells (lining the bronchiole) was for the first time reported in 2005.[4] In contrast, mouse ESC-derived cardiomyocytes were isolated and characterised in the early 90s[5] and refinements to their production and use have made great progress.[6–8] Availability of suitable donor cells from ESC or adult tissues has triggered early research on cell-based tissue repair in the heart,[9] the brain,[10] and the liver.[11] Given the similarity of challenges faced by researchers endeavouring to design cell therapies for these organs and for the lung, we will be drawing examples from these earlier areas of research.

2. Visual Detection of Engrafted Cells — Mechanisms Underlying Misleading Results

(See also Chapter 5)

Perhaps the most widely employed method to detect engraftment of transplanted cells is to introduce a donor population which is genetically marked to be easily identifiable post-engraftment. Male bone marrow-derived murine stem cells injected into female mice models of lung injury have been detected via fluorescent *in situ* hybridisation of the Y chromosome in lung explants.[12–15] These cells displayed the morphology of the alveolar epithelium and expressed tissue specific markers, demonstrating that cells originating from the bone marrow could integrate and differentiate into lung tissue. Murine bone marrow-derived stem cells genetically modified to constitutively express eGFP or LacZ have also been detected in the murine host lung following lung injury and displayed morphology of the alveolar epithelial cells.[16,17] Although not reported in these studies, functional recovery was suggested by the demonstration that donor cells were recruited more vigorously to damaged lungs and made a significant contribution (up to 20%) of the alveolar epithelium. However, a chimeric mouse model, wherein the bone marrow was depleted and replaced with transplanted cells that harboured an eGFP construct driven by an alveolar epithelium-specific promoter, failed to produce any GFP fluorescing cells in the alveolar epithelium, even after lung injury.[18,19] A detailed analysis of the generation of false positives in models wherein the donor tissue constitutively expressed a visual marker demonstrated that marked cells were either juxtaposed over true alveolar epithelium or retained stem cell morphology.[18,20] These results not only draw into question whether bone marrow-derived stem cells can contribute to the repair of a damaged lung, but also challenge the underlying notion that bone marrow-derived stem cells differentiate to alveolar epithelium *in vivo*.

Although not addressed in these studies, another source of false-positive observations is the occurrence of cell fusion. A progenitor cell can easily be construed to have differentiated into a functional cell type as a result of fusion to an existing cell in the host organ. Following reports that stem cells residing in one tissue may have the capacity to produce differentiated cell types for other tissues (a phenomenon often referred to as transdifferentiation or stem cell plasticity), unequivocal evidence of spontaneous fusion *in vitro* suggested that this mechanism might also underlie "transdifferentiation" events *in vivo*. Findings in cell therapy studies aiming at the repair of several organs including liver, heart and lung have been challenged by the claims of cell fusion. These have been recently reviewed for the heart,[21] the liver,[22] and the lung.[23]

Although cell fusion is presented here as an additional factor in the analysis of cellular engraftment, it can be viewed as a potential aid to cell therapy. Notably, cell fusion seems to take place with particularly high incidence in the liver. The combined cell, or heterokaryon, contains new genetic material which can provide a selective advantage over resident hepatocytes that stimulate proliferation and liver repopulation. Therefore genetic deficiencies might be a specific target for cell fusion therapies.[24]

However, methodology to eliminate false positives due to cellular fusion was shown in a study of liver cell therapy employing xenogenic transplantation.[25] Human hepatocytes were transplanted into mice spleen.

Human and mice hepatocytes were subsequently detected in the liver using mouse and human-specific pan-centromeric probes. Double fluorescing cells which also overlapped with DAPI labelling were products of fusion while single fluorescing; DAPI-labelled cells were not. Furthermore, the mice were deficient for the liver specific enzyme Fah (fumarylacetoacetate hydrolase), and thus human hepatocytes were discerned by the presence of human FAH.

In an example drawn from the field of cardiac cell therapy aimed at explicit demonstration that bone marrow (BM)-derived cells can acquire cardiomyogenic phenotypes by transdifferentiation, researchers constructed an experimental design wherein male donor cells from three types of genetically engineered mice strains were transplanted to injured female host hearts.[26] eGFP was either constitutively expressed in all transplanted bone marrow cells (β-actin promoter-eGFP), or turned on only in cells that had adopted a cardiac myocyte fate (cardiac α-MHC promoter-eGFP). In the third design, c-myc was expressed and targeted to the nuclei in donor cells upon differentiation into cardiac myocytes (cardiac α-MHC promoter-c-myc-nuc-Act). The donor origin of cells injected into the heart was thus determined by expression of eGFP, by nuclear akt tagged with c-myc, and by the presence of the Y chromosome. That transdifferentiation and engraftment did not occur but cell fusion was demonstrated by the presence of only one set of sex chromosomes and by total DNA content. As formalin-fixed cardiac tissue has high background fluorescence, the possibility of false positives detected in tissue immunofluorescence was alleviated through the use of primary antibodies directly conjugated with fluorescent quantum dots. At the surface, this appears to be an exhaustive approach. However, the presented data lacked key controls showing that the cardiomyocyte-specific eGFP and c-myc tag was not readily expressed in the bone marrow cells when transplanted. Also, to eliminate the problem of autofluorescence, DAB immunostaining and light microscopy with adequate negative controls should have been included.

Thus, the cardiomyogenic potential of adult BM cells after directly transplantation into ischemically injured heart remains a controversial issue. To address this question, several populations of BM cells (low-density mononuclear, c-kit-enriched, and highly enriched lineage (–)

c-kit(+)) were obtained from adult transgenic mice ubiquitously expressing eGFP, and injected around the infarcted myocardiums of nontransgenic mice.[27] The ability of transplanted BM cells to develop intracellular calcium transients in response to membrane depolarisation *in situ* was studied. After 9–10 days the mice were killed and the hearts removed, perfused, loaded with the calcium-sensitive fluorophore rhod-2, and subjected to two-photon laser scanning fluorescence microscopy (TPLSM) to monitor action potential-induced Ca(2+) ion transients in eGFP-expressing donor-derived cells and non-expressing host cardiomyocytes. Bone marrow-derived cells were located in clusters throughout the scar or peri-infarct zone. However, whereas spontaneous and electrically evoked calcium transients were found to occur synchronously in host cardiomyocytes along the graft–host border and in areas remote from the infarct, they were absent in all of the >3,000 imaged BM-derived cells. Functional, electrophysiological coupling with host cardiomyocytes and calcium transients that are synchronous with those of neighbouring host cardiomyocytes are a key feature of transplanted, *bona fide* donor cardiomyocytes.[28] Therefore, it was concluded that engrafted BM-derived cells lack attributes of functioning cardiomyocytes, calling into question the concept that adult BM cells can give rise to substantive cardiomyocyte regeneration within the infarcted heart.

There are still other mechanisms over and above cell fusion whereby the donor cell would express proteins specific to the host organ without actually constituting functional cell engraftment. For example, recent *in vitro* co-culture studies of human mesenchymal stem cells (MSC) with rat cardiomyocytes analyzed by light, fluorescence, and scanning electron microscopy revealed intercellular contacts between both cell types via nanotube formation and transient exchange of cytosolic elements.[29] The transport of cytosolic material had no specific direction but resulted in the expression of human-specific myosin in MSC. Cell-to-cell connections by nanotube formation was also described in co-culture experiments of cardiomyocytes with endothelial cells.[30] These observations suggest another, fusion-independent mechanism by which the expression of some cardiomyocyte markers can be induced in non-myogenic cells. Potentially, the transport of intracellular elements to MSC (and possibly other donor cells) might induce their differentiation. However it is more likely to be a mechanism leading to false-positive results.

Phagocytosis was also described as a mechanism by which the immunoreactivity to reporter genes and myocyte markers can be induced in donor cells independent of their cardiomyogenic differentiation.[31] In this study a sophisticated lineage tracing approach was used to test the cardiomyogenic differentiation potential of cardiac explant-derived cells (EDCs). EDCs are a proliferation-competent fraction of the adult heart tissue purported to be the resident stem cell, homologous to the type II pneumocytes of the lung epithelium.[32–34] They emerge as round, "phase contrast bright", clonogenic and expandable cells after some weeks in cell culture, and express GATA-4 mRNA, a transcription factor known to be expressed during cardiomyogenic differentiation. Following transplantation in experimental models of myocardial infarction, differentiation of EDCs into cardiac myocytes accompanied by an improvement in cardiac function has been claimed.[32,34]

However, Shenje and co-workers employed a binary, conditional, cardiac-restricted transgenic reporter system, the double heterozygous MLC2v-Cre/ZEG reporter mouse, for cardiac-specific lineage tracing of EDCs.[31] In this model, Cre recombinase is expressed in cells which differentiate into ventricular cardiomyocytes. In consequence, a loxP-flanked β-geo transgene (LacZ gene) is deleted and GFP expression is activated exclusively in ventricular myocytes. On the other hand, all non-myocytes, in principle, are targeted to express LacZ. Cardiac explants from MLC2v-Cre/ZEG reporter mice were cultured and EDCs were successfully derived *in vitro*. Interestingly, GFP immunoreactivity and epifluorescence were identified in EDCs, suggesting their cardiomyogenic differentiation; but Cre recombinase-based deletion of the LacZ gene did not occur, bringing myogenic differentiation into question. Ultrastructural analysis by transmission electron microscopy revealed structures suggestive of active phagocytes in EDCs. Thus, the presence of GFP as well as vimentin and sarcomeric actinin protein (cardiac markers) appeared to result from phagocytotic activity rather than from myogenic differentiation. Although phagocytosis offers a ready explanation for the presence of GFP epiflourescence without recombination-based LacZ gene deletion, the unexpected absence of LacZ activity in EDCs, observed in parallel, is challenging. Missing LacZ activity in ZEG reporter mouse lines has previously been described in the liver and lungs, suggesting it may be an occasional phenomenon related to the poor expression of prokaryotic reporter genes in some eukaryotic cell types.[35] However, EDCs also failed to differentiate into functional cardiomyocytes three weeks after transplantation into injured hearts (shown by the lack of intracellular calcium transients). In summary, this and related studies documenting the absence of cardiomyogenic differentiation of MSC, despite some "myocyte-specific" gene transcription,[36,37] showed that: (1) transcriptional induction of a limited number of target genes is insufficient evidence of lineage induction; and (2) the use of sophisticated reporter gene lines for lineage tracing can generate misleading results. Therefore, combined analytical methodology such as qRT-PCR, immunohistochemistry, electron microscopy, and functional *in situ* analysis are needed to avoid misrepresentation of functional cell engraftment.

In vivo differentiation of BM-derived cells into lung epithelial cells in animal models of lung injury has also been reported. But concerns as to whether fusion, rather than transdifferentiation, was the mechanism behind these observations were raised.[23] To test the hypothesis that "damage" signals released from disrupted lung tissue can induce expression of lung-related genes in BM-derived cells, experiments were performed *in vitro* in a manner designed to explicitly exclude cell fusion events.[38] In the presence of disrupted lung tissue (but not in medium and liver tissue controls) BM-derived haematopoietic progenitor cells differentiated into suspension and adherent populations with dendritic cell- and Langerhans cell-like characteristics. Cytokeratins 7 and 18 and surfactant protein B mRNA expression (markers of pneumonocytes), were upregulated or induced in the dendritic cell-like population only. These experiments provide evidence for a fusion or cell–cell contact-independent mechanism capable of inducing expression of specific lineage markers in BM-derived cells, possibly in response to soluble signals. The study supports the notion that supporting *in vivo* assessment of cell engraftment and differentiation with *in vitro* analysis leads to a clearer picture. However, in light of the above discussion on the overrated potential of stem cells to differentiate into cardiomyocytes, the expression of a limited number of lineage-restricted genes is not sufficient to claim faithfully the formation of a fully functional, tissue-specific cell type.

3. *In Vivo* Monitoring of Cell Engraftment and Assessment of Functional Improvement

(See also Chapter 15)

In disease models of the lung, functional improvement is most easily recorded *in vivo* by demonstration of prolonged and increased survival. However, in a model of engineered lung tissue, functional improvement could be assessed *in vitro* through demonstration that transplanted grafts has integrated with the host vasculature, thereby permitting gas exchange through the grafted tissue.[2] Murine foetal pulmonary cells seeded in a three-dimensional plug of Matrigel and collagen developed alveolar-like structures that were infiltrated with host vasculature after transplantation into the anterior abdominal wall of host animals. To demonstrate that host vasculature had been recruited to supply the transplanted tissue, recipient mice were injected through the tail vein with FITC conjugate dextran. Subsequent immunohistochemical staining of the engrafted tissue demon strated the presence of dextran-labelled vessels connected to the host vasculature. To determine whether immature endothelial cells present within the foetal pulmonary cell preparation contributed to vessel formation, the donor tissue was pre-labelled with a fluorescent dye. Merged images of the cell-tracking dye and the FITC-dextran demonstrated that donor-derived vasculature interconnected with host vasculature, or developing host blood vessels recruited donor-derived endothelial cells.

Cell therapy for the purpose of treating type 1 diabetes is unique in that there is generally no need for cellular engraftment to achieve functional recovery. The beta cell that responds to glucose in the bloodstream with the release of insulin can function wherever there is an adequate blood supply. For this reason, islet transplantation is not discussed here. However, another avenue of cellular therapy is the transplantation of pancreatic primordia with the idea that this immature tissue will mature into functional insulin secreting islets *in situ*.[39] In this case, engraftment of the transplanted tissue must occur in a manner that permits maturation of the islets, and functional engraftment can be determined *in vitro* through the detection of insulin and insulin-producing beta cells from the graft. In one example, pig pancreatic primordial was transplanted into the mesentery of macaque models of type 1 diabetes.[40] Long-term engraftment of pig beta cells was detected in the liver, pancreas, and mesenteric lymph nodes using immunohistochemistry, RT-PCR and *in situ* hybridisation for porcine proinsulin mRNA, and electron microscopy.

Similar to the transplantation of functional islets, bone marrow transplantation has an easier functional output. If sufficient time has elapsed after transplantation to rule out detection of transplanted peripheral blood or haematopoietic progenitors, all bone marrow-derived cells of donor origin detected in the peripheral blood or in the marrow should derive from functionally engrafted bone marrow stem cells. Detection of donor cells can be accomplished by flow cytometry or PCR of bone marrow or peripheral blood samples.[41] Similarly, functional engraftment of hepatocytes in a liver cell therapy model can be determined simply by the expression of secreted components such as serum albumin and plasma C3a which originated from the donor.[25]

Other sophisticated methods have been developed to demonstrate functional improvement *in vivo* or *in situ* by cardiac cell therapy. It is relatively straightforward to understand how cellular transplantation can improve cardiac function if a significant number of donor cells survive and directly participate in a functional syncytium via electrophysiological and mechanical coupling with the host myocardium. As outlined above, Rubart and Field have developed a method based on two-photon molecular excitation laser scanning microscopy to monitor calcium transients in isolated, perfused rodent hearts *in situ*.[42] The method was used to show that calcium transients in GFP-labelled donor cardiomyocytes were synchronous with and had kinetics indistinguishable from those of neighbouring host cardiomyocytes,[28] providing unequivocal evidence that *bona fide* cardiomyocytes do functionally integrate into recipient hearts. Subsequently, the method has been applied to assess functional integration of several other donor cell types transplanted to injured hearts, including haematopoietic stem cells,[43] skeletal myoblasts,[44] cardiac explant-derived cells[31] and adult bone marrow cells.[27]

However, assuming that *bona fide* cardiomyocytes could functionally integrate and replace loss of contractile tissue, improvement of heart function in animal models is currently hampered by the low efficiency of donor cardiomyocyte implantation.[21,45] This is illustrated in studies using Magnet Resonance Imaging (MRI) and echocardiography to repeatedly measure overall heart function (i.e. left ventricular ejection fraction) for several weeks in ischemia-damaged mouse or rat hearts receiving injection of human ESC-derived cardiomyocytes. At termination, immunohistology-based morphometric analysis was performed to determine donor cell numbers and their tissue distribution.

Laflamme and co-workers found that transplantation of cardiomyocytes blocked the progressive decrease in cardiac function at four weeks post-infarction, the latest point in time analyzed.[46] In contrast, van Laake and colleagues[47] followed the long-term survival of cardiomyocytes for up to 12 weeks. Similarly to the rat study, MRI assessment showed an attenuated and progressive decrease in cardiac function at four weeks post-infarction. These findings are exciting as the cardioprotective effect appeared to be cardiomyocyte-dependent. However, quantitative morphometry revealed an unexpectedly small graft size in both studies (i.e. average long-term transplantation efficiency of only 2.3% was documented in mice), casting doubt over the substantial contribution of donor cardiomyocytes to heart function by physiological coupling to the host heart. Subsequently, in the mouse study, significant functional improvement observed at four weeks post-MI completely ceased upon reassessment at 12 weeks. While the mechanism(s) underlying the transient benefit is presently unclear, it seems likely that for long-term improvement, larger grafts consisting of functional and physiological integrated cardiomyocytes would be required.

Another problem relates to the suitability of rodent models to assess human cardiomyocyte function. It is questionable whether human ESC-derived cardiomyocytes would functionally couple long-term to adult rodent cardiomyocytes beating at rates typical of intact rodent hearts (~400–600 beats per minute versus ~60–100 in human). While sustained slow pacing is known to induce heart failure in large mammals, prolonged rapid pacing could potentially underlie the loss of graft efficacy at the

12-week time point observed in the functional mouse study by van Laake and co-worker.[47]

Although the low engraftment efficiency of donor cardiomyocytes, at present, is inadequate to significanlty increase the pump performance of damaged hearts, the functional coupling of relatively small donor cell numbers (i.e. engraftment of 3,000–20,000 mouse embryonic cardiomyocytes per mouse heart; Roell *et al.*[48]) to the host myocardium has the potential to reduce life-threatening ventricular dysrhythmias, which are the main cause of sudden death in patients after myocardial infarction. This has recently been demonstrated[48] by applying a novel method to track electrophysiological coupling of donor cells that have been engineered to express the fluorescent calcium ion indicator GCaMP2 in mouse hearts *in vivo*.

However, morphometric quantification of donor cell engraftment is cumbersome and time consuming. Ideally, the best methodology should be a quantitative assessment of donor cell survival in living animals combined with functional monitoring of organ performance. A comparative study to assess the heart function in rodent models of acute myocardial infarction revealed that MRI imaging was superior to other modalities such as positron emission tomography (PET) and single-photon emission computed tomography (SPECT), due to its greater spatial resolution and ability to detect necrotic myocardium directly.[49] MRI also has the advantage for concurrent monitoring both heart function and implantation of iron oxide-labelled donor cells in living animals. It has been shown that MRI provides a relative high resolution detection of cell location in the implanted organ. On the other hand, the MRI signal correlates poorly with the number of live donor cells; the iron oxide label poorly reflects no changes such as donor cell death or cell proliferation *in vivo*. By contrast, bioluminescence imaging, which can be used to track engineered donor cells expressing firefly luciferase, correlates well with cell integrity. This has been demonstrated by monitoring luminescence intensity *in vivo* followed by post mortem immunohistochemistry of human ESC transplanted to the heart, which were double-labelled with luciferase and GFP.[50] These data suggest that parallel use of MRI and bioluminescence imaging, is an ideal methodology to monitor and optimise cell transplantation in the heart and other organs.

4. Conclusion and the Way Forward

This brief review of the progress made in assessing *in vitro* and *in vivo* cell engraftment to the lung, heart and others highlights the following issues. First, detection of visually labelled cells as a criterion for engraftment in models of cell therapy can lead to false positive results. In fact, the use of fluorescent tracers in transplanted tissue has led to false positive results due to inappropriate immunohistochemistry protocols and cell fusion events, and other avenues of sharing cytoplasmic material between adjacent cells. It is therefore important to adopt experimental strategies to limit the possible occurrence of these challenges. Second, a truly accurate assessment of engraftment requires a parallel measure of functional improvement. However, this review provides examples that demonstrate that hasty conclusions of functional improvement, even if assessed with sophisticated, state-of-the-art methodology in the presence of adequate controls, could lead to false hopes. This could avoid false expectations of cell therapy and risk to patients' lives.

References

1. van Haaften, T. and Thebaud, B. Adult bone marrow-derived stem cells for the lung: implications for pediatric lung diseases. *Pediatr Res* **59**(4 Pt 2), 94R–99R (2006).
2. Mondrinos, M.J., *et al. In vivo* pulmonary tissue engineering: contribution of donor-derived endothelial cells to construct vascularization. *Tissue Eng* **14**(3), 361–368 (2007).
3. Nichols, J.E. and Cortiella, J. Engineering of a complex organ: progress toward development of a tissue-engineered lung. *Proc Am Thorac Soc* **5**(6), 723–730 (2008).
4. Coraux, C., *et al.* Embryonic stem cells generate airway epithelial tissue. *Am J Respir Cell Mol Biol* **32**(2), 87–92 (2005).
5. Wobus, A.M., Wallukat, G., and Hescheler, J. Pluripotent mouse embryonic stem cells are able to differentiate into cardiomyocytes expressing chronotropic responses to adrenergic and cholinergic agents and Ca^{2+} channel blockers. *Differentiation* **48**(3), 173–182 (1991).
6. Zweigerdt, R., *et al.* Generation of confluent cardiomyocyte monolayers derived from embryonic stem cells in suspension: a cell source for new therapies and screening strategies. *Cytotherapy* **5**(5), 399–413 (2003).

7. Schroeder, M., *et al*. Differentiation and lineage selection of mouse embryonic stem cells in a stirred bench scale bioreactor with automated process control. *Biotechnol Bioeng* **92**(7), 920–933 (2005).
8. Xu, X.Q., *et al*. Highly enriched cardiomyocytes from human embryonic stem cells. *Cytotherapy* **10**(4), 376–389 (2008).
9. Murry, C.E., Field, L.J., and Menasche, P. Cell-based cardiac repair: reflections at the 10-year point. *Circulation* **112**(20), 3174–3183 (2005).
10. Yu, D. and Silva, G.A. Stem cell sources and therapeutic approaches for central nervous system and neural retinal disorders. *Neurosurg Focus* **24**(3–4), E11 (2008).
11. Enns, G.M. and Millan, M.T. Cell-based therapies for metabolic liver disease. *Mol Genet Metab* **95**(1–2), 3–10 (2008).
12. Krause, D.S., *et al*. Multi-organ, multi-lineage engraftment by a single bone marrow-derived stem cell. *Cell* **105**(3), 369–377 (2001).
13. Theise, N.D., *et al*. Radiation pneumonitis in mice: a severe injury model for pneumocyte engraftment from bone marrow. *Exp Hematol* **30**(11), 1333–1338 (2002).
14. Ortiz, L.A., *et al*. Mesenchymal stem cell engraftment in lung is enhanced in response to bleomycin exposure and ameliorates its fibrotic effects. *Proc Natl Acad Sci USA* **100**(14), 8407–8411 (2003).
15. Rojas, M., *et al*. Bone marrow-derived mesenchymal stem cells in repair of the injured lung. *Am J Respir Cell Mol Biol* **33**(2), 145–152 (2005).
16. Kotton, D.N., *et al*. Bone marrow-derived cells as progenitors of lung alveolar epithelium. *Development* **128**(24), 5181–5188 (2001).
17. Jiang, Y., *et al*. Pluripotency of mesenchymal stem cells derived from adult marrow. *Nature* **418**(6893), 41–49 (2002).
18. Chang, J.C., *et al*. Evidence that bone marrow cells do not contribute to the alveolar epithelium. *Am J Respir Cell Mol Biol* **33**(4), 335–342 (2005).
19. Kotton, D.N., Fabian, A.J., and Mulligan, R.C. Failure of bone marrow to reconstitute lung epithelium. *Am J Respir Cell Mol Biol* **33**(4), 328–334 (2005).
20. Shannon, J.M., Mason, R.J., and Jennings, S.D. Functional differentiation of alveolar type II epithelial cells *in vitro*: effects of cell shape, cell-matrix interactions and cell–cell interactions. *Biochim Biophys Acta* **931**(2), 143–156 (1987).
21. Rubart, M. and Field, L.J. Cell-based approaches for cardiac repair. *Ann N Y Acad Sci* **1080**, 34–48 (2006).

22. Xu, Y.Q. and Liu, Z.C. Therapeutic potential of adult bone marrow stem cells in liver disease and delivery approaches. *Stem Cell Rev* **4**(2), 101–112 (2008).
23. Herzog, E.L. and Krause, D.S. Engraftment of marrow-derived epithelial cells: the role of fusion. *Proc Am Thorac Soc* **3**(8), 691–695 (2006).
24. Popp, F.C., *et al.* Therapeutic potential of bone marrow stem cells for liver diseases. *Curr Stem Cell Res Ther* **1**(3), 411–418 (2006).
25. Bissig, K.D., *et al.* Repopulation of adult and neonatal mice with human hepatocytes: a chimeric animal model. *Proc Natl Acad Sci USA* **104**(51), 20507–20511 (2007).
26. Rota, M., *et al.* Bone marrow cells adopt the cardiomyogenic fate *in vivo*. *Proc Natl Acad Sci USA* **104**(45), 17783–17788 (2007).
27. Scherschel, J.A., *et al.* Adult bone marrow-derived cells do not acquire functional attributes of cardiomyocytes when transplanted into peri-infarct myocardium. *Mol Ther* **16**(6), 1129–1137 (2008).
28. Rubart, M., *et al.* Physiological coupling of donor and host cardiomyocytes after cellular transplantation. *Circ Res* **92**(11), 1217–1224 (2003).
29. Plotnikov, E.Y., *et al.* Cell-to-cell cross-talk between mesenchymal stem cells and cardiomyocytes in co-culture. *J Cell Mol Med* **12**(5A), 1622–1631 (2008).
30. Koyanagi, M., *et al.* Cell-to-cell connection of endothelial progenitor cells with cardiac myocytes by nanotubes: a novel mechanism for cell fate changes? *Circ Res* **96**(10), 1039–1041 (2005).
31. Shenje, L.T., *et al.* Lineage tracing of cardiac explant derived cells. *PLoS ONE* **3**(4), e1929 (2008).
32. Messina, E., *et al.* Isolation and expansion of adult cardiac stem cells from human and murine heart. *Circ Res* **95**(9), 911–921 (2004).
33. Steele, A., *et al.* Stem-like cells traffic from heart *ex vivo*, expand *in vitro*, and can be transplanted *in vivo*. *J Heart Lung Transplant* **24**(11), 1930–1939 (2005).
34. Smith, R.R., *et al.* Regenerative potential of cardiosphere-derived cells expanded from percutaneous endomyocardial biopsy specimens. *Circulation* **115**(7), 896–908 (2007).
35. Novak, A., *et al.* Z/EG, a double reporter mouse line that expresses enhanced green fluorescent protein upon Cre-mediated excision. *Genesis* **28**(3–4), 147–155 (2000).
36. Belema Bedada, F., *et al.* Activation of myogenic differentiation pathways in adult bone marrow-derived stem cells. *Mol Cell Biol* **25**(21), 9509–9519 (2005).

37. Schulze, M., *et al.* Mesenchymal stem cells are recruited to striated muscle by NFAT/IL-4-mediated cell fusion. *Genes Dev* **19**(15), 1787–1798 (2005).
38. Field-Corbett, C. and O'Dea, S. Soluble signals from mechanically disrupted lung tissue induce lung-related gene expression in bone marrow-derived cells *in vitro*. *Stem Cells Dev* **16**(2), 231–242 (2007).
39. Hammerman, M.R. Xenotransplantation of pancreatic and kidney primordia — where do we stand? *Transpl Immunol* **21**(2), 93–100 (2009).
40. Rogers, S.A., *et al.* Long-term engraftment following transplantation of pig pancreatic primordia into non-immunosuppressed diabetic rhesus macaques. *Xenotransplantation* **14**(6), 591–602 (2007).
41. Narayan, A.D., *et al.* Human embryonic stem cell-derived hematopoietic cells are capable of engrafting primary as well as secondary fetal sheep recipients. *Blood* **107**(5), 2180–2183 (2006).
42. Rubart, M., *et al.* Two-photon molecular excitation imaging of Ca^{2+} transients in Langendorff-perfused mouse hearts. *Am J Physiol Cell Physiol* **284**(6), C1654–C1668 (2003).
43. Murry, C.E., *et al.* Haematopoietic stem cells do not transdifferentiate into cardiac myocytes in myocardial infarcts. *Nature* **428**(6983), 664–668 (2004).
44. Rubart, M., *et al.* Spontaneous and evoked intracellular calcium transients in donor-derived myocytes following intracardiac myoblast transplantation. *J Clin Invest* **114**(6), 775–783 (2004).
45. Zweigerdt, R. The art of cobbling a running pump — will human embryonic stem cells mend broken hearts? *Semin Cell Dev Biol* **18**(6), 794–804 (2007).
46. Laflamme, M.A., *et al.* Cardiomyocytes derived from human embryonic stem cells in pro-survival factors enhance function of infarcted rat hearts. *Nat Biotechnol* **25**(9), 1015–1024.
47. van Laake, L.W., *et al.* Human embryonic stem cell-derived cardiomyocytes and cardiac repair in rodents. *Circ Res* **102**(9), 1008–1010 (2008).
48. Roell, W., *et al.* Engraftment of connexin 43-expressing cells prevents post-infarct arrhythmia. *Nature* **450**(7171), 819–824 (2007).
49. Thomas, D., *et al.* Noninvasive assessment of myocardial viability in a small animal model: comparison of MRI, SPECT, and PET. *Magn Reson Med* **59**(2), 252–259 (2008).
50. Li, Z., *et al.* Comparison of reporter gene and iron particle labeling for tracking fate of human embryonic stem cells and differentiated endothelial cells in living subjects. *Stem Cells* **26**(4), 864–873 (2008).

Chapter 15
In Vivo Imaging for Cell Therapy

Richard Newton and Guang-Zhong Yang

The development of cell therapies in animal models and patients requires robust *in vivo* imaging techniques to track the administration, migration, integration, division and fate of administered stem cells and to monitor any morphological and physiological effects on the damaged tissue after therapy. This chapter outlines the well-established imaging modalities such as CT (computed tomography), MRI (magnetic resonance imaging), PET (positron emission tomography), SPECT (single photon emission computed tomography), and EBUS (endobronchial ultrasound) and discusses recent advances in biophotonic technologies including OCT (optical coherence tomography), DOI (diffuse optical imaging), and CLSM (confocal laser scanning microscopy) and their integration with endoscopy for *in situ*, *in vivo* cellular level imaging. Comparisons are made of sensitivity and spatial resolution for detecting stem cells *in vivo*, as well as imaging the lung in small animal models and patients. We also outline the challenges of labelling stem cells *in vivo* and the role of responsive agents to visualise the local physiological conditions.

Keywords: Lung imaging; CT; MRI; PET; SPECT; optical imaging; confocal laser microscopy; optical coherence tomography; fluorescence bronchoscopy.

Outline

of cell-based therapy. In addition to anatomy, MRI can provide physiological and functional information such as blood flow and blood volume.[10] Haemodynamic changes are also used in functional MRI (fMRI) to observe neuronal activity in the central nervous system.[11] With MR spectroscopy, metabolic parameters can be sought.[12]

Conventional MRI, however, has had limited success with lung imaging due to low proton density of the lung, magnetic susceptibility effects and respiratory movement. A significant recent development has been that of new inhaled hyperpolarised gases that act as contrast agents to create HP-MRI.[13,14] The detail of regional microstructure is perhaps still not as refined as with HRCT, but functional measurement such as regional ventilation[15] and Va/Q mapping[16] become possible. HP noble gas isotopes are technically challenging to make and are only available at specialist centres,[17] but include ^{3}He and ^{129}Xenon, the former producing better spatial resolution and signal-to-noise ratio,[14] but the latter allows measurement of chemical diffusion[18] from the lung airspaces due to its increased density and solubility.

In most clinical applications of MRI, intravenous contrast agents function as extracellular or sometimes intracellular substances to provide better image contrast. In order to visualise the fate of stem cells and their progeny *in vivo*, they must be similarly contrasted against the background host tissue with contrast agents. Contrast agents interact with surrounding hydrogen atoms and affect their relaxation times back to alignment. They affect the T1 (spin-lattice) and T2 (spin-spin) relaxation times, T1 and T2 providing different weighted images and each being affected differently by particular contrast agents. T1 agents (lanthanides) are paramagnetic metals such as manganese, or more usually gadolinium, which have weak, transiently inducible magnetism.[19] This is in contrast with ferromagnetic materials like iron, nickel and cobalt, which are the basis for T2 contrast agents. T2 agents are generally preferred because the local magnetic field disruption is greater,[20] and if cell disruption or cell death occurs, the iron can be eliminated through normal recycling mechanisms and avoid the impression of a viable stem cell. Gadolinium toxicity provokes concern due to possible dechelation or stem cell death.[21] The major drawback of T2 agents, though, is that they generate negative (dark) contrast, unlike the positive (white) images of T1 agents. This makes it difficult to discriminate between other hypo-intense areas such as scar tissue and blood that may be present in patients or lung disease animal models already. However, positive contrast with T2 contrast agent can be created by examining the off-resonance effects.[22]

The ideal agent provides strong and unambiguous contrast for visualisation of potentially small cell numbers surrounded by other sources of endogenous hypo/hyper intensity. It is non-toxic and does not affect cell behaviour, and if internalised pre-implantation it has a long half-life and is minimally affected by dilution caused by stem cell mitosis. It also reliably binds to, or is incorporated within the stem cell to avoid leakage or transfer to host cells. Most widely described is the method of *ex vivo* pre-implantation internalisation of strong T2 contrast agent like iron oxide particles.[23] Different sized iron-oxide particles are used: MPIOs (micro-sized paramagnetic iron oxide particles), SPIOs (small super-paramagnetic iron-oxide particles) and USPIOs (ultra-small paramagnetic iron oxide particles). Figure 1 shows MRI sections of swine hearts in a model of myocardial infarction after the injection of SPIOS (Feridex) labelled mesenchymal stem cells into the myocardium, with spreading of the signal over one week suggestive of MSC migration and possible division.[24]

Phagocytes spontaneously take up contrast agents,[25] but stem cells are more stubborn and require techniques such as transfection (co-incubation of stem cells with contrast agent and protamine sulphate or poly-L-lysine for 24 hours),[26] magnetoelectroporation (MEP)[27] involving the application of small pulsed voltages to perturb the cell membrane, or lipofection.[28] These techniques can have marginal effects on cell survival, proliferation and differentiation, though.[29] Unlike iron particles, fluorine nanoparticles are readily phagocytosed by stem cells[30] but ^{19}F MRI and MR Spectroscopy require specialised coils and hardware.

An alternative is to use contrast agents that respond to the local physiological and metabolic properties of the cell, and thus only provide contrast in mature or differentiated engrafted cells. This avoids the pitfalls of an expired iron-labelled stem cell appearing viable due to iron embedded in macrophages.[31] For example, cells can be transfected with a beta-galactosidase gene as well as a galactose chelated gadolinium contrast.[32] Until the enzyme cleaves the galactose, the gadolinium is shielded from affecting the local field of the cellular water protons and the T1 relaxation

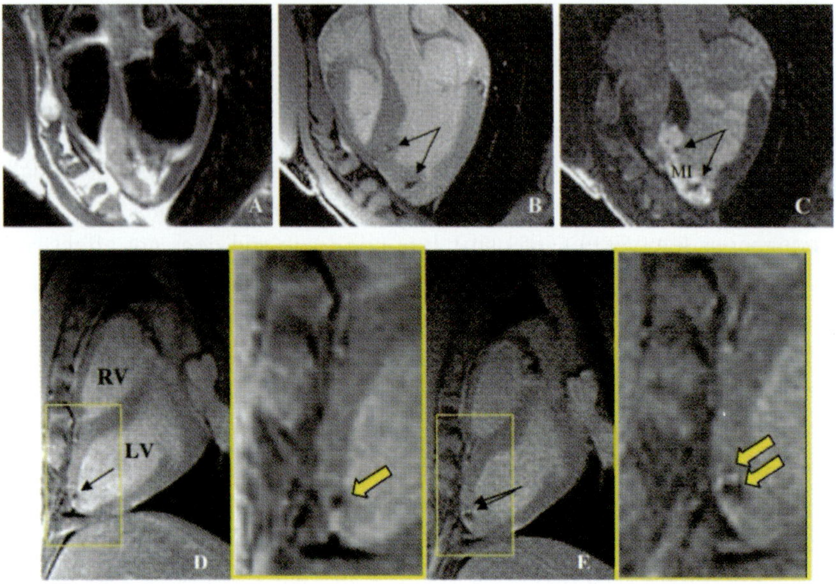

Figure 1: *In vivo* MRI scans of swine hearts with an induced myocardial infarction followed by myocardial injection of Feridex-labelled mesenchymal stem cells (A–D) at 24 hours; and (E) displaying spread and decreasing delineation of the borders of the injected area after one week. The hyperintense area in C demonstrates the infarcted area, and the ovoid hypointense areas show the mesenchymal stem cells.[24] From Kraitchman *et al.* (2003).[24] Reproduced with permission from LWW.

time remains unmodulated. Harnessing the lipase activity in viable cells to unblock aliphatic side chains chelating a contrast agent remains another option,[33] or even enabling stem cells to manufacture their own contrast agent by stimulating overproduction of transferrin receptor or ferritin which incorporates extracellular matrix ferrous ions into the cell.[34]

2.2. *Stem cell biodistribution with PET and SPECT*

PET is a nuclear medicine technique to evaluate the spatial distribution of biochemical or metabolic activity. A biologically active molecule of physiological interest (such as a glucose analogue, water or ammonia) is labelled with a positron-emitting radioisotope (such as ^{18}F, ^{11}C or ^{13}N) to form a radiopharmaceutical tracer which is injected or inhaled. Most commonly in PET for oncological purposes, the glucose analogue fluorodeoxyglucose (FDG) is labelled with ^{18}F and injected, and the most metabolically active cells (such as within metastatic deposits or perhaps actively dividing stem cells) preferentially take up this tracer, and after almost irreversible phosphorylation of FDG, the radioactive tracer becomes locked in the cytoplasm. Positron emission decay (beta decay) then occurs, as an unstable neutron in the ^{18}F nucleus decays into a proton, emitting a neutron, a positron (opposite charge to an electron) and an anti-neutrino. After a few millimetres, the positron merges with an electron to yield two high-energy 511KeV photons (gamma rays) travelling in almost opposite directions. A quantitative map of tracer distribution is made by recording the direction of these paired photons.

Due to nanomolar sensitivity, PET/SPECT can measure processes at low concentrations, and small animal imaging systems with <2 mm spatial resolution have been devised.[35] In research and clinical practice, the logical progression has been to merge nuclear imaging technology with MRI or CT to create SPECT/CT and PET/MRI. This co-registration of images provides the highly sensitive quantitative location of the active cells superimposed on an accurate anatomical map. Radioisotopes can be injected, or for functional assessment of the lung they can be inhaled. For example, ^{81m}Kr has a half-life of 13.1 seconds and can be used for regional ventilation imaging.[36]

PET and SPECT are now used to track stem cells. For real-time observation of stem cell biodistribution, implanted cells can be labelled via incubation with relatively long-lived isotopes such as ^{111}N for SPECT[37] and ^{64}Cu for PET,[38] and then free radioisotopes washed off prior to implantation. However, similar to the labelling of stem cells for MR imaging, efflux of radiotracer from cells over time is a problem.[39] Further, the incubation process may impair proliferation and function and the imaging technique provides no information on the viability or function of the cells. The solution is to use electroporation or cationic lipid-based transfection agents to stably transfect stem cells with a strong viral promoter sequence (such as from cytomegalovirus, CMV[40]) and a reporter gene. Three classes of reporter gene are discussed by Acton *et al.*,[41] which include:

- Enzymes: This involves transfection with enzymes such as yeast cytosine deaminases or HSV1-tk (mutant herpes simplex virus type-1 thymidine kinase). HSV1-tk produces HSV1-TK enzyme which then phosphorylates the radioactive reporter probe ^{18}F-FHBG (^{18}F labelled fluoro-3-hydroxymethylbutylguanine) in the cytoplasm, locking in the tracer to the stem cell. This technology has been used for looking at the engraftment of rat cardiac myocytes (H9c2 cells).[42]
- Transporters: The human norepinephrine transporter gene and the sodium-iodine symporter (NIS) gene encode membrane glycoprotein receptors. The NIS symporter transports sodium and iodine into thyroid cells in particular.[43] The expression of the NIS symporter can be imaged with ^{123}I for SPECT, and ^{131}I and ^{125}I for PET.[44,45] However, it has been suggested that to prevent efflux of radioactive iodine from transfected cells, co-expression of thyroperoxidase to trap the iodine is required.[46] With a cardiac-specific promoter, the NIS reporter gene has been used with PET imaging to show the outcome of stem cell implantation in a transgenic mouse.[47] Furthermore, excellent histological correlation with SPECT imaging has been shown with metastatic lung tumour deposits resulting from NIS transfected tumour cells and the ability to detect lesions of <3 mm.[48]
- Receptors: An alternative is for transfection with the somatostatin subtype 2 (SSTr2) gene or the dopamine D2R reporter gene. The D2R receptor migrates to the cell surface and binds radioactive ligands such as ^{18}F-FESP (^{18}F-fluoroethylspiperone). This radiotracer provides a quantitative measure of D2R expression. However, the endogenous agonist can affect binding and also, if ^{18}F-FESP binds, the cAMP levels rise with physiological consequences,[49] which has necessitated the development of mutant D2R strains.

An antibiotic resistance gene is usually included to demonstrate successful transfection prior to stem cell transplant, and the highest reporter gene expression clones can also be selected. After engraftment, there is a constitutional expression of the reporter gene to monitor its survival.

Conversely, a dead stem cell will have no gene expression, no uptake of reporter probe and therefore no signal. An additional benefit of using reporter genes is that a therapeutic gene can be added to the transfection construct, allowing simultaneous gene therapy and cell tracking. For example, skeletal myoblast transplant has been used to deliver VEGF (vascular endothelial growth factor) which may have a role in improving the local blood supply.[50]

PET/SPECT has high sensitivity for detecting engrafted functioning cells but multimodality is likely to be the future, perhaps with MRI for delineation of tissue anatomy or using the multimodality fusion reporter system.[51] With the latter, a gene is transfected expressing a tri-fusion protein (a red fluorescent protein, a renilla luciferase and the HSV1-TK enzyme). The fluorescence can be used for studying the cells *ex vivo* or for pre-implantation cell sorting; the bioluminescence can be used for *in vivo* imaging and the PET/SPECT for quantitative accuracy.

2.3. *Endobronchial ultrasound*

Pleural imaging and the evaluation of effusions is often performed with conventional ultrasound, and since its bronchoscopic integration (EBUS), the modality can now image deep inside the lung, highlighting the layers of the bronchial tree as well as mediastinal and hilar lymph nodes beyond.

Radial probe EBUS consists of a small diameter (1–2 mm) rotary transducer ultrasound probe that is inserted down a sheath lining the working channel of a conventional bronchoscope. It can image to the subsegmental bronchus, with 5 cm penetration and around 1 mm spatial resolution. The guide sheath can be left in its advanced position for biopsy or brushings (or potentially stem cell delivery). An integrated linear probe bronchoscope has also been developed to provide images that are easier to interpret, and also has a doppler mode for studying blood flow through vessels.

3. Biophotonics *in Vivo*

A key challenge for cell therapy research in the *in vivo* lung is to monitor biochemical and molecular biological activity and pathways in engrafted tissue and its surrounding native tissue. Optical imaging covers a plethora of techniques that use combinations of bioluminescence, fluorescence, absorption and reflectance created from light in the visible (400–700 nm) or near infrared (NIR) spectrum (700–900 nm) as mechanisms to provide a contrasted image.[52] The contrast and information gleaned can be altered with molecular imaging techniques that use biomarker probes to image particular intracellular targets and pathways.

Generally, optical imaging has a very high sensitivity for measuring optical contrast agents and reporter molecules at picomolar or even femtomolar concentrations. The downside is poor penetration, particularly at visible wavelengths.[53] Generally, light absorption and scattering decrease with increasing wavelength and so NIR light has better penetration (centimetres) than visible light (millimetres).[54] Above 900 nm, water absorption interferes with signal to background ratio.

3.1. *Diffuse optical imaging*

DOI (diffuse optical imaging) measures the absorption and the elastically scattered light, when visible or NIR light is transmitted between two points with tissue in the middle.[55] At these wavelengths several physiologically interesting molecules (chromophores) absorb and scatter photons.[56] For example oxyhaemoglobin (HbO) and deoxyhaemoglobin (HHb) absorb different frequencies and provide an indicator of tissue oxygenation and local haemodynamics,[57] a principle which is used in the pulse oximeter.

Good resolution at depth remains a challenge as the absorption coefficient (μa — absorption events per unit length) is much less than the scatter coefficient (μs) so signals over a few millimetres are dominated by scattered light. The scatter coefficient is affected by the refractive index of intracellular and extracellular fluids and subcellular components like nuclei or mitochondria. The distinction is blurred between diffuse optical topography where 2D images of a plane parallel to the source and detector have limited depth information, and diffuse optical tomography where 3D images are created from multiple sources and detectors widely spaced over the surface of the object.

DOI can be allied with molecular imaging[57] in which contrast agents bind to specific genes or proteins and perhaps stem cell labelling can be harnessed. Molecular beacons such as quantum dots acting as fluorophores[58] or other agents utilising bioluminescence and fluorescence can be used. For these techniques to become useful within the lung, future improvements require better sources (helped by new laser diodes), increased penetration and faster image acquisition, and better understanding of optimal wavelength combinations for particular applications. DOI has been used for breast cancer diagnostics[59] and joint imaging[60] but in the lung, it may ultimately be limited by the need for small probes either side of the tissue and has so far only been used in the lung in *in vivo* mice lung tumour models.[61]

3.2. *Optical coherence tomography*

Optical coherence tomography (OCT) was devised in 1990[62] and generates a 2D or 3D cross-sectional image in a manner similar to radar or B-mode ultrasonography,[63] by beaming near infra-red light (750–1300 nm wavelength) at tissue. Most light is scattered, creating a glare that OCT eliminates, allowing only coherent light to be measured by the interferometer. It is an interferometric technique using broadband (broad frequency range) superbright LEDs or extremely short pulse lasers, and, being higher frequency than ultrasound, can image through air and does not suffer from a trade-off between depth penetration and resolution. It images to a depth of 2 or 3 mm, which is three times more than its nearest rival, confocal laser scanning microscopy (CLSM). Whilst its submicron resolution[64] is inferior to CLSM (see Section 3.3), it is superior to MRI or ultrasonography.

OCT's radial image probes and longitudinal image probes are good for smaller and larger diameter lumens, respectively, and transverse image probes provide the best depth of field.[52] They are available as an optic fibre to pass down an endoscope channel. The role of OCT has been tested in the follow-up of patients who have undergone local resection of non-invasive transitional cell carcinoma of the bladder,[65] mucositis in chemotherapy patients, intravascular imaging, and in the visualisation of the biliary tract.[66] OCT has identified Barrett's metaplasia in the distal oesophagus, transmural inflammation in inflammatory bowel disease, and has distinguished hyperplastic from adenomatous polyps in the colon.[67] In the lung, the terminal bronchiole can be reached and the fine resolution of OCT delineates epithelium, mucous glands and ducts, cartilage and perichondrium and closely mirrors the microstructure on conventional histology.[68] The lamina propria and submucosa is not quite as well demarcated and the refractive indices of blood vessels and smooth muscle are similar beneath the epithelium, and thus these structures are difficult to differentiate. However, OCT enables accurate differentiation between normal bronchial epithelium, dysplastic epithelium,[69] carcinoma-in-situ and invasive carcinoma using the increasing thickness of the epithelium and increasing visibility of the nuclei (see Fig. 2).[70] The basement membrane and upper submucosa is well-delineated due to significant scattering, but very obviously breached in the case of invasive carcinoma.

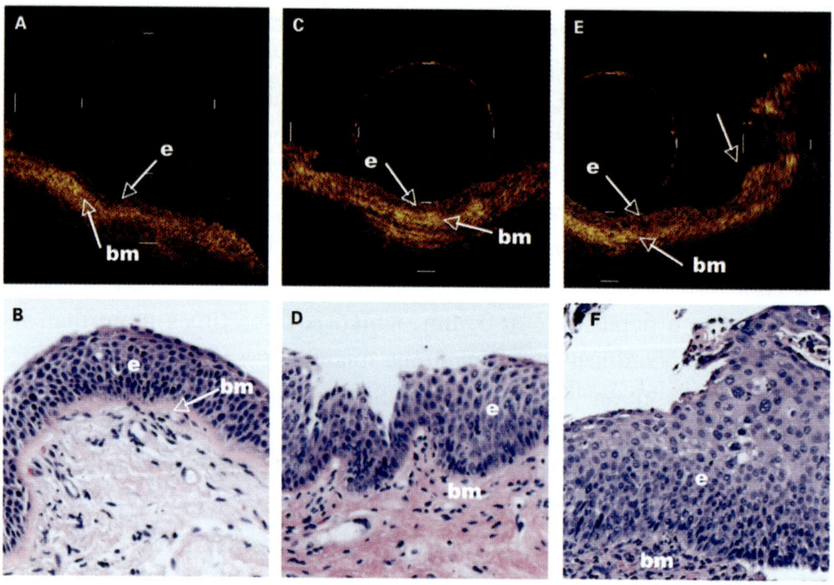

Figure 2: Optical coherence tomography views of the bronchial epithelium (e) on basement membrane (*bm*), demonstrating increasing dysplasia (A, C and E) compared with corresponding normal haematoxylin and eosin staining below (B, D and F).[70] With permission from Lam *et al.* (2008).[70] Reproduced with permission of the American Association for Cancer Research via Copyright Clearance Center.

Different grades of dysplasia are not currently differentiable, but if morphometric measurements can provide an answer, better spatial resolution will be required.

OCT has also been used to quantify the regional changes *in vivo* to rabbit bronchial epithelium in a smoke inhalation model.[71] It may be that doppler capability can be added for haemodynamic information in the lung.[72] Another hurdle is the large amount of data created from the intrinsically high spatial resolution, enabling little more than highly selected areas to be optically imaged at the present.

3.3. *Confocal fluorescence laser scanning microscopy (CLSM)*

Marvin Minsky developed CLSM in 1955,[73] and as a desktop instrument it became popular in the 1980s to perform high resolution, blur-free fluorescence microscopy and immunohistochemistry on excised tissues. A laser beam of blue light (440–500 nm) is shone via a pinhole aperture to reduce out-of-focus light from above and below the focal plane, focussed via an objective lens, and the image is then captured through a dichromatic mirror. By scanning across the tissue, sequential images with a small depth of field are amalgamated to provide high quality 3D images that are indispensable in identifying cellular and subcellular microstructures. These depth-resolved optical slices include reflected light, autofluorescence, and exogenous fluorescence from intravenous or topical contrast agents.

More recently, miniaturisation of CLSM has enabled it to be used as an *in vivo* device to take the microscope to the tissue and provide virtual histology. Optic fibres are employed, providing both a focussed narrow point light source and a detection pinhole. There are two main commercially available systems and they are mostly being used in conjunction with endoscopy in the gastrointestinal tract.[74] The Cellvizio-GI (Mauna Kea Technologies, Paris) is a flexible mini-probe system, which is inserted down the endoscope biopsy channel to create probe-based confocal laser endomicroscopy (pCLE). There is also a 1.4 mm pCLE for reaching the alveolus (AlveoFlex, Mauna Kea Technologies) where the fluorescence collected by the fibre bundle is divided for imaging (80%)

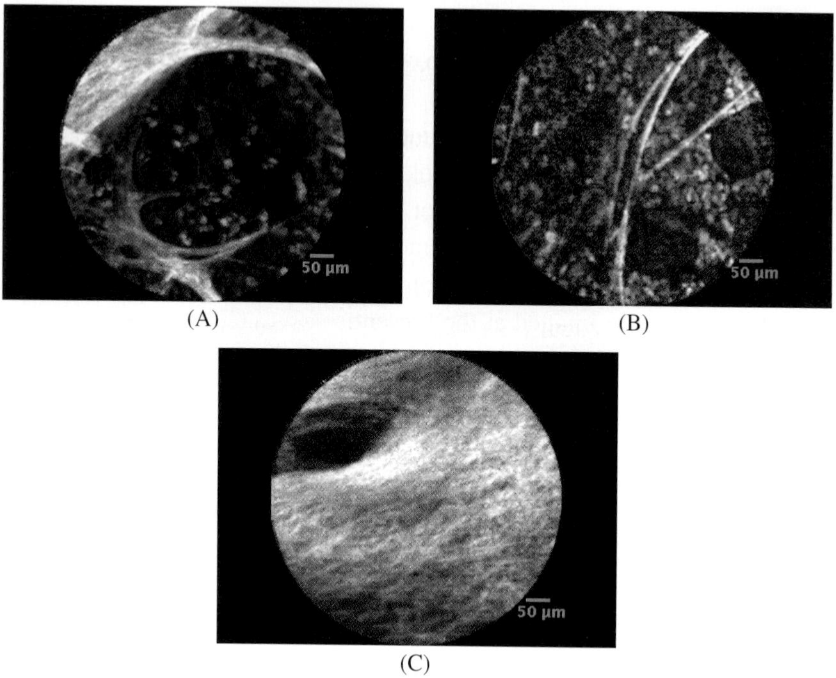

Figure 3: The AlveoloFlex pBCLE demonstrating (A and B) alveolar structure and macrophages (courtesy of Dr. Kyle Hogarth at the University of Chicago and Dr. Raspaud at the Clinique Pasteur in Toulouse) and the autofluorescing elastin component of the basement membrane in the bronchus (courtesy of Dr. Sterman at the University of Pennsylvania Medical Center in Philadelphia). With permission from Mauna Kea Technologies.

and spectroscopy (20%) by a beam splitter. Figures 3A and 3B show the Alveoflex demonstrating the *in vivo, in situ* views of terminal airways with alveolar walls and sacs evident, scattered with fluorescing macrophages, and Fig. 3C shows the autofluorescence of the elastin component of the basement membrane of the more proximal airways. Pentax and Optiscan have developed a system with a confocal microscope integrated into the distal end of an endoscope, creating an endomicroscope that is held in direct contact with the gastrointestinal mucosa. At the press of a button on the console the image depth is advanced 4 μm along the z-axis to focal planes anywhere from 0–250 μm. The benefits postulated include 1) immediate diagnosis and in some cases treatment and

2) reducing the numbers of conventional biopsies, cutting the financial costs and risks of perforation and bleeding.[75,76]

Thiberville *et al.* demonstrated the potential of a narrower 1 mm prototype Cellvizio 400/S dual-fibred CLSM probe used down the channel of a bronchoscope without topical or intravenous contrast, imaging from 0–50 μm below the surface with a lateral resolution of 5 μm.[77] They demonstrated that the microautofluorescence signal varied between healthy patients and corresponded closely with the intensity of the auto-fluorescence bronchoscopy signal. It was also suggested from *ex vivo* work that the microautofluorescence seemed to be caused by the elastin fibres in the more superficial layers of the subepithelial bronchial wall and that five distinct elastin patterns were evident as the imaging progressed distally through the bronchial tree:

1. A dense homogeneous pattern lacking identifiable cross-fibres in the anterior tracheal wall;
2. Tightly compacted cross-linked fibres at the carina of lobar bronchi close to gland openings;
3. At the pars membrana of the large bronchi, a multilayer of thin, perpendicularly aligned fibres, that probably correspond to the lamina densa lying above the lamina reticularis;
4. A loose, web-like pattern at the origins of the segmental bronchi;
5. A ring-like pattern on distal non-cartilagious bronchiolar airways covered with longitudinal fibres.

Furthermore, the team demonstrated that the microautofluorescence was decreased or absent in dysplastic areas or areas with CIS, and showed specific differences in the CLSM images in sarcoidosis and Mounier Khun syndrome that were not observable under white light bronchoscopy. Whilst bronchial epithelium has been poorly characterised, the CLSM has been used to identify single nuclei and cell to cell borders in *in vivo* CLSM with topically applied contrast agents like fluorescein.[78] In the upper gastrointestinal tract, non-erosive reflux disease of the distal oesophagus has been diagnosed with a high degree of sensitivity and specificity in patients with reflux symptoms, based around the degree of dilatation of the intercellular spaces and the number of capillary loops in the upper third of the mucosal layer on CLSM. The pCLE technique has recently been shown to have 99% negative predictive value in the diagnosis of endoscopically invisible neoplasia in the setting of Barrett's oesophagus (intestinal metaplasia) using four quadrant optical biopsies taken from the distal oesophagus.[75] The endomicroscope has also been used to identify Barrett's oesophagus in at-risk patient groups, and an endomicroscopic classification for the detection of associated neoplasia has been proposed, looking at vessel and crypt architecture. In 2005, using acriflavine endomicroscopy, helicobacter pylori, a bacterium associated with peptic ulcer disease and gastric cancer, was identified as characteristic white dots with an attached flagellum within the superficial gastric epithelium.[79] Even flowing individual erythrocytes can be visualised[80] and it can only be a matter of time before this powerful tool has more data on lung imaging and can ultimately be used for assessment of the engraftment of lung stem cells.

However, there are potential obstacles to CLSM. The miniprobe is not easy to manipulate and requires perpendicular apposition to the tissue with the correct pressure. Many operators use a short, transparent mucosectomy cap on the endoscope tip to aid stabilisation and targeting. It also needs to slide across the tissue gently and smoothly to mosaic anything other than a very small window and this may prove damaging to new stem cells. Surface secretions can impair the CLSM view, and the white light view from the endomicroscope is covered when the instrument has to contact the tissue for the CLSM function, which may reduce target accuracy. Additionally, the high magnification of CLSM means that cardiorespiratory movements can become troublesome for the operator. An important engineering consideration has been the linking of point-based optical biopsy to large area tissue visualisation. Image mosaicing is employed, a well-developed subject in computer vision that stitches together sequential images with partial overlapping to create a seamless panoramic wide field of view.[81]

3.4. *Endoscopic integration*

The realisation that a third of pre-invasive bronchial carcinomas cannot be visualised under standard white light bronchoscopy drove investigation into fluorescence bronchoscopy. Early procedures used drug-induced fluorescence from pre-procedural administration of photosensitisers such as haematoporphyrin derivatives or delta-aminolevulinic acid (ALA), but high costs and potential side effects have led to a concentration on contrast-free autofluorescence. When blue light (380–460 nm) is shone at normal bronchial mucosa, structural fluorophores such as collagen and elastin, and cellular metabolic fluorophores such as nicotinamide adenine dinucleotide (NAD) and flavins, emit fluorescent light. This emission peaks at 520 nm (green), with a minor peak around 630 nm (red), with the autofluorescence yield being ten times greater from the subepithelial tissues than from the epithelium. The transition in smokers' lungs from normal epithelium to squamous metaplasia, dysplasia, carcinoma-in-situ and invasive carcinoma represents a ten-fold decrease in the local overall autofluorescence, but the suppression of green autofluorescence is proportionately more than that of red autofluorescence, changing the green to red ratio from 5:3 to 2:3. This is due to a decrease in the amount of collagen in the elastin, an increase in the number of epithelial cell layers and the microvascular blood volume, and a decrease in the concentration of cellular fluorophores in premalignant/malignant cells. The result is that many abnormal areas appear mild red and whilst autofluorescence is relatively sensitive for detecting pre-invasive carcinoma and can quickly scan large areas, complementary techniques like conventional or optical biopsy are needed to improve the specificity.

Conventional white light bronchoscopy uses the full visible wavelength range (400–700 nm) to produce a red-green-blue image. HR-NBI (high resolution narrow band imaging) uses higher magnification video endoscopy to show vascular networks by using filters to restrict the incident light into two narrow bands of different wavelength range (blue around 415 nm and green around 540 nm) whilst simultaneously increasing the relative intensity of the blue band. As blue is particularly well absorbed by haemoglobin, the 415 nm band highlights the superficial capillary networks and the 540 nm image channel analyses the collecting vessels better, as longer wavelengths penetrate further. NBI has been used to improve the identification rate of squamous cell carcinomas of the oesophagus, and to better predict tumour depth invasion.[82] HR-NBI may also detect lung squamous intraepithelial neoplasia by enhancing the microvascular architecture in the bronchi.[83]

More recently, Olympus have introduced prototypes of an endocytoscope, which is essentially a 3 mm diameter endoscope that can be inserted down a biopsy channel of a normal endoscope or bronchoscope and magnifies 450 times, visualising to the single cell level and nucleus. Most tissue characterisation has been on the gastrointestinal tract[84] but studies are being done to characterise the full bronchial tree. 0.5% methylene blue is used as a contrast agent, but there is some evidence that it may have deleterious effects on DNA.[85]

Lung applications are not currently documented for chromoendoscopy, but it is a promising technique and somewhat re-emergent in gastroenterology since the advent of high-resolution endoscopes, and may have a role in the assessment of bronchial epithelium. It involves the topical application of stains or pigments to improve tissue localisation and characterisation of epithelial attributes during endoscopy.[86] These pigments include:

- absorptive/vital stains that highlight abnormal mucosa after absorption into cytoplasm. Examples include crystal violet, methylene blue, toluoidine blue and Lugol's solution/dark brown agent (glycogen-containing normal squamous epithelium is stained dark brown but abnormal multilayer metaplastic/dysplastic mucosa remains unstained);
- contrast stains (e.g. indigo carmine), highlighting mucosal topography;
- reactive stains that identify changes in pH but are likely to remain important only to hydrochloric acid-secreting areas of the gastrointestinal tract.

The stains are generally cheap and safe, and the only additional equipment needed is a reusable spray catheter to create a fine mist (e.g. Olympus PW-5L).

4. Photogenic Small Animal Models and Cell Delivery

Imaging reporter gene expression via PET, SPECT or MRI has already been described. However, reporter genes can also encode enzymes that create biophotonic molecular tags that can highlight transfected stem cells.[87] GFP (green fluorescent protein from jellyfish) transfection requires light to be shone upon the transfected tissue, but bioluminescent molecular tags can act as an internal biological light source. For example, luciferase transfection from the firefly forces a tranfected cell to oxidise the pigment luciferin and create light. This has the advantage that different luciferase enzymes from varying organisms are specific for distinct substrates, which when split create specific wavelength photons for the simultaneous use of more than one marker.[88]

Mesenchymal stem cells have been transfected with luciferase and then shown to migrate from their femoral artery injection site to a damaged area of tibial muscle, with some trapping in the lung microvasculature.[89] The light produced is then detected by a charge-coupled device camera (CCD) such as the IVIS system (Xenogen) (see Fig. 4). Much work has been done to develop fluorescent proteins that avoid known deleterious effects on cells,[90] including the creation of enhanced green, cyan and yellow fluorescent proteins.[91,92] A significant improvement awaited is the development of fluorescent proteins with bright, near-infrared spectra to create greater tissue penetration.[93]

The most effective method of stem cell administration remains to be established (see Chapter 13). In general, intravenous delivery is simpler than intra-arterial, but the concentration of cells reaching the target site is often limited. Intra-arterial administration via coronary arteries has been used extensively in cardiac stem cell research. It is likely that targeted organ injections will provide the most accurate mechanism and the highest concentration of cells in the target region, and may benefit from MRI guidance with systems similar to those used in X-ray fluoroscopic delivery approaches.[94] Steerable catheters have been used to inject stem cells into the ventricles, with hypointensities created on MRI by the needle. To create enhanced identification of the needle tip, active catheter tracking is possible by incorporating various radiofrequency receiver antennas into the catheter tip,[95] to provide a high-signal area after processing of the images.[95] However, all these catheter tips provoke concern over the differing degrees of heating, which have the potential for damage to patient tissues and the stem cells being administered.

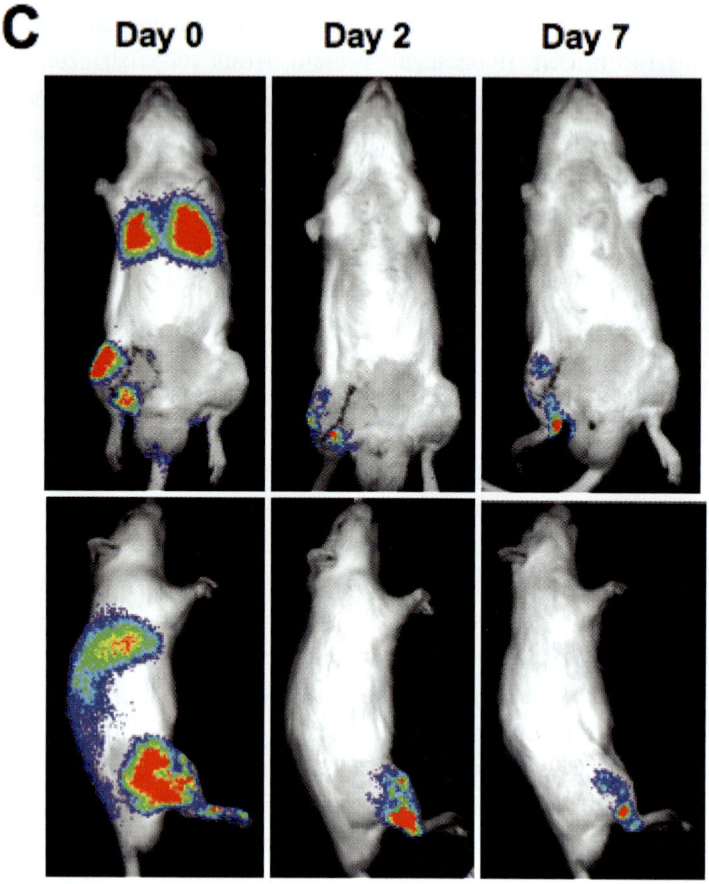

Figure 4: *In vivo* luciferase imaging of injected stem cells from luciferase/LacZ double TG LEW rats that have had prior induction of right tibial muscle degeneration. Injections were in either right femoral artery (top panels) or tibial muscle (lower panels) at days 0, 2 and 7. Reprinted from Hara *et al.* (2008).[89] With permission from Elsevier Science.

5. Conclusion and the Way Forward

In this chapter, we have discussed some of the major imaging modalities that are important for cell therapy. It is worth noting that there are many other techniques that may be of importance for cell therapy imaging in future. For example, photoacoustic tomography (PAT) has a background in materials analysis, and harnesses the photoacoustic effect. A single laser pulse with a wavelength in the nanosecond range induces a wideband ultrasonic emission in the target tissue due to the transient thermoelastic expansion of the tissue. It has been used experimentally to image the cerebral cortex[96] and blood vessels.[97] As another example, Terahertz pulsed imaging (TPI) uses frequencies between NIR and microwave (3 mm to 30 μm) and has been used to characterise *ex vivo* basal cell carcinomas[98] and breast tumours.[99] However, the more developed techniques like MRI are likely to occupy the limelight for the near future.

Currently there is no single imaging approach that can observe all the relevant aspects of *in vivo* cell therapy. Applied to stem cell therapies in the lung, the optimal modality will depend on the question being posed. HRCT and MRI will create 3D recreations of bronchial and parenchymal microstructure before and after cell therapy with far better temporal and spatial resolution than most biophotonic techniques. With appropriately labelled stem cells, MRI and nuclear imaging will monitor the tracking and fate of the administered cells together with photogenetic labelling. However, the avoidance of ionizing radiation and problematic magnetic fields makes the biophotonic techniques appealing, particularly when it is considered that the equipment is smaller, thus enabling endoscopic integration, and that they have excellent potential for interrogation of tissue function. However, the acid test remains whether any intervention affects cell and organ function. Multimodality is likely to be the future, for example with MRI providing the background anatomical detail for delivery, superimposed with PET supplying information about cell tracking, and multiple molecular properties detailing viability and function elucidated simultaneously with different colours through biofluorescence.

In the future, intravascular stem cell delivery will continue, but in the lung more locally targeted techniques will also be employed. This may be via image-guided percutaneous transthoracic injection or bronchoscopic administration, the latter benefiting from integrated imaging systems such as EBUS, CLSM, OCT and NBI. Minimally invasive surgical techniques such as video-assisted thoracic surgery (VATS), and robotic surgery with systems such as the da Vinci surgical system (Intuitive Surgical, California) and newly emerging platforms such as the i-Snake (imaging-sensing navigated and kinematically enhanced) robot may play a role in precise tissue dissection and stem cell delivery. These robotic systems may improve the accuracy of delivery by compensating for cardiorespiratory movement either by feature-based motion tracking[100] or through the concept of perceptual docking[101] for effective human–robot interaction and haptic feedback with dynamic active constraints.[102] MRI-adapted catheters can provide excellent resolution for stem cell delivery,[95] and indeed MRI-compatible surgical robots have been developed.[103] However, the manoeuvrability, size, and image quality produced by the technology being developed in the field of biophotonics means that integration into the surgical robots of the future must be the next logical forward step for stem cell delivery.

References

1. Frangioni, J.V. New technologies for human cancer imaging. *J Clin Oncol* **26**(24), 4012–4021 (2008).
2. Seo, Y., Mari, C., and Hasegawa, B.H. Technological development and advances in single-photon emission computed tomography/computed tomography. *Semin Nucl Med* **38**(3), 177–198 (2008).
3. Hansell, D.M. High-resolution computed tomography in the evaluation of fibrosing alveolitis. *Clin Chest Med* **20**(4), 739–760, viii (1999).
4. Townsend, D.W. Positron emission tomography/computed tomography. *Semin Nucl Med* **38**(3), 152–166 (2008).
5. Pichler, B.J., Wehrl, H.F., Kolb, A., and Judenhofer, M.S. Positron emission tomography/magnetic resonance imaging: the next generation of multi-modality imaging? *Semin Nucl Med* **38**(3), 199–208 (2008).
6. Wilson, B.C. Detection and treatment of dysplasia in Barrett's esophagus: a pivotal challenge in translating biophotonics from bench to bedside. *J Biomed Opt* **12**(5), 051401–051422 (2007).
7. Passier, R., van Laake, L.W., and Mummery, C.L. Stem-cell-based therapy and lessons from the heart. *Nature* **453**(7193), 322–329 (2008).
8. Barker, N., van Es, J.H., Kuipers, J., Kujala, P., van den Born, M., Cozijnsen, M., *et al.* Identification of stem cells in small intestine and colon by marker gene Lgr5. *Nature* **449**(7165), 1003–1007 (2007).

9. Mansfield, P. Snapshot magnetic resonance imaging (Nobel lecture). *Angew Chem Int Ed Engl* **43**(41), 5456–5464 (2004).
10. Yang, G.Z. Exploring *in vivo* blood flow dynamics. *IEEE Eng Med Biol Mag* **17**(3), 64–72 (1998).
11. Dolan, R.J. Neuroimaging of cognition: past, present, and future. *Neuron* **60**(3), 496–502 (2008).
12. Hudsmith, L.E. and Neubauer, S. Detection of myocardial disorders by magnetic resonance spectroscopy. *Nat Clin Pract Cardiovasc Med* **5**(Suppl 2), S49–S56 (2008).
13. Albert, M.S., Cates, G.D., Driehuys, B., Happer, W., Saam, B., Springer, C.S., Jr., *et al.* Biological magnetic resonance imaging using laser-polarized 129Xe. *Nature* **370**(6486), 199–201 (1994).
14. Fain, S.B., Korosec, F.R., Holmes, J.H., O'Halloran, R., Sorkness, R.L., and Grist, T.M. Functional lung imaging using hyperpolarized gas MRI. *J Magn Reson Imaging* **25**(5), 910–923 (2007).
15. Salerno, M., Altes, T.A., Brookeman, J.R., de Lange, E.E., and Mugler, J.P. 3rd. Dynamic spiral MRI of pulmonary gas flow using hyperpolarized (3)He: preliminary studies in healthy and diseased lungs. *Magn Reson Med* **46**(4), 667–677 (2001).
16. Rizi, R.R., Baumgardner, J.E., Ishii, M., Spector, Z.Z., Edvinsson, J.M., Jalali, A., *et al.* Determination of regional VA/Q by hyperpolarized 3He MRI. *Magn Reson Med* **52**(1), 65–72 (2004).
17. Hersman, F.W., Ruset, I.C., Ketel, S., Muradian, I., Covrig, S.D., Distelbrink, J., *et al.* Large production system for hyperpolarized 129Xe for human lung imaging studies. *Acad Radiol* **15**(6), 83–92 (2008).
18. Driehuys, B., Cofer, P.G., Pollaro J., Mackel, J.B., Hedlund, L.W., and Johnson, G.A. Imaging alveolar-capillary gas transfer using hyperpolarized 129Xe MRI. *Proc Natl Acad Sci USA* **103**(48), 18278–18283 (2006).
19. Aime, S., Cabella, C., Colombatto, S., Geninatti Crich, S., Gianolio, E., and Maggioni, F. Insights into the use of paramagnetic Gd(III) complexes in MR-molecular imaging investigations. *J Magn Reson Imaging* **6**(4), 394–406 (2002).
20. Heyn, C., Bowen, C.V., Rutt, B.K., and Foster, P.J. Detection threshold of single SPIO-labeled cells with FIESTA. *Magn Reson Med* **53**(2), 312–320 (2005).

21. Dousset, V., Tourdias, T., Brochet, B., Boiziau, C., and Petry, K.G. How to trace stem cells for MRI evaluation? *J Neurol Sci* **265**(1–2), 122–126 (2008).
22. Cunningham, C.H., Arai, T., Yang, P.C., McConnell, M.V., Pauly, J.M., and Conolly, S.M. Positive contrast magnetic resonance imaging of cells labeled with magnetic nanoparticles. *Magn Reson Med* **53**(5), 999–1005 (2005).
23. Bulte, J.W. and Kraitchman, D.L. Monitoring cell therapy using iron oxide MR contrast agents. *Curr Pharm Biotechnol* **5**(6), 67–84 (2004).
24. Kraitchman, D.L., Heldman, A.W., Atalar, E., Amado, L.C., Martin, B.J., Pittenger, M.F., *et al. In vivo* magnetic resonance imaging of mesenchymal stem cells in myocardial infarction. *Circulation* **107**(18), 2290–2293 (2003).
25. Saleh, A., Wiedermann, D., Schroeter, M., Jonkmanns, C., Jander, S., and Hoehn, M. Central nervous system inflammatory response after cerebral infarction as detected by magnetic resonance imaging. *NMR Biomed* **17**(4), 163–169 (2004).
26. Frank, J.A., Miller, B.R., Arbab, A.S., Zywicke, H.A., Jordan, E.K., Lewis, B.K., *et al.* Clinically applicable labeling of mammalian and stem cells by combining superparamagnetic iron oxides and transfection agents. *Radiology* **228**(2), 480–487 (2003).
27. Walczak, P., Kedziorek, D.A., Gilad, A.A., Lin, S., and Bulte, J.W. Instant MR labeling of stem cells using magnetoelectroporation. *Magn Reson Med* **54**(4), 769–774 (2005).
28. Daldrup-Link., H.E., Rudelius, M., Oostendorp, R.A., Jacobs, V.R., Simon, G.H., Gooding, C., *et al.* Comparison of iron oxide labeling properties of hematopoietic progenitor cells from umbilical cord blood and from peripheral blood for subsequent *in vivo* tracking in a xenotransplant mouse model XXX. *Acad Radiol* **12**(4), 502–510 (2005).
29. Kostura, L., Kraitchman, D.L., Mackay, A.M., Pittenger, M.F., and Bulte, J.W. Feridex labeling of mesenchymal stem cells inhibits chondrogenesis but not adipogenesis or osteogenesis. *NMR Biomed* **17**(7), 513–517 (2004).
30. Partlow, K.C., Chen, J., Brant, J.A., Neubauer, A.M., Meyerrose, T.E., Creer, M.H., *et al.* 19F magnetic resonance imaging for stem/progenitor cell tracking with multiple unique perfluorocarbon nanobeacons. *FASEB J* **21**(8), 1647–1654 (2007).
31. Cahill, K.S., Germain, S., Byrne, B.J., and Walter, G.A. Non-invasive analysis of myoblast transplants in rodent cardiac muscle. *Int J Cardiovasc Imaging* **20**(6), 5938 (2004).

32. Louie, A.Y., Huber, M.M., Ahrens, E.T., Rothbacher, U., Moats, R., Jacobs, R.E., *et al.* *In vivo* visualization of gene expression using magnetic resonance imaging. *Nat Biotechnol* **18**(3), 321–325 (2000).
33. Himmelreich, U., Aime, S., Hieronymus, T., Justicia, C., Uggeri, F., Zenke, M., *et al.* A responsive MRI contrast agent to monitor functional cell status. *Neuroimage* **32**(3), 1142–1149 (2006).
34. Deans, A.E., Wadghiri, Y.Z., Bernas, L.M., Yu, X., Rutt, B.K., and Turnbull, D.H. Cellular MRI contrast via coexpression of transferrin receptor and ferritin. *Magn Reson Med* **56**(1), 519 (2006).
35. Park, S.J., Rogers, W.L., and Clinthorne, N.H. Design of a very high-resolution small animal PET scanner using a silicon scatter detector insert. *Phys Med Biol* **52**(15), 4653–4657 (2007).
36. Stavngaard, T. and Mortensen, J. Assessment of ventilation inhomogenity with Krypton SPECT and planar imaging. *Clin Physiol Funct Imaging* **25**(2), 106–112 (2005).
37. Brenner, W., Aicher, A., Eckey, T., Massoudi, S., Zuhayra, M., Koehl, U., *et al.* ^{111}In-labeled CD34+ hematopoietic progenitor cells in a rat myocardial infarction model. *J Nucl Med* **45**(3), 512–518 (2004).
38. Adonai, N., Nguyen, K.N., Walsh, J., Iyer, M., Toyokuni, T., Phelps, M.E., *et al.* *Ex vivo* cell labeling with 64Cu pyruvaldehyde-bis (N4-methyl-thiosemicarbazone) for imaging cell trafficking in mice with positron-emission tomography. *Proc Natl Acad Sci USA* **99**(5), 3030–3035 (2002).
39. Zhou, R., Thomas, D.H., Qiao, H., Bal, H.S., Choi, S.R., Alavi, A., *et al.* *In vivo* detection of stem cells grafted in infarcted rat myocardium. *J Nucl Med* **46**(5), 816–822 (2005).
40. Chung, S., Andersson, T., Sonntag, K.C., Bjorklund, L., Isacson, O., and Kim, K.S. Analysis of different promoter systems for efficient transgene expression in mouse embryonic stem cell lines. *Stem Cells* **20**(2), 139–145 (2002).
41. Acton, P.D. and Zhou, R. Imaging reporter genes for cell tracking with PET and SPECT. *Q J Nucl Med Mol Imaging* **49**(4), 349–360 (2005).
42. Wu, J.C., Inubushi, M., Sundaresan, G., Schelbert, H.R., and Gambhir, S.S. Optical imaging of cardiac reporter gene expression in living rats. *Circulation* **105**(14), 1631–1634 (2002).
43. Chung, J.K. Sodium iodide symporter: its role in nuclear medicine. *J Nucl Med* **43**(9), 1188–1200 (2002).

44. Shin, J.H., Chung, J.K., Kang, J.H., Lee, Y.J., Kim, K.I., Kim, C.W., *et al.* Feasibility of sodium/iodide symporter gene as a new imaging reporter gene: comparison with HSV1-tk. *Eur J Nucl Med Mol Imaging* **31**(3), 425–432 (2004).
45. Niu, G., Gaut, A.W., Ponto, L.L., Hichwa, R.D., Madsen, M.T., Graham, M.M., *et al.* Multimodality noninvasive imaging of gene transfer using the human sodium iodide symporter. *J Nucl Med* **45**(3), 445–449 (2004).
46. Huang, M., Batra, R.K., Kogai, T., Lin, Y.Q., Hershman, J.M., Lichtenstein, A., *et al.* Ectopic expression of the thyroperoxidase gene augments radioiodide uptake and retention mediated by the sodium iodide symporter in non-small cell lung cancer. *Cancer Gene Ther* **8**(8), 612–618 (2001).
47. Kang, J.H., Lee, D.S., Paeng, J.C., Lee, J.S., Kim, Y.H., Lee, Y.J., *et al.* Development of a sodium/iodide symporter (NIS)-transgenic mouse for imaging of cardiomyocytespecific reporter gene expression. *J Nucl Med* **46**(3), 479–483 (2005).
48. Marsee, D.K., Shen, D.H., MacDonald, L.R., Vadysirisack, D.D., Lin, X., Hinkle, G., *et al.* Imaging of metastatic pulmonary tumors following NIS gene transfer using single photon emission computed tomography. *Cancer Gene Ther* 11(**2**), 121–127 (2004).
49. MacLaren, D.C., Toyokuni, T., Cherry, S.R., Barrio, J.R., Phelps, M.E., Herschman, H.R., *et al.* PET imaging of transgene expression. *Biol Psychiatry* **48**(5), 337–348 (2000).
50. Suzuki, K., Murtuza, B., Smolenski, R.T., Sammut, I.A., Suzuki, N., Kaneda, Y., *et al.* Cell transplantation for the treatment of acute myocardial infarction using vascular endothelial growth factor-expressing skeletal myoblasts. *Circulation* **104**(12 Suppl 1), I207–I212 (2001).
51. Ray, P., De, A., Min, J.J., Tsien, R.Y., and Gambhir, S.S. Imaging tri-fusion multimodality reporter gene expression in living subjects. *Cancer Res* **64**(4), 1323–1330 (2004).
52. Azar, F.S. and Intes, X. *Translational Multimodality Optical Imaging* (Aretech House, Boston, London, 2008).
53. Contag, C.H., Contag, P.R., Mullins, J.I., Spilman, S.D., Stevenson, D.K., and Benaron, D.A. Photonic detection of bacterial pathogens in living hosts. *Mol Microbiol* **18**(4), 593–603 (1995).

54. Weissleder, R. and Ntziachristos, V. Shedding light onto live molecular targets. *Nat Med* **9**(1), 123–128 (2003).
55. Gibson, A.P., Hebden, J.C., and Arridge, S.R. Recent advances in diffuse optical imaging. *Phys Med Biol* **50**(4), R1–R43 (2005).
56. Mourant, J.R., Freyer, J.P., Hielscher, A.H., Eick, A.A., Shen, D., and Johnson, T.M. Mechanisms of light scattering from biological cells relevant to noninvasive optical-tissue diagnostics. *Appl Opt* **37**(16), 586–593 (1998).
57. Culver, J.P., Durduran, T., Furuya, D., Cheung, C., Greenberg, J.H., and Yodh, A.G. Diffuse optical tomography of cerebral blood flow, oxygenation, and metabolism in rat during focal ischemia. *J Cereb Blood Flow Metab* **23**(8), 911–924 (2003).
58. Salata, O. Applications of nanoparticles in biology and medicine. *J Nanobiotechnology* **2**(1), 3 (2004).
59. Leff, D.R., Warren, O.J., Enfield, L.C., Gibson, A., Athanasiou, T., Patten, D.K., *et al.* Diffuse optical imaging of the healthy and diseased breast: a systematic review. *Breast Cancer Res Treat* **108**(1), 9–22 (2008).
60. Hillman, E.M., Hebden, J.C., Schweiger, M., Dehghani, H., Schmidt, F.E., Delpy, D.T., *et al.* Time resolved optical tomography of the human forearm. *Phys Med Biol* **46**(4), 1117–1130 (2001).
61. Koenig, A., Herve, L., Josserand, V., Berger, M., Boutet, J., Da Silva, A., *et al. In vivo* mice lung tumor follow-up with fluorescence diffuse optical tomography. *J Biomed Opt* **13**(1), 011008 (2008).
62. Huang, D., Swanson, E.A., Lin, C.P., Schuman, J.S., Stinson, W.G., Chang, W., *et al.* Optical coherence tomography. *Science* **254**(5035), 1178–1181 (1991).
63. Testoni, P.A. and Mangiavillano, B. Optical coherence tomography in detection of dysplasia and cancer of the gastrointestinal tract and bilio-pancreatic ductal system. *World J Gastroenterol* **14**(42), 6444–6452 (2008).
64. Drexler, W., Morgner, U., Kartner, F.X., Pitris, C., Boppart, S.A., Li, X.D., *et al. In vivo* ultrahigh-resolution optical coherence tomography. *Opt Lett* **24**(17), 1221–1223 (1999).
65. Tearney, G.J., Brezinski, M.E., Southern, J.F., Bouma, B.E., Boppart, S.A., and Fujimoto, J.G. Optical biopsy in human urologic tissue using optical coherence tomography. *J Urol* **157**(5), 1915–1919 (1997).

66. Seitz, U., Freund, J., Jaeckle, S., Feldchtein, F., Bohnacker, S., Thonke, F., *et al.* First *in vivo* optical coherence tomography in the human bile duct. *Endoscopy* **33**(12), 1018–1021 (2001).
67. Brand, S., Poneros, J.M., Bouma, B.E., Tearney, G.J., Compton, C.C., and Nishioka, N.S. Optical coherence tomography in the gastrointestinal tract. *Endoscopy* **32**(10), 796–803 (2000).
68. Yang, Y., Whiteman, S., Gey, van Pittius D., He, Y., Wang, R.K., and Spiteri, M.A. Use of optical coherence tomography in delineating airways microstructure: comparison of OCT images to histopathological sections. *Phys Med Biol* **49**(7), 1247–1255 (2004).
69. Whiteman, S.C., Yang, Y., Gey, van Pittius D., Stephens, M., Parmer, J., and Spiteri, M.A. Optical coherence tomography: real-time imaging of bronchial airways microstructure and detection of inflammatory/neoplastic morphologic changes. *Clin Cancer Res* **12**(3 Pt 1), 813–818 (2006).
70. Lam, S., Standish, B., Baldwin, C., McWilliams, A., leRiche, J., Gazdar, A., *et al. In vivo* optical coherence tomography imaging of preinvasive bronchial lesions. *Clin Cancer Res* **14**(7), 2006–2011 (2008).
71. Brenner, M., Kreuter, K., Ju, J., Mahon, S., Tseng, L., Mukai, D., *et al. In vivo* optical coherence tomography imaging of preinvasive bronchial lesions. *Clin Cancer Res* **13**(3), 034001 (2008).
72. Mason, C., Markusen, J.F., Town, M.A., Dunnill, P., and Wang, R.K. Doppler optical coherence tomography for measuring flow in engineered tissue. *Biosens Bioelectron* **20**(3), 414–423 (2004).
73. Amos, W.B. and White, J.G. How the confocal laser scanning microscope entered biological research. *Biol Cell* **95**(6), 335–342 (2003).
74. Goetz, M. and Kiesslich, R. Confocal endomicroscopy: *in vivo* diagnosis of neoplastic lesions of the gastrointestinal tract. *Anticancer Res* **28**(1B), 353–360 (2008).
75. Pohl, H., Rosch, T., Vieth, M., Koch, M., Becker, V., Anders, M., *et al.* Miniprobe confocal laser microscopy for the detection of invisible neoplasia in patients with Barrett's oesophagus. *Gut* **57**(12), 1648–1653 (2008).
76. Meining, A., Saur, D., Bajbouj, M., Becker, V., Peltier, E., Hofler, H., *et al.* *In vivo* histopathology for detection of gastrointestinal neoplasia with a portable, confocal miniprobe: an examiner blinded analysis. *Clin Gastroenterol Hepatol* **5**(11), 1261–1267 (2007).

77. Thiberville, L., Moreno-Swirc, S., Vercauteren, T., Peltier, E., Cave, C., and Bourg Heckly, G. *In vivo* imaging of the bronchial wall microstructure using fibered confocal fluorescence microscopy. *Am J Respir Crit Care Med* **175**(1), 22–31 (2007).
78. Koenig, F., Knittel, J., and Stepp, H. Diagnosing cancer *in vivo. Science* **292**(5520), 1401–1403 (2001).
79. Kiesslich, R., Goetz, M., Burg, J., Stolte, M., Siegel, E., Maeurer, M.J., *et al.* Diagnosing Helicobacter pylori *in vivo* by confocal laser endoscopy. *Gastroenterology* **128**(7), 2119–2123 (2005).
80. Meining, A., Bajbouj, M., and Schmid, R.M. Confocal fluorescence microscopy for detection of gastric angiodysplasia. *Endoscopy* **39**(Suppl 1), E145 (2007).
81. Becker, V., Vercauteren, T., von Weyhern, C.H., Prinz, C., Schmid, R.M., and Meining A. High-resolution miniprobe-based confocal microscopy in combination with video mosaicing (with video). *Gastrointest Endosc* **66**(5), 1001–1007 (2007).
82. Reddymasu, S.C. and Sharma, P. Advances in endoscopic imaging of the esophagus. *Gastroenterol Clin North Am* **37**(4), 763–774 (2008).
83. Shibuya, K., Hoshino, H., Chiyo, M., Iyoda, A., Yoshida, S., Sekine, Y., *et al.* High magnification bronchovideoscopy combined with narrow band imaging could detect capillary loops of angiogenic squamous dysplasia in heavy smokers at high risk for lung cancer. *Thorax* **58**(11), 989–995 (2003).
84. Anandasabapathy, S. Endoscopic imaging: emerging optical techniques for the detection of colorectal neoplasia. *Curr Opin Gastroenterol* **24**(1), 64–69 (2008).
85. Olliver, J.R., Wild, C.P., Sahay, P., Dexter, S., and Hardie, L.J. Chromoendoscopy with methylene blue and associated DNA damage in Barrett's oesophagus. *Lancet* **362**(9381), 373–374 (2003).
86. Stevens, P.D., Lightdale, C.J., Green, P.H., Siegel, L.M., Garcia-Carrasquillo, R.J., and Rotterdam, H. Combined magnification endoscopy with chromoendoscopy for the evaluation of Barrett's esophagus. *Gastrointest Endosc* **40**(6), 747–749 (1994).
87. Conchello, J.A., and Lichtman, J.W. Optical sectioning microscopy. *Nat Methods* **2**(12), 920–931 (2005).

88. Wood, K.V., Lam, Y.A., Seliger, H.H., and McElroy, W.D. Complementary DNA coding click beetle luciferases can elicit bioluminescence of different colors. *Science* **244**(4905), 700–702 (1989).
89. Hara, M., Murakami, T., and Kobayashi, E. *In vivo* bioimaging using photogenic rats: fate of injected bone marrow-derived mesenchymal stromal cells. *J Autoimmun* **30**(3), 163–171 (2008).
90. Vintersten, K., Monetti, C., Gertsenstein, M., Zhang, P., Laszlo, L., Biechele, S., *et al.* Mouse in red: red fluorescent protein expression in mouse ES cells, embryos, and adult animals. *Genesis* **40**(4), 241–246 (2004).
91. Tao, W., Evans, B.G., Yao, J., Cooper, S., Cornetta, K., Ballas, C.B., *et al.* Enhanced green fluorescent protein is a nearly ideal long-term expression tracer for hematopoietic stem cells, whereas DsRed-express fluorescent protein is not. *Stem Cells* **25**(3), 670–678 (2007).
92. Hadjantonakis, A.K., Macmaster, S., and Nagy, A. Embryonic stem cells and mice expressing different GFP variants for multiple non-invasive reporter usage within a single animal. *BMC Biotechnol* **2**, 11 (2002).
93. Shcherbo, D., Merzlyak, E.M., Chepurnykh, T.V., Fradkov, A.F., Ermakova, G.V., Solovieva, E.A., *et al.* Bright far-red fluorescent protein for whole-body imaging. *Nat Methods* **4**(9), 741–746 (2007).
94. Sherman, W., Martens, T.P., Viles-Gonzalez, J.F., and Siminiak, T. Catheter-based delivery of cells to the heart. *Nat Clin Pract Cardiovasc Med* **3**(Suppl 1), S57–S64 (2006).
95. Karmarkar, P.V., Kraitchman, D.L., Izbudak, I., Hofmann, L.V., Amado, L.C., Fritzges, D., *et al.* MR-trackable intramyocardial injection catheter. *Magn Reson Med* **51**(6), 1163–1172 (2004).
96. Yang, X. and Wang, L.V. Monkey brain cortex imaging by photoacoustic tomography. *J Biomed Opt* **13**(4), 044009 (2008).
97. Hoelen, C.G., de Mul, F.F., Pongers, R., and Dekker, A. Three-dimensional photoacoustic imaging of blood vessels in tissue. *Opt Lett* **23**(8), 648–650 (1998).
98. Wallace, V.P., Fitzgerald, A.J., Pickwell, E., Pye, R.J., Taday, P.F., Flanagan, N., *et al.* Terahertz pulsed spectroscopy of human basal cell carcinoma. *Appl Spectrosc* **60**(10), 1127–1133 (2006).
99. Fitzgerald, A.J., Wallace, V.P., Jimenez-Linan, M., Bobrow, L., Pye, R.J., Purushotham, A.D., *et al.* Terahertz pulsed imaging of human breast tumors. *Radiology* **239**(2), 533–540 (2006).

100. Stoyanov, D., Mylonas, G.P., Deligianni, F., Darzi, A., and Yang, G.Z. Soft-tissue motion tracking and structure estimation for robotic assisted MIS procedures. *Med Image Comput Comput Assist Interv Int Conf Med Image Comput Comput Assist Interv* **8**(Pt 2), 139–146 (2005).
101. Yang, G.Z., Mylonas, G.P., Kwok, K.W., and Chung, A. Perceptual docking for robotic control. In: Dohi, T., Sakuma, I., Liao, H. (eds.), *Medical Imaging and Augmented Reality* (Springer Berlin/Heidelberg, 2008).
102. Mylonas, G.P., Darzi, A., and Yang, G.Z. Gaze-contingent control for minimally invasive robotic surgery. *Comput Aided Surg* **11**(5), 256–266 (2006).
103. Sutherland, G.R., Latour, I., and Greer, A.D. Integrating an image-guided robot with intraoperative MRI: a review of the design and construction of neuroArm. *IEEE Eng Med Biol Mag* **27**(3), 59–65 (2008).

Chapter 16
Haematopoietic Transplantation and Pulmonary Chimerism

Benjamin T. Suratt

The discovery of donor-derived cells incorporating and apparently differentiating into human pulmonary tissue following haematopoietic stem cell transplantation (HSCT) was first reported in 2003, and subsequent studies on this phenomenon using ever more sophisticated techniques have described a wide range of findings. Thus far, it has been suggested that epithelial, endothelium, and interstitial tissues of the lung may be chimeric following HSCT, but debate about the extent and even existence of such chimerism continues. Although this phenomenon has only been described in humans in the incidental setting of transplantation, the implication is that such a process occurs in the absence of such conditions, relying on *endogenous* circulating marrow-derived stem cells to repair or remodel the lung. Whether these reports represent some other source of chimerism (such as maternal–foetal) or even cellular fusion, the incorporation of donor-derived cells into the lung, should it represent not only antigenic but functional conversion to lung tissue and be inducible at sufficient frequency, would present broad therapeutic possibilities. Furthermore, the study of this phenomenon may yield insights into disease pathogenesis that present even further therapeutic avenues, beyond the use of stem cells themselves.

Keywords: Pulmonary chimerism; haematopoietic stem cell transplantation; epithelium; endothelium; fibroblasts.

Outline

1. Introduction

The term "pulmonary chimerism" describes the finding of circulating non-autologous cells that have incorporated into the lung and "become" tissue-specific structural cells, including epithelium, endothelium, or interstitium (e.g. fibroblasts). This finding has been described after allogeneic haematopoietic stem cell transplantation (HSCT), lung allografting, and several other clinical settings. Whether, in the case of HSCT, this process reflects donor stem cell differentiation or transdifferentiation, fusion between donor and host cells, or perhaps simply artefact of the detection techniques remains an open question, and reports of this phenomenon have been met with both excitement and scepticism. Clinical settings such as HSCT provide a unique opportunity to examine the behaviour of adult stem populations in humans because, similarly to many animal experiments, they provide a "model" of stem cell tracking due to differing detectable markers that allow the distinction of host- and donor-derived cells in tissues. This chapter will focus on the evidence for pulmonary chimerism following HSCT, as the majority of human data reported fall in this category, and this literature will be discussed in the greater context of other described settings of such chimerism, as well as the growing body of animal modelling data that dissects this phenomenon.

2. A Brief Review of HSCT Concepts and Techniques

The clinical use of haematopoietic stem cell transplantation is predicated on the finding that infused whole bone marrow, either autologous or syngeneic, can readily repopulate the entire lymphohaematopoietic system of an animal following radioablation.[1] Clinical use of HSCT was pioneered by E. Donnall Thomas and colleagues in leukaemia patients in 1959,[2] following work exploring this technique in canine models. Early experiences were restricted to syngeneic (identical) twins as donor and recipient, until allogeneic transplantation became feasible in the late 1960s following the discovery of the major histocompatibility complex (MHC) and its role in cellular immunity and rejection. In the 1970s, Thomas and others reported the first allogeneic bone marrow transplants in human leukocyte antigen (HLA)-matched siblings,[3] and since this original patient series, the use of both autologous and allogeneic HSCT has grown dramatically, with more than 45,000 transplants per year currently performed worldwide.[4] HSCT continues to be used primarily in the treatment of malignancy, but its use has expanded to include a wide range of non-malignant haematologic and immune diseases as well.

Historically, HSCT techniques consisted of therapeutic "conditioning" or ablation of the recipient's haematopoietic system using total body irradiation (TBI), followed by the intravenous infusion of donor bone marrow cells to engraft the host and recapitulate the haematopoietic system. Over the past 20 years, the process of HSCT has evolved to include complex conditioning regimens, multiple sources of haematopoietic stem cells, and the use of multiple pharmacologic agents to enhance haematopoietic engraftment or, in the case of allogeneic HSCT, suppress graft versus host immune response (GVHD). Commonly used conditioning regimens include TBI in conjunction with an alkylating agent (such as cyclophosphamide), radiation-free regimens (such as carmustine, etoposide, cytarabine, and melphalan or "BEAM"), and even some reduced-intensity regimens that are incompletely myeloablative (such as are used in "mini" allogeneic HSCT). All conditioning regimens are associated with some degree of regimen-related toxicity, with the lung as the most frequently injured organ.

Although HSCT is still commonly referred to as "bone marrow transplant", marrow is now relatively rarely used as a source of HSC, having been superseded by peripheral blood HSC collected by leukaphoresis from HLA-matched siblings or a matched unrelated donor (MUD). HSC, typically selected or quantified based on the CD34 cell surface marker, may also be derived from umbilical cord blood immediately following birth. Although its use in adults is still limited, cord blood, which is typically banked, provides a critical resource for patients lacking an HLA-matched donor, as its use is less often associated with GVHD and thus allows a greater degree of HLA-mismatching in HSCT.[5] All three sources of HSC (bone marrow, peripheral blood and cord blood) differ somewhat in their overall cellular composition and the phenotype of the CD34+ cells present, and hence their behaviours differ following HSCT.[4]

As much as animal modelling using HSCT provides an excellent method for the investigation of HSC trafficking and plasticity, clinical HSCT provides a relevant "model" for this work in humans, particularly given the availability of large banks of stored biopsy and autopsy specimens related to the care of these patients.

3. Techniques for the Detection of Pulmonary Chimerism Following HSCT

Numerous techniques have been developed in animal models for detecting the incorporation of donor-derived cells into the lungs of recipient animals (chimerism) following HSCT. These range from the most common, such as the use of marrow from transgenic donor animals that ubiquitously and constitutively express a reporter protein (e.g. green fluorescent protein), to more complex approaches, such as the use of wild type marrow transplanted into knockout mice lacking specific proteins (e.g. CFTR),[6] or donor marrow transgenic for a lung-specific promoter-driven reporter (e.g. Sp-C promoter-driven GFP)[7] into wild type mice. In the search for adult stem cell engraftment of human tissues, obviously, none of these approaches for detection are feasible. Thus, researchers have been left to study the incidental occurrence of this approach to "modelling", that arises clinically from allogeneic haematopoietic stem cell transplantation, as well as lung allografting. Such an examination is feasible only in instances in which 1) differing donor- and host-associated cell markers are reliably detectable; 2) discrimination between haematopoietic and non-haematopoietic cells is possible; and 3) sufficient utilisable lung tissue from either biopsy or autopsy is available for study.

To date, at least six groups have published studies examining clinical samples for evidence of respiratory tissue chimerism following HSCT (see Table 1) and, in all but two, detection of donor cells has relied on the sex chromosomes as the identifying marker. This approach employs the use of *in situ* hybridisation (ISH) of both X and Y chromosomes, and is thus predicated on the examination of tissue samples from sex-mismatched HSCT patients (i.e. female patients who have received male HSC, or less commonly, the reverse). XY ISH has most often been performed in conjunction with immunostaining of relevant tissue-specific markers, such as cytokeratin (for epithelium), performed either sequentially on the same slide, or independently on adjacent slides cut from the same tissue block. Given the loss of tissue-specific peptide markers (e.g. cytokeratin) that accompanies most approaches to XY ISH (which requires fairly aggressive antigen retrieval techniques that can result in destruction of cell surface epitopes utilised for immunostaining), most authors have taken the approach of performing immunostaining of the tissue section first when attempting to correlate cell tissue-type (e.g. epithelium) and derivation (donor or recipient) on the same slide. Staining is typically followed by photomicroscopic mapping of the section before ISH is performed, after which the section is again mapped, and the corresponding images from immunostaining and ISH are then compared or overlaid (see Fig. 1). Such an approach is superior to serial sectioning, in which hybridised nuclei in one section may not accurately associate with the immunostained cytoplasm/membrane seen in adjacent sections, but suffers from significant loss of tissue morphology. Attempts at simultaneous dual-immunostaining/ISH in HSCT patient samples have been limited by loss of tissue-specific markers and severe degradation of overall tissue morphology.[8] Recently, advanced techniques for dual-immunostaining/ISH without significant tissue degradation have been developed for mouse tissues.[6,9,10] Such an approach may be feasible in human samples, but may require special tissue preparation, thus potentially limiting its utility in the examination of archival tissue blocks.

Controversy surrounding the results of XY ISH studies tends to centre on the possibility of erroneous inclusion of leukocytes, which, following HSC engraftment, are typically greater than 95% donor in origin at the time points examined.[11,12] Such a possibility is due in large part to the

Table 1: Studies examining clinical samples for evidence of respiratory tissue chimerism following HSCT.

Paper	Technique/ Markers	Pt #	Diagnosis at Xplant	Age at Xplant (y)	Conditioning	HSC Source	HLA Match	Sex	GVHD	Time after Xplant	Sample Source	Lung Pathology	% Donor-Derived Epith	Endo	Fibro
Davies	XY ISH/IS	1	ALL	NR	NR	BM	6/6 related	F	NR	10y	Nasal	N/A	0.0	N/A	NR
	Pancyto	2	CML	NR	NR	BM	6/6 MUD	F	NR	8y	Brush	N/A	0.0	N/A	NR
		3	ALL	NR	NR	BM	6/6 related	F	NR	15y	"	N/A	0.0	N/A	NR
		4	ALL	NR	NR	BM	6/6 MUD	F	NR	10y	"	N/A	0.0	N/A	NR
		5	ALL	NR	NR	BM	6/6 related	F	NR	2y	"	N/A	0.0	N/A	NR
		6	CML	NR	NR	BM	6/6 related	F	NR	1y	"	N/A	0.0	N/A	NR
Suratt	XY ISH/IS	1	HD	24	TBI/Mel/ATG	CB	5/6 MUD	F	No	200d	Autopsy	DAD	2.5	35.7	NR
	Pancyto/CD31/	2	Breast	32	Bu/Mel/ATG	CB	6/6 MUD	F	No	50d	Autopsy	DAD	0.0	0.0	NR
	CD45	3	NHL	40	TBI/Mel/ATG	CB	5/6 MUD	F	Yes	462d	Surgical	OB	8.0	42.3	NR
		4	CML	34	Bu/Mel/ATG	BM/PBSC	6/6 related	F	Yes	225d†	Autopsy	OB	0.0	0.0	NR
Kleeberger	LCM/STR	8	AC	13	NR	NR	NR	NR	NR	246d	TBBX	Aspergillus	0.0	ND	NR
	Pancyto/CD45/	9	AA	5	NR	NR	NR	NR	NR	75d	TBBX	Aspergillus	0.0	ND	NR
	CD68	10	CLL	57	NR	NR	NR	NR	NR	313d	TBBX	Pneumonitis	0.0	ND	NR
Mattsson	XY ISH/IS	1	RCC	47	Flu/TBI/ATG*	PBSC	6/6 MUD	F	Yes	60d	Autopsy	NR	2.5–3.0	ND	NR
	Pancyto/SpA/CD68	2	HCC	58	Flu/TBI*	PBSC	6/6 related	F	Yes	480d	Autopsy	NR	2.0–6.0	ND	NR
Albera	XY ISH/IS	9	Lymph	38	NR	BM	NR	F	NR	38d	Autopsy	Chronic injury	"+"	"+"	NR
	SpB/SpC/Pancyto/	10	Lymph	34	NR	BM	NR	F	NR	51d	Autopsy	Chronic injury	"+"	"+"	NR
	SMA/CD45	11	Lymph	35	NR	BM	NR	F	NR	65d	Autopsy	Chronic injury	"+"	"+"	NR
Zander	XY ISH/IS	1	AML	44	NR	PBSC	NR	F	NR	15d	TBBX	DAD	0.0	ND	NR
	SpB/EMA	2	HD	41	NR	PBSC	NR	F	NR	53d	TBBX	Pneumonitis	1.8	ND	NR
		3	NHL	20	NR	PBSC	NR	F	NR	23d	TBBX	DAD	0.0	ND	NR
		4	CML	54	NR	BM	NR	F	NR	314d	TBBX	DAD	0.0	ND	NR
Brocker	LCM/STR vs. ISH	13	AML	41	NR	NR	NR	M	Yes	450d	Surgical	OB	ND	ND	"+"
	CD45/CD68/SMA	14	OMF	30	NR	NR	NR	F	Yes	3y	Explant	OB	ND	ND	6.4–13.2

AA = Aplastic anaemia; AC = agranulocytosis; ALL = acute lymphoblastic leukaemia; AML = acute mylogenous leukaemia; ATG = antithymocyte globulin; BM = bone marrow; Bu = busulfan; CB = cord blood; CML = chronic myelogenous leukaemia; DAD = diffuse alveolar damage; EMA = epithelial membrane antigen; Endo = endothelial; Epith = epithelial; FISH = fluorescent *in situ* hybridisation; Flu = fludarabine; GVHD = graft versus host disease; HCC = hepatocelular carcinoma; HD = Hodgkin's disease; HSC = haematopoietic stem cell; IS = immunostaining; Lymph = lymphoma NOS; Mel = melphalan; MUD = matched unrelated donor; N/A = not applicable; ND = note done; NHL = non-Hodgkin's lymphoma; NR = not reported; OB = obliterative bronchiolitis; OMF = osteomyelofibrosis; Pancyto = pancytokeratin; PBSC = peripheral blood stem cells; RCC = renal cell carcinoma; SpA = pro-surfactant A protein; SpB = pro-surfactant B protein; SpC = pro-surfactant C protein; TBBX = transbronchial lung biopsy; TBI = total body irradiation; Xplant = transplant; † = no haematopoietic engraftment; * = non-myeloablative regimen.

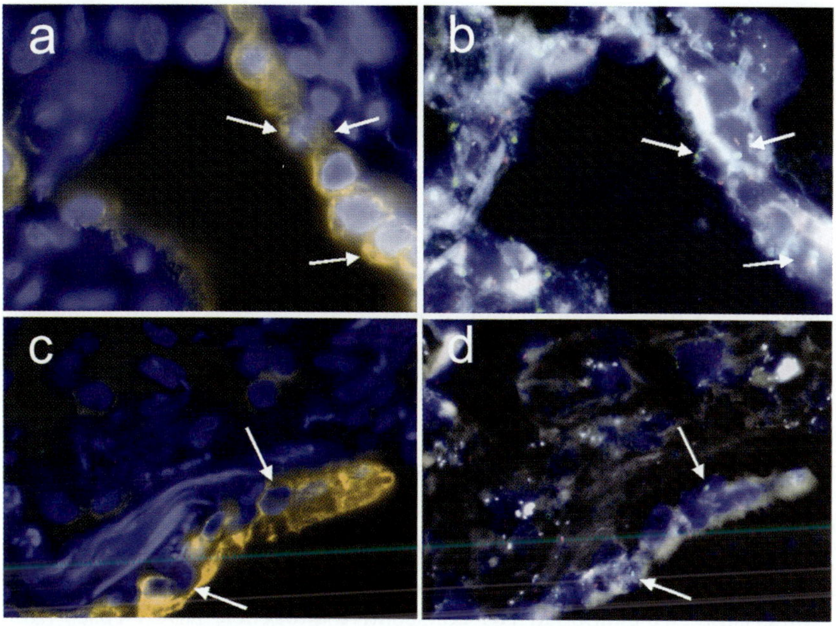

Figure 1: Examination of lung specimens from female HSCT recipients using immunostaining with epithelial cell-specific antibody (cytokeratin), followed by fluorescent ISH for the sex chromosomes. (a and c) Staining of specimens with epithelial cell-specific antibody (cytokeratin, yellow) and nuclear dye (DAPI, blue) showing distribution of epithelium in the alveolar wall (a) and terminal bronchiole (b). Arrows indicate same cells in paired images (a and b, c and d). (b and d) Same specimens labelled by *in situ* hybridisation of X (red) and Y (green) chromosomes, showing presence of donor-derived (male, green) epithelial cells (arrows) in the alveolar walls and distal airways. Reprinted from Suratt *et al.* (2003).[11] © The American Thoracic Society.

universal finding of injury and/or infection in lung biopsies or autopsy specimens of HSCT patients (see Table 1). This is accompanied by extensive tissue infiltration of leukocytes, which may be extremely difficult to identify morphologically. Several authors have attempted to exclude leukocytes using immunostaining for leukocyte-specific antigens such as CD45.[8,11–13] Yet, without multiplanar confocal imaging, which is not feasible in the absence of simultaneous dual-immunostaining/ISH, the possibility remains that leukocyte contamination may account for a significant portion of the donor-derived cells identified in these studies. A further consideration, though apparently not nearly as common, is the occurrence of cell fusion and the formation of polyploidy. Such a situation could potentially give rise to cells containing donor marker chromosomes, and thus might be misinterpreted as being donor-derived through differentiation. Thorough studies in mice have suggested that the incidence of this in the lung is extremely low following HSCT,[14,15] but it remains a potential concern in human studies. Efforts to control for this have focussed on direct examination of ploidy as part of ISH, but again this approach is dependent on the investigator's ability to evaluate completely the cell nucleus, for instance through the use of confocal microscopy.

Beyond the technical aspects, the use of XY ISH presents an additional limitation: only sex-mismatched HSCT cases are appropriate for examination. This, in addition to the considerations of tissue availability and usability, rapidly narrows the number of patient samples appropriate for study. For example, in our study of human pulmonary chimerism,[11] we examined 161 consecutive allogeneic HSCT cases from a six-year period to find 46 instances of male HSC transplantation into female recipients, and of these only four patients had usable lung tissue samples. Similar difficulties were reported by Albera and colleagues,[8] who screened >200 HSCT patients from two centres to obtain three appropriate lung samples for study.

One alternate approach reported is the use of short tandem repeat (STR) PCR. This approach relies on genetic polymorphisms that exist at high levels in the general population to define markers for both donor and recipient cell origin, and thus does not have the constraint of requiring sex-mismatched HSCT. The use of PCR, however, requires that the cells examined be retrieved from archived tissue samples using laser microdissection. This is essential in order to collect cells of specific type (e.g. bronchial epithelium), and, perhaps more importantly, to avoid leukocyte contamination. In the two published reports employing this approach to examine the lung,[13,16] the authors used laser microdissection of immunostained tissue sections, followed by PCR of a highly polymorphic STR marker within the human β-actin-related pseudogene, H-β-Ac-psi-2, in their examination of both HSCT and lung allograft patient samples (see Fig. 2). Using quantitative PCR and calibration curves similar to those used to determine mixed leukocyte chimerism in

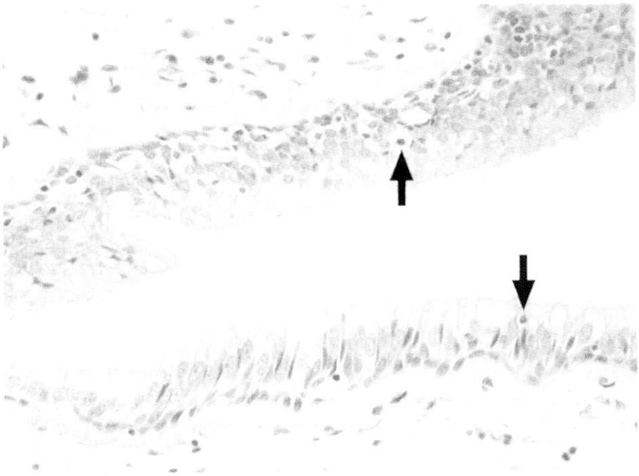

(A)

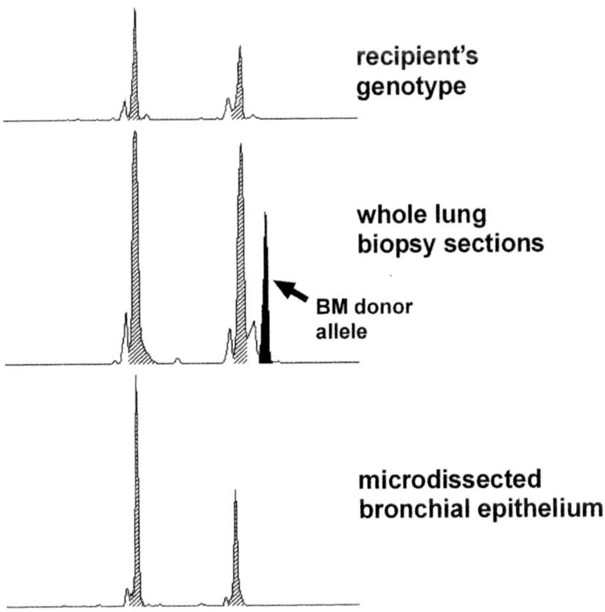

(B)

Figure 2: Short tandem repeat (STR) PCR analysis of bronchi from recipients of bone marrow transplantation. (A) Representative bronchial epithelium from one patient who had received therapeutic bone marrow transplantation. Note the infiltration of single lymphocytes into the epithelial layer (arrows). (B) Electropherograms demonstrating the absence of chimerism within microdissected bronchi (see text for details). Reprinted from Kleeberger *et al.* (2003).[13] With permission from the American Society for Investigative Pathology.

HSCT, the authors were also able to estimate the degree of tissue chimerism present. The limitations of this approach include a requirement for greater quantities of archival tissue compared to ISH (with a higher detection threshold), as well as the availability of pre-transplant patient tissue samples (for derivation of a calibration curve). It is important to recognise, also, that this approach is perhaps even more susceptible to error than ISH, due to leukocyte contamination or cell fusion, as it relies critically on accurate laser microdissection as the only safeguard against such error.

4. Evidence for Pulmonary Chimerism Following HSCT

As with reports of experimental animal models of pulmonary chimerism following HSCT, the findings of human studies vary widely (Table 1). Following the initial report of Davies and colleagues,[17] who found no evidence of donor-derived upper respiratory epithelium from nasal brushes of HSCT patients, the primary focus of investigators has been on pulmonary epithelial chimerism following HSCT. Starting with our group's report in 2003,[11] which suggested levels of epithelial chimerism of up to 8%, several other investigators have reported the presence of donor-derived epithelial cells in the airways and alveoli of HSCT recipients. Among other findings of the collected reports, the occurrence of chimerism may be related to tissue injury,[11,13] and appears to predominate in the alveoli more than the airways.[8,11,12] These findings are consistent with animal experiments suggesting that a threshold of injury must be reached before chimerism occurs,[18,19] and that epithelial chimerism more commonly occurs in the alveoli than the airways.[19–22] Comparable work has been reported from another clinical "model" of human lung chimerism examining recipients of lung transplant.[8,13,16,23–26] Much of this work has been presented by the same groups reporting findings in HSCT patients (often in the same publications), and comes to a similar array of conclusions, but with a wider range of epithelial chimerism rates (0 to 26%).

As shown in Table 1, although some have found significant epithelial engraftment of the lung,[8,11,12] others have found vanishingly little,[25] or none at all[13] following HSCT. This is consistent with the findings of various animal models,[10] and gives rise to speculation regarding the source of such variability. As may be appreciated from the data presented in Table 1, the patients examined in the seven reports to date vary widely in a number of potentially important clinical characteristics, among them transplant conditioning regimen, HSCT graft source, duration of HSC engraftment at sampling, and the degree/mechanism of tissue injury present in the lung samples.

Conditioning regimen and severity of tissue injury have previously been shown to alter levels of lung chimerism following HSCT in mouse models.[18,19] In particular, the use of total body irradiation (TBI) has been associated with greater levels of chimerism, thus, one might expect patients conditioned with TBI to have higher levels of chimerism.[19] This was seen in our series, in which the two patients with measurable chimerism had received TBI conditioning, whilst a third patient who had not received TBI (but who had haematopoietic engraftment) showed no evidence of chimerism.[11] Other factors related to tissue injury may be operative as well, but there exists no clear correlation when examining the cumulative patient series, at least on the basis of reported lung histological changes (see Table 1). In animal models, the source of transplanted HSC also appears to be relevant, as the engraftment of different tissues of the lung may be dependent on the population of cells infused during transplantation.[27] Examination of this variable in the reported human subjects is limited, but no trend is evident (Table 1). Similarly, time since transplant varies significantly among the reported patients (15d to over 3y) and has been hypothesised to affect levels of tissue chimerism;[11] yet there is again no consistent correlation between pulmonary chimerism and this variable across the entire cohort (Table 1).

Non-clinical variables in the reported literature are likely to have had significant effects on the results as well. These include the means of lung tissue sampling and the chimerism detection techniques used. Lung tissue may be sampled in one of three ways: transbronchial biopsy (TBBX) during bronchoscopy (typically very small — a few millimetres — with significant architectural distortion), surgical biopsy (larger — 1–2 centimetres — with preserved architecture), and autopsy (large, but often with some degree of tissue degradation). Reports that have relied on TBBX samples[13,25] have shown very little evidence of chimerism, and in at least one report[13] the authors comment on the possibility that extremely small biopsy size may have limited their ability to detect donor-derived cells (for instance alveolar tissue could not be examined). The use of autopsy specimens has also been associated with difficulties due to loss tissue morphology and failure of antigen retrieval.[8] As discussed in the previous section, aspects of detection technique also are likely to alter results. In the case of ISH, for instance, it has been shown that antigen (XY) retrieval may be markedly different between leukocytes and epithelial cells in the same sample.[28] Thus, retrieval conditions must be tailored to the cell-type of interest, and donor marker detection normalised to the overall detection rate of host marker (e.g. %XY/%XX), or the findings may appear falsely low or negative. Lastly, one of the major concerns raised by the disparate reported findings is that reports of donor-derived cell incorporation may be altogether erroneous, and in fact represent artefacts of the techniques used. This possibility will be discussed in greater detail in the next section.

Chimerism of lung tissues other than the epithelium has been reported as well following human HSCT (see Table 1). The appearance of donor-derived *endothelial* cells, in particular, was quite significant in our series (36–42%),[11] and was also noted in the report of Albera and colleagues.[8] A more recent report examining sex-mismatched lung allografts by ISH[26] found a much lower incidence of endothelial chimerism (1.3–2.1%), but the meaning of this discrepancy is unclear. Several reasons may be suggested: 1) the mechanism of HSCT (infusion or donor cells that pass through the lung) and the particular cell populations infused (marrow and cord blood-derived cells in particular)[29] may facilitate engraftment of the endothelium; 2) the lungs examined in the lung transplant study were only mildly injured while endothelial chimerism may require more severe injury; and 3) lungs from this study were sampled only by transbronchial biopsy, with limited tissue available for examination. Although the presence of donor-derived *interstitial* tissue in the lung has been demonstrated in a number animal studies,[30–32] it has only been described in one human study to date.[16] In this study, the authors examined two HSCT patients with obliterative bronchiolitis, a fibrosing process of the small airways, and found donor-derived fibroblasts in the lesions of both patients (examined by STR), accounting for 6–13% of these cells in one patient sample when examined by ISH. Similar results were seen in lung transplant patients in the same study.

5. Possible Origins of Pulmonary Chimerism

The excitement that has surrounded reports of human pulmonary chimerism following HSCT arises due to the view that this finding may reflect the ability of circulating adult stem cells to not only incorporate but *differentiate* into elements of human lung tissue. However, substantial controversy has arisen as to the mechanisms behind this phenomenon, and hence its biological significance and therapeutic potential. Certainly, as suggested in numerous reports, there are other potential aetiologies, both biological and methodological, to explain these findings.

One possibility is the occurrence of so-called maternal–foetal chimerism,[33] in which it is believed that foetal cells transit to the maternal circulation and may embed in maternal tissues, including the lung. Studies have suggested that this phenomenon may occur following blood transfusions as well.[34,35] One recent report describes the occurrence of lung chimerism in tissue from 10/38 women examined who had borne male children.[36] Although the male cells described were found largely in the alveolar septae and airspaces of the samples (suggesting they were leukocytes), the authors did not use immunohistology, and cell types could therefore not be determined. Of the reported HSCT patients described in Table 1, transfusion status is unknown (but suspected to be fairly prevalent given the clinical setting), and only two reports (Suratt *et al.*,[11] and Zander *et al.*[25]) include data on pregnancies (two patients had male children; patients 4 and 2, in the respective series). However, studies using STR should distinguish foetal- or blood donor-derived cells from HSC donor-derived cells, and none have been reported.[37] Thus, although tissue chimerism related to pregnancy or transfusion cannot be completely ruled out in many of the reported studies, it appears unlikely to explain their findings.

Cell fusion, in which two or more cells fuse to form one cell, typically with a polyploid nucleus, is another possible explanation for the findings of "donor-derived" cells in the lung, and has received growing attention in relation to the reported chimerism studies. Although data on cellular fusion in the lung is limited, mouse models have either shown no evidence of this process[14] or that it may occur at extremely low rates and only in the setting of repeated injury to the lung.[15] This second study is, nevertheless, relevant, as the setting of recurrent lung injury applies to most human data reported. None of the clinical studies using ISH, however, have found significant levels of polyploidy (though this may not be sufficient to rule out fusion);[15] and in our study we found that despite multiple HSC infusions in one patient (#4), no chimerism was evident, suggesting that at least cell fusion during initial pulmonary vascular transit may not account for the chimerism phenomenon.

Lastly, as discussed in Section 3, serious consideration must be given to the possibility that methodological limitations and artefact may influence the findings of chimerism studies. Just as issues of tissue sample size and differential antigen retrieval may lead to an under-appreciation of chimerism, failure to effectively exclude leukocytes may lead to falsely positive results. This may occur due to use of insufficiently specific antigens and reagents to define cell type (as has been suggested may occur with cytokeratin staining),[18] failure to adequately remove leukocytes from the tissue analyzed (as may be seen with laser microdissection for STR), or inability to recognise leukocytes in the full tissue thickness (as may occur with ISH, particularly without confocal microscopy). Currently it is impossible to determine to what degree such error may account for false positive findings in the reported studies, but it should be noted that as detection techniques used in animal models have improved in the past five years, the reported rate of lung tissue chimerism in these studies has dropped substantially. This has led some to suspect that although chimerism may indeed occur, it may occur at such vanishingly low rates that, in the absence of successful techniques to increase these rates, structural engraftment of the lung through HSCT or other variants of adult stem cell infusion may yield limited therapeutic prospects.[38] Furthermore, it is critical to note that no study of archival tissue can demonstrate evidence of cellular function. Thus, irrespective of the true degree of pulmonary chimerism, it has not been shown that such chimerism results in functionally differentiated cells of the lung in humans, and such data is scarce in murine models.

6. Conclusions and Future Perspectives

The discovery of donor-derived cells incorporating and apparently differentiating into human pulmonary tissue following HSCT or lung allografting has relevance to both the therapeutic and even broader biological realms. Although this phenomenon has only been described in humans in the incidental setting of transplantation (and perhaps transfusion or pregnancy), the implication is that such a process occurs in the absence of such conditions, relying on *endogenous* circulating marrow-derived stem cells to either directly repair or remodel the lung or to act more indirectly to these ends through replenishment of tissue-resident stem cell niches.[38] As discussed elsewhere in this volume (see Chapter 5), animal models have yielded conflicting findings pertaining to the mechanisms and even relevance of such structural chimerism. Although this has recently to some degree dampened enthusiasm for the therapeutic use of marrow-derived stem cells, it remains unknown whether the development of techniques that might increase pulmonary engraftment of such cells would render this phenomenon clinically relevant. The use of cytokine manipulation, specific stem cell populations, or genctic alteration of these cells may yet yield enhanced tissue engraftment.

Beyond the concept of circulating stem cells as a sort of "therapeutic spackle", one must also consider the possibility that the processes evidenced by pulmonary chimerism may to some degree underlie human disease states such as obliterative bronchiolitis, pulmonary fibrosis, or even pulmonary hypertension. Furthermore, it is possible that certain diseases may in part reflect a failure of marrow-derived stem cells to properly respond, as has been suggested by recent human studies of both COPD[39] and ARDS.[40] It is therefore likely that even if direct strategies to use infused marrow-derived adult stem cells fail, efforts to exploit our understanding of the endogenous process may lead to more effective therapies to alter disease progression through manipulations of this axis. Moreover, the therapeutic effects of mesenchymal stem cells, although likely to be independent of significant structural engraftment, appear to be a promising new avenue for cellular therapies (see Chapter 5).

Although controversy continues to surround reports of human chimerism following both HSCT and lung allografting, it is important to recognise that whether the reports represent some other source of chimerism (such as maternal–foetal) or even cellular fusion, the incorporation of donor-derived cells into the lung, should it represent not only antigenic but functional conversion to lung tissue and be inducible at sufficient frequency, would present broad therapeutic possibilities. Furthermore, the study of this phenomenon may yield insights into disease pathogenesis that present even further therapeutic avenues, beyond the use of stem cells themselves. As clinical trials of cellular therapies progress, it will be critical to use our rapidly advancing technical expertise honed by animal studies to investigate further the behaviour and effects of adult stem cells in humans, as it is evident that much is yet to be learned.

References

1. Lorenz, E., Uphoff, D., Reid, T.R., and Shelton, E. Modification of irradiation injury in mice and guinea pigs by bone marrow injections. *J Natl Cancer Inst* **12**, 197–201 (1951).
2. Thomas, E.D., Lochte, H.L., Jr., Cannon, J.H., Sahler, O.D., and Ferrebee, J.W. Supralethal whole body irradiation and isologous marrow transplantation in man. *J Clin Invest* **38**, 1709–1716 (1959).
3. Thomas, E.D., *et al.* One hundred patients with acute leukemia treated by chemotherapy, total body irradiation, and allogeneic marrow transplantation. *Blood* **49**, 511–533 (1977).
4. Copelan, E.A. Hematopoietic stem-cell transplantation. *New Engl J Med* **354**, 1813–1826 (2006).
5. Wagner, J.E., *et al.* Transplantation of unrelated donor umbilical cord blood in 102 patients with malignant and nonmalignant diseases: influence of CD34 cell dose and HLA disparity on treatment-related mortality and survival. *Blood* **100**, 1611–1618 (2002).
6. Loi, R., Beckett, T., Goncz, K.K., Suratt, B.T., and Weiss, D.J. Limited restoration of cystic fibrosis lung epithelium *in vivo* with adult bone marrow-derived cells. *Am J Resp Crit Care Med* **173**, 171–179 (2006).

7. Kotton, D.N., Fabian, A.J., and Mulligan, R.C. Failure of bone marrow to reconstitute lung epithelium. *Am J Resp Cell Mol Biol* **33**, 328–334 (2005).
8. Albera, C., *et al.* Repopulation of human pulmonary epithelium by bone marrow cells: a potential means to promote repair. *Tissue Eng* **11**, 1115–1121 (2005).
9. Trotman, W., Beckett, T., Goncz, K.K., Beatty, B.G., and Weiss, D.J. Dual Y chromosome painting and *in situ* cell-specific immunofluorescence staining in lung tissue: an improved method of identifying donor marrow cells in lung following bone marrow transplantation. *Histochem Cell Biol* **121**, 73–79 (2004).
10. Krause, D.S. Bone marrow-derived lung epithelial cells. *Proceedings of the Am Thorac Soc* **5**, 699–702 (2008).
11. Suratt, B.T., *et al.* Human pulmonary chimerism after hematopoietic stem cell transplantation. *Am J Resp Crit Care Med* **168**, 318–322 (2003).
12. Mattsson, J., Jansson, M., Wernerson, A., and Hassan, M. Lung epithelial cells and type II pneumocytes of donor origin after allogeneic hematopoietic stem cell transplantation. *Transplantation* **78**, 154–157 (2004).
13. Kleeberger, W., *et al.* Increased chimerism of bronchial and alveolar epithelium in human lung allografts undergoing chronic injury. *Am J Pathol* **162**, 1487–1494 (2003).
14. Harris, R.G., *et al.* Lack of a fusion requirement for development of bone marrow-derived epithelia. *Science* **305**, 90–93 (2004).
15. Herzog, E.L., *et al.* Lung-specific nuclear reprogramming is accompanied by heterokaryon formation and Y chromosome loss following bone marrow transplantation and secondary inflammation. *FASEB J* **21**, 2592–2601 (2007).
16. Brocker, V., *et al.* Fibroblasts of recipient origin contribute to bronchiolitis obliterans in human lung transplants. *Am J Resp Crit Care Med* **173**, 1276–1282 (2006).
17. Davies, J.C., *et al.* Bone marrow stem cells do not repopulate the healthy upper respiratory tract. *Pediatr Pulmonol* **34**, 251–256 (2002).
18. Herzog, E.L., Van Arnam, J., Hu, B., and Krause, D.S. Threshold of lung injury required for the appearance of marrow-derived lung epithelia. *Stem Cells* **24**, 1986–1992 (2006).
19. Kotton, D.N., *et al.* Bone marrow-derived cells as progenitors of lung alveolar epithelium. *Development* **128**, 5181–5188 (2001).

20. Krause, D.S., *et al.* Multi-organ, multi-lineage engraftment by a single bone marrow-derived stem cell. *Cell* **105**, 369–377 (2001).
21. Grove, J.E., *et al.* Marrow-derived cells as vehicles for delivery of gene therapy to pulmonary epithelium. *Am J Resp Cell Mol Biol* **27**, 645–651 (2002).
22. Abe, S., Lauby, G., Boyer, C., Rennard, S.I., and Sharp, J.G. Transplanted BM and BM side population cells contribute progeny to the lung and liver in irradiated mice. *Cytotherapy* **5**, 523–533 (2003).
23. Bittmann, I., *et al.* Cellular chimerism of the lung after transplantation. An interphase cytogenetic study. *Am J Clin Pathol* **115**, 525–533 (2001).
24. Spencer, H., *et al.* Transbronchial biopsies provide longitudinal evidence for epithelial chimerism in children following sex mismatched lung transplantation. *Thorax* **60**, 60–62 (2005).
25. Zander, D.S., Cogle, C.R., Theise, N.D., and Crawford, J.M. Donor-derived type II pneumocytes are rare in the lungs of allogeneic hematopoietic cell transplant recipients. *Ann Clin Lab Sci* **36**, 47–52 (2006).
26. Murata, H., *et al.* Endothelial cell chimerism associated with graft rejection after human lung transplantation. *Transplantation* **85**, 150–154 (2008).
27. Krause, D.S. Bone marrow-derived cells and stem cells in lung repair. *Proc Am Thorac Soc* **5**, 323–327 (2008).
28. Zander, D.S., *et al.* Bone marrow-derived stem-cell repopulation contributes minimally to the type II pneumocyte pool in transplanted human lungs. *Transplantation* **80**, 206–212 (2005).
29. Mimeault, M. and Batra, S.K. Concise review: recent advances on the significance of stem cells in tissue regeneration and cancer therapies. *Stem Cells* **24**, 2319–2345 (2006).
30. Epperly, M.W., Guo, H., Gretton, J.E., and Greenberger, J.S. Bone marrow origin of myofibroblasts in irradiation pulmonary fibrosis. *Am J Resp Cell Mol Biol* **29**, 213–224 (2003).
31. Hashimoto, N., Jin, H., Liu, T., Chensue, S.W., and Phan, S.H. Bone marrow-derived progenitor cells in pulmonary fibrosis. *J Clin Invest* **113**, 243–252 (2004).
32. Phillips, R.J., *et al.* Circulating fibrocytes traffic to the lungs in response to CXCL12 and mediate fibrosis. *J Clin Invest* **114**, 438–446 (2004).
33. Bianchi, D.W., Zickwolf, G.K., Weil, G.J., Sylvester, S., and DeMaria, M.A. Male fetal progenitor cells persist in maternal blood for as long as 27 years postpartum. *Proc Natl Acad Sci USA* **93**, 705–708 (1996).

34. Lee, T.H., Paglieroni, T., Ohto, H., Holland, P.V., and Busch, M.P. Survival of donor leukocyte subpopulations in immunocompetent transfusion recipients: Frequent long-term microchimerism in severe trauma patients. *Blood* **93**, 3127–3139 (1999).
35. Utter, G.H., *et al.* Leukoreduction of blood transfusions does not diminish transfusion-associated microchimerism in trauma patients. *Transfusion* **46**, 1863–1869 (2006).
36. Koopmans, M., *et al.* Chimerism occurs in thyroid, lung, skin and lymph nodes of women with sons. *J Reprod Immunol* **78**, 68–75 (2008).
37. Mengel, M., Lehmann, U., and Kreipe, H. Chimerism after solid organ transplantation. *Am J Transplant* **5**, 3021 (2005).
38. Weiss, D.J., Kolls, J.K., Ortiz, L.A., Panoskaltsis-Mortari, A., and Prockop, D.J. Stem cells and cell therapies in lung biology and lung diseases. *Proc Am Thorac Soc* **5**, 637–667 (2008).
39. Palange, P., *et al.* Circulating haemopoietic and endothelial progenitor cells are decreased in COPD. *Eur Resp J* **27**, 529–541 (2006).
40. Burnham, E.L., *et al.* Increased circulating endothelial progenitor cells are associated with survival in acute lung injury. *Am J Resp Crit Care Med* **172**, 854–860 (2005).

Chapter 17
Stem Cells and Lung Cancer

Arielle C. Lebrenne, Shahriar Islam
and Malcolm R. Alison

Lung cancer is the most common cause of death from cancer. It is estimated that every 15 minutes in the UK, one person dies of lung cancer[1] and in both the UK and USA more people die of it than any other type of cancer,[2,3] it being responsible for about one quarter of all deaths from cancer. Its high mortality rate is also reflected on a global scale, with lung cancer accounting for more than 1 million deaths per year.[4] Lung cancer is usually sub-divided into three types: small cell lung cancer (SCLC), non-small cell lung cancer (NSCLC), and mesothelioma a rare type of cancer that affects the pleura. Stem cells and cancer are inextricably linked; the perceived wisdom is that the process of carcinogenesis initially affects normal stem cells or their closely related progenitors, and then at some point, neoplastic stem cells are generated that propagate and ultimately maintain the process. Many, if not all cancers contain a population of self-renewing stem cells; the so-called *cancer stem cells* (CSCs) that are entirely responsible for sustaining the tumour as well as giving rise to proliferating but progressively differentiating cells that contribute to the cellular heterogeneity typical of many solid tumours. Thus tumours, like normal cell populations, may have a hierarchical structure. Adherents of the CSC hypothesis believe that the bulk of the tumour is therefore not the clinical problem, and so the identification of CSCs and the factors that regulate their behaviour are likely to have an enormous bearing on the way we treat neoplastic disease in the future. This chapter summarises 1) the histogenesis and molecular pathogenesis of lung tumours that probably take origin from normal pulmonary stem cells; 2) the evidence for the existence of CSCs in neoplastic lung tissue; and 3) illustrates some of the cellular pathways, often related to stem cell behaviour, that are frequently aberrant in lung cancer and may represent druggable targets.

Keywords: Stem cells; bronchioalveolar duct junction (BADJ); lung cancer; CD133; SP; drug resistance.

Outline

1. Stem Cells and the Origins of Lung Tumours

Progress in identifying lung stem cells has been impeded by the slow turnover of airway and alveolar epithelium. The consensus view is that there is no single multipotential stem cell for the lung, but rather there are regiospecific stem cell zones in the proximal and distal lung (Fig. 1) (see Chapter 4). In the mouse trachea, basal cells scattered in the ducts of submucosal glands expressing high levels of CK5 that were label-retaining cells (LRCs) after a period of BrdU labelling during tracheal injury, are likely stem/progenitor cells for this pseudostratified epithelium.[5] Other tracheal basal cells are also present, probably acting as progenitor cells, and expression of p63 appears important for their self-maintenance.[6] The mouse bronchial tree is lined by a number of cell types including basal cells, ciliated cells and non-ciliated cells (serous, goblet and Clara cell secretory protein (CCSP)-expressing (CE) cells). Here, both CE cells and basal cells appear to be able to act as transit amplifying/progenitor cells, and when CE cells are selectively ablated, a major subset of basal cells (basal cells identified by the binding of the lectin *Griffonia simplicifolia isolectin* B_4 (GSI-B_4) upregulate CK14 and are able to regenerate the entire epithelium.[7] In the bronchiolar epithelium, stem cell function appears to be the property of rare pollutant-resistant CE cells, co-localised with pulmonary

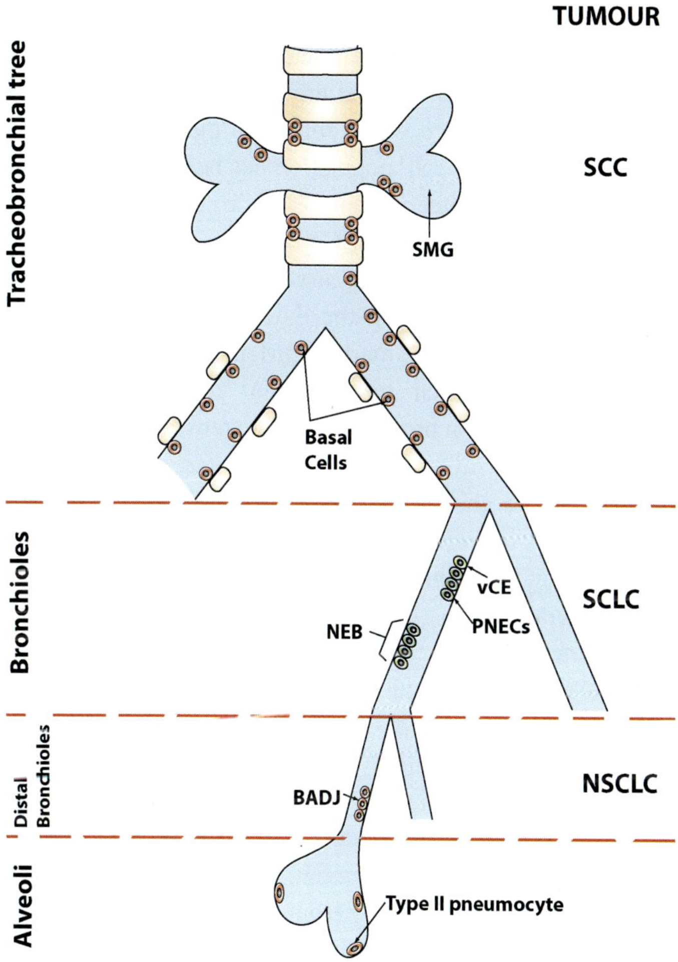

Figure 1: Probable locations of stem/progenitor cells in the respiratory tree and the tumours that might arise from them. BADJ, bronchioalveolar duct junction; NEB, neuroepithelial body; NSCLC, non small cell lung cancer; PNECs, pulmonary neuroepithelial cells; SCC, squamous cell carcinoma; SCLC, small cell lung cancer; SMG, submucosal gland; vCE, variant (CCSP) expressing cell.

neuroendocrine cells (PNECs), normally located in cell clusters termed *neuroepithelial bodies* (NEBs); PNECs can only act as progenitors for more PNECs.[8] Pollutant resistance of CE cells appears to be related to a deficiency in the phase I drug metabolizing enzyme CYP450 2F2. At the bronchioalveolar duct junction (BADJ), Giangreco *et al.*[9] have identified other pollutant-resistant CE cells, not associated with NEBs, that probably serve a stem cell function in the terminal bronchioles; CE cells that were also LRCs were located within three cell diameters of the BADJ. A non-haematopoietic ($CD45^-$) SP population has been isolated from murine lung and, like NEB-associated variant CE (vCE) cells, these cells showed expression of CCSP and absence of CYP450 2F2; a frequency of <0.9% amongst intrapulmonary conducting airway epithelia is consistent with a low abundance of stem cells.[10] At the BADJ are rare CE cells that also express the type II pneumocyte marker, surfactant protein C (SP-C), suggestive of both bronchiolar and alveolar differentiation potential, deserving of the appellation bronchioalveolar stem cells (BASCs).[11] In the mouse, these BASCs also express CD34 and Sca-1, but not the haematopoietic or endothelial markers CD45 and CD31.[11] In the alveoli, type II pneumocytes are widely believed to be progenitor cells, undergoing hyperplasia in response to the loss of the squamous type I pneumocytes, giving rise to type I cells as well as self-renewing. Type II cells are characteristically recognised by the presence of intracellular lamellar bodies and the production of surfactant proteins.

The phenotypic diversity of lung tumours, related to their location within the pulmonary tree, is highly likely to be partly a reflection of these complex region-specific variations in stem cell nature.[11] As Kim and colleagues[11] observed in murine models, it seems that only certain specific locations can foster tumour development, there being a proximal to distal distribution pattern; moving distally from the trachea these are squamous cell carcinomas (SCCs), small cell lung carcinomas (SCLCs) and bronchioloalveolar carcinomas (BACs)/adenocarcinomas (Fig. 1). We cannot be quite so certain of the origins of human lung cancers. The notion that there may be pro-oncogenic stem cell niches is reinforced by the fact that in murine models where there has been global knockdown of a tumour suppressor gene (e.g. *p53*) or upregulation of a proto-oncogene under the regulation of a widely expressed lung-specific promoter, in essence a "field cancerisation" effect, there has not been a similarly wide distribution of tumours produced. For example, *K-RAS* mutations are very common in human lung cancer, but murine models with widespread *K-ras* mutations only result in atypical adenomatous hyperplasia and eventually adenocarcinoma in the bronchioalveolar region.[12] This seems to be related to expansion of self-renewing, multipotent BASCs in the BADJ.[11] Thus, although extensive field cancerisation has been observed in the human tracheal–bronchial tree with an identical *TP53* codon 245 transversion found at 7/10 sites in both lungs, presumed to derive from a single mutated clone,[14] any synchronous carcinoma development would seem to depend on the location of the various stem cell niches.

2. Histogenesis and Molecular Pathogenesis of Lung Tumours

2.1. *The development of SCLCs*

Human SCLCs localise to mid-level bronchioles, and express a number of neuroendocrine markers including calcitonin gene-related peptide (CGRP) normally expressed by PNECs in NEBs. As such, the consensus view is that SCLCs have origin in PNECs.[12] Murine models that delete both *Rb* and *p53* demonstrate hyperplasia specifically in the microenvironment of NEBs with resulting metastatic SCLCs.[15] The proposed relationship between PNECs and SCLC is further strengthened by the fact that the Hedgehog and Notch-delta pathways both have roles in PNEC growth and SCLC progression.[16,17]

2.2. *The development of NSCLCs*

Squamous cell carcinomas (SCCs) generally occur in the proximal airways and appear to develop in a step-wise fashion beginning with basal cell hyperplasia, progressing to squamous metaplasia, dysplasia, carcinoma *in situ* and finally invasive SCC[18] (Fig. 2a). An origin from CK14-positive basal cells found either in the submucosal gland ducts or intercartilagenous boundaries is consistent with their increased numbers in mouse models of SCC, and the fact that they occur in the same locations where SCCs arise.[19]

Adenocarcinomas sometimes co-express airway and alveolar traits, e.g. CCSP and SP-C, and together with the observed expansion of BASCs prior to tumour formation in mouse models, the evidence strongly

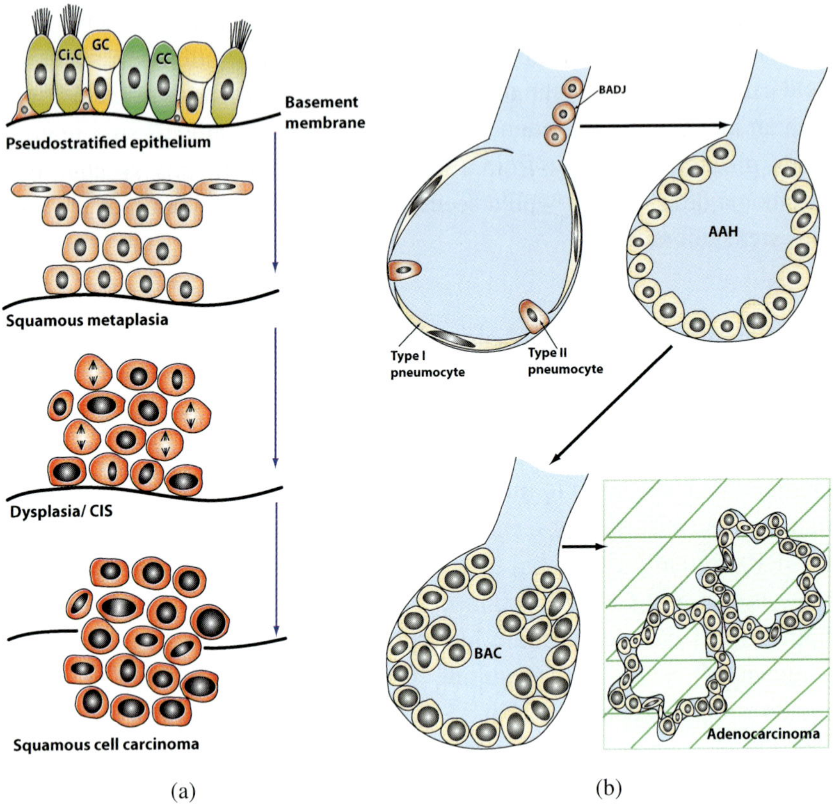

Figure 2: Simplified cartoons illustrating the major recognisable stages in the histological development of (a) squamous cell carcinoma (SCC) and (b) adenocarcinoma, presumed to reflect the multistage nature of lung cancer development. For detailed morphological descriptions the reader is referred to Ref. 18. The development of SCC begins in the upper airways with basal cell hyperplasia, proceeding to squamous metaplasia, increasing dysplasia resulting in carcinoma *in situ* (CIS), and finally invasive squamous carcinoma. Peripheral adenocarcinoma may have its origin in BASCs at the BADJ, and proceeds through atypical adenomatous hyperplasia (AAH), bronchioloalveolar carcinoma (BAC), finally to invasive adenocarcinoma. AAH is thought to be equivalent to an adenoma in the adenoma-carcinoma sequence. CC, clara cell; CiC, ciliated cell; GC, goblet cell.

suggests that many adenocarcinomas arise from BASCs at the BADJ,[11,20] although an origin from Clara cells or type II pneumocytes is also possible. Likewise, many CCSP or SP-C promoter-driven murine models of lung adenocarcinoma, including mutated epidermal growth factor (EGF) receptor[21] and activated *K-ras* (G12D),[13] produce mixed phenotype tumours located close to terminal bronchioles resembling human bronchioalveolar carcinomas (BACs), again suggesting that the cell of origin of adenocarcinoma and BAC resides at or near the BADJ. The Polycomb repressor Bmi1, often overexpressed in NSCLC, is required for normal stem cell maintenance and tumorigenicity in many tumours, functioning, for example, by blocking expression of $p19^{arf}$ that would normally sequester mdm2 thereby raising p53 levels. In the *K-ras* initiated model of lung adenocarcinoma, loss of Bmi1 inhibits tumorigenesis by blocking prior expansion of BASCs.[22]

The molecular regulation of BASCs at the BADJ has come under close scrutiny. In *K-ras* (G12V) activated lung tumorigenesis in the mouse, loss of the stress activated protein kinase (SAPK) p38α, also known as MAPK14 and required for lung branching morphogenesis,[23] results in hypercellularity at the BADJ with increased numbers of CCSP/SP-C expressing cells, rendering the mice highly sensitive to *K-ras*-induced lung tumorigenesis.[24] Loss of p38α was associated with a loss of differentiation signals such as the transcription factors C/EBPα and HNF3β, but an increase in proliferation signals including EGFR and cyclin D1. The relevance of p38α to human lung cancer was clear, since compared to normal lungs, the human tumours only expressed about one third of the levels of p38α. Furthermore, mutation of $p27^{Kip1}$ and loss of $p18^{Ink4c}$ are also associated with increased numbers of BASCs in mouse lung.[25,26] P38α negatively regulates the cell cycle activity by inhibiting c-Jun N-terminal kinase (JNK)-c-Jun pathway, and p38α deficiency in newborn $Mapk^{\Delta/\Delta}$ mice have a very distorted alveolar structure.[27]

Class I phosphatidylinositol 3-kinases (PI3Ks) phosphorylate phosphatidylinositol 4,5-bisphosphate (PIP2) to produce phosphatidylinositol 3,4,5-trisphosphate (PIP3). PI3K is required for the initial events in Ras-induced tumour formation. In two mouse models of *K-ras*-induced lung cancer, BASC expansion and tumour progression were retarded by pharmacological inhibition of PI3K, but enhanced by genetic inactivation of *Pten*, a negative regulator PI3K.[28] Phosphatase and tensin homologue deleted on chromosome ten (PTEN) removes phosphate groups from PIP3, PIP3 being essential for the activation of the serine/threonine kinase known as Akt (also know as protein kinase B); in turn this leads to

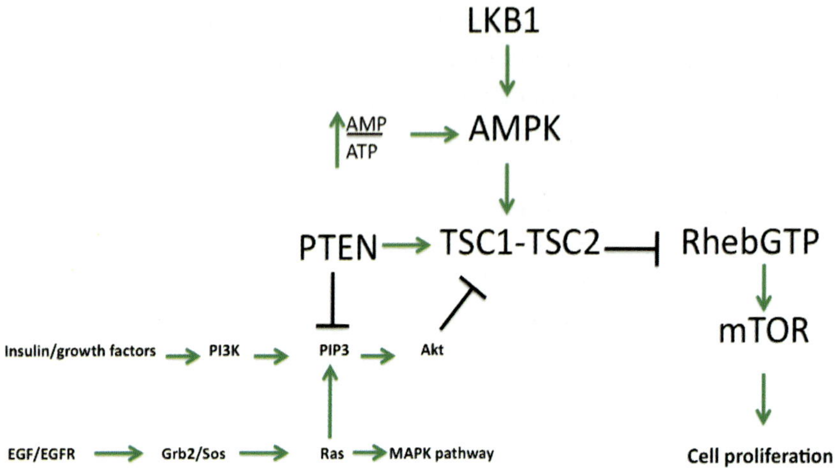

Figure 3: The relationship between three signalling pathways involved in the disease syndromes Peutz Jeghers (LKB1), Cowden's disease (PTEN) and tuberous sclerosis (TSC1-TSC2) that essentially do not cause malignancy or pulmonary disease, but nevertheless illustrate interactions between molecules frequently implicated in lung cancer.[32] LKB1 acts as a sentinel, halting cell division in times of energy stress (overriding growth factor signalling) through phosphorylation of AMPK (5′-AMP-activated kinase), that in turn activates the tuberous sclerosis complex (TSC1-TSC2). Mutations in TSC1 or TSC2 impair the GTPase activity of the complex for Rheb (Ras homologue enriched in brain), leading to inappropriate mTOR signalling. Thus, targeting mTOR would seem to be potentially beneficial to lung cancer patients with not only activation of growth factor signalling pathways but also with inactivating mutations/loss of *LKB1*, *PTEN* or *TSC* genes.

upregulation of mammalian target of Rapamycin (mTOR) that is associated with cell cycle progression (Fig. 3). In a similar vein, conditional inactivation of *Pten* in SP-C expressing cells can result in spontaneous lung tumours.[29] Moreover, *Pten* conditional deletion models for a variety of organs show upregulation of stem/progenitor cells with activation of mTOR, suggesting that mTOR is an important therapeutic target.[30] Examining CCSP/SP-C double positive murine BASCs, it has been suggested that micro RNAs (miRNAs) play a significant role in self-renewal and that their dysregulated expression is a major event in cancer progression.[31]

The study of three genetic syndromes, Peutz-Jeghers syndrome (gene affected *LKB1*), Cowden's disease (*PTEN*) and tuberous sclerosis

(*TSC1/TSC2*), all of which share parallel phenotypes of primarily non-malignant and non-pulmonary disease, has nevertheless provided an important insight into a common pathway activated in a significant proportion of lung cancers[32] (Fig. 3). A tumour suppressor gene *LKB1*, also known as *STK11* (serine/threonine kinase11), largely responsible for Peutz-Jeghers syndrome, is inactivated in 30% of human lung cancer.[32] In times of energy stress (high AMP/ATP), LKB1 negatively regulates mTOR via TSC1-TSC2, thus blocking cell proliferation, overriding growth factor (e.g. EGF) signalling (Fig. 3). In animal models the homozygous loss of *Lkb1* in combination with *K-ras* mutation produces considerably more aggressive lung tumours than those produced by mutant K-ras alone.[32] Clearly our increasing knowledge regarding the regulation of BASCs, seemingly the cells of origin of many NSCLCs, is going to have a major impact on the development of new therapies for lung cancer, particularly those that target mTOR and related pathways.

Most adenocarcinomas develop in a step-wise manner, with the first recognisable precursor lesion being atypical adenomatous hyperplasia (Fig. 2B), AAH (a clonal lesion), that progresses to BAC and then to adenocarcinoma; the fact that AAHs can express markers of Clara cells and type II pneumocytes reinforces the belief that AAH arises from BASCs and AAH is the forerunner of adenocarcinoma.[33] In general, the number and severity of mutational load (e.g. EGFR mutation)[34] is seen to increase with each successive stage, though strangely no major differences in MMP expression were seen between non-invasive BACs and invasive adenocarcinomas,[35] though as observed for many tumours, the MMP expression is largely confined to the desmoplastic cancer-associated fibroblasts.[36]

2.3. *Molecular pathology of established tumours*

A detailed discussion of the genetic abnormalities in lung cancer is beyond the remit of this chapter. A number of reviews are available for the interested reader.[37–42] In SCLC, the common tumour suppressor genes inactivated are *TP53*, *RB* and *FHIT* (fragile histidine triad gene) along with overexpression of BCL-2. In NSCLC, inactivation of *INK4A*, *TP53* and to a lesser extent *RB* and *FHIT*, are common, along with EGFR overexpression and *K-RAS* mutation. In a detailed analysis of 623 genes potentially related to cancer in lung cancer, other genes implicated in lung cancer pathogenesis included those encoding c-erbB-4, ephrin receptors and KDR (VEGFR2) along with NF1, ATM and LRP1B (LDL receptor-related protein 1B).[43] Attempts have been made to classify human lung cancers based on association with stages of murine lung development, on the premise that murine orthologs of human genes upregulated in lung cancer would be downregulated with time during mouse lung development.[44] As expected, overexpressed genes in lung tumours had a significant likelihood of having a mouse lung developmental profile that, in the mouse, decreased with time. In particular, there was an association between placement along the mouse lung development trajectory and patient survival.

Two immortalised bronchial epithelial cell lines exposed to chemical carcinogens had increased levels of cytosine-DNA methyltransferase I (DNMT1), the major methyltransferase, associated with promoter hypermethylation of a number of genes, particularly of the cadherin family.[45] Stable knockdown of DNMT1 reversed transformation, suggesting that demethylation therapy could be a primary prevention strategy for lung cancer in smokers. Similarly, promoter hypermethylation of *Cadm1* (also known as *TSLC1*, for tumour suppressor in lung cancer 1), a member of the Ig superfamily of cell adhesion molecules (CAMs), is seen in tumour cells isolated from *c-myc* and *c-raf* double transgenic mice;[46] gene expression could be restored with the demethylating agent, 5-aza-2'-deoxycytidine. Hypermethylation of the *TSLC1* promoter seems to be the second hit in many human cases of NSCLC with loss of heterozygosity.[47] There is little information on hereditary cancer syndromes associated with increased lung cancer incidence. The best known examples are Li Fraumeni syndrome; smokers who carry a *TP53* mutation are more at risk than non-smokers with the same mutation, and carriers of *RB* mutations have a higher lifetime risk of lung cancer.[48]

KEAP1 gene mutations have been identified in a number of lung cancer patients, therefore conferring a cytoprotective advantage to the mutated cells since the transcription factor Nrf2 is now available to upregulate cytoprotective genes.[49] Thus, inhibition of Nrf2 maybe useful in patients with activation of Nrf2. TTF-1 (thyroid transcription factor-1) is required for lung development and regulates expression of surfactant proteins;[50] in stage 1 adenocarcinoma, TFF-1 expression defines a favourable prognostic group.[51] A genome-wide association study has identified region 5p15.33 as a lung cancer susceptibility locus, implicating the *TERT* (human telomerase reverse transcriptase) gene.[52] A group of conserved genes known as homeobox (*HOX*) genes are important regulators of development, and in humans there are four *HOX* gene clusters. They are highly expressed in the developing lung, but levels of most *HOX* genes decline with advancing maturation. Since the neoplastic process often recapitulates ontogeny, *HOX* genes are increasingly implicated in neoplastic development. Upregulation of *HOX* genes is seen in SCLC,[53] and in NSCLC there is even increased expression of *HOX* genes not seen in the developing lung.[54] Lack of response to the negative growth influences of TGFβ are also apparent in human SCLC, possibly related to the mutagenicity of benzo[a]pyrene, a component of cigarette smoke causing defective TβRII expression,[55] and CpG hypermethylation of the TβRII promoter is seen in NSCLC.[56] Inhibiting Wnt signalling results in increased apoptosis in human NSCLC, seemingly related to downregulation of the human inhibitor of apoptotic proteases, survivin.[57] Activating mutations in *CTNNB1* (β-catenin) are seen in a significant proportion of human adenocarcinomas.[58]

3. Cancer Stem Cells

For many years it has been apparent that stem cells feature in processes as diverse as wound healing, metaplasia and cancer. Two-stage models of skin cancer in rodents carried out in the 1950s strongly suggested that cancers had their origins in long-lived epidermal stem cells, but the idea that cancers themselves might have malignant cancer stem cells (CSCs) is only just gaining widespread acceptance, despite the fact that Hamburger and Salmon, in 1997,[59] using colony formation in soft agar as a surrogate stem cell assay, found that for many human tumours, only 1 in a 1,000 to 1 in 5,000 cells was able to form a macroscopic colony.

Cancer could arise from the dedifferentiation of mature cells that have retained the ability to divide, or it could result from the "maturation arrest" of immature stem cells.[60] The idea of "blocked ontogeny" has gained wide acceptance, and we now believe that the arrested differentiation of tissue-based stem cells or their immediate progenitors is closely linked to the development of not only teratocarcinomas and haematological malignancies, but also carcinomas. Some of the most frequent cancers occur in tissues with a high cell turnover such as the skin and the epithelial lining of the gastrointestinal tract. It is argued, not unreasonably in our view, that in these tissues the stem cells are the only cells with sufficient lifespan to acquire the requisite number of genetic abnormalities for malignant transformation. Additionally, with a self-renewal mechanism already in place, seemingly fewer alterations are required to change normal stem cells into CSCs. In the now classical two-stage model of mouse skin carcinogenesis, severely delaying the interval between 7,12-dimethylbenz(a)anthracene (DMBA) initiation and the application of the phorbol ester promoter had no bearing on subsequent tumour yield — strongly suggestive of an origin in a long-lived cell, an epidermal stem cell.[61]

3.1. *Markers of CSCs*

Considerable effort is being expended in the search for "markers" of stem cells, with every expectation that many of the molecules expressed by normal stem cells will also be found in their malignant counterparts. Collectively these molecules appear to be involved in maintaining "stemness" (transcription factors such as Oct-4 and Nanog), ensuring adhesion to the niche and involvement in cytoprotection.[62] For example, in the human epidermis, superior colony forming ability has been found in cells selected on the basis of expression of the hemidesmosomal integrin $\alpha6$, which partners $\beta4$ to attach cells to the basement membrane component, laminin V,[63] while Jones and Watt noted that selection of basal keratinocytes based upon high expression of the $\beta1$-integrin enriched for colony-forming ability.[64] Likewise, in the murine mammary gland, either the β-1 integrin (CD29) or the α-6 integrin (CD49f) has been used to isolate multipotential stem cells, each capable of generating entire mammary glands after *in vivo* transplantation.[65,66]

Many putative stem cells have acquired the ability to withstand cytotoxic insults through either efficient enzyme-based detoxification systems or the ability to rapidly export potentially harmful xenobiotics. In 1996, Goodell *et al.* reported on a new method for the isolation of HSCs based on the ability of HSCs to efflux a fluorescent dye,[67] an ability that was inhibited by verapamil. Cells are subjected to Hoechst 33342 dye staining and fluorescence-activated cell sorting (FACS) analysis; those that actively efflux the Hoechst dye appear as a distinct population of cells on the side of the profile, hence the name "side population" (SP). The SP phenotype of HSCs in mice and humans is largely determined by the expression of a protein known as the ABCG2 transporter (ATP-binding cassette [ABC] subfamily G member 2, also known as BCRP1).[68,69] The ABC superfamily of membrane transporters is one of the largest protein classes known, and is characterised by expression of an ATP-binding cassette region functioning to hydrolyse ATP to support energy-dependent substrate exportation against steep concentration gradients across membranes, principally from the intracellular cytoplasm to the extracellular space.[70] Various websites detail the 49 human ABC transporters, which are organised into seven families, A–G (see http://www.nutrigene.4t.com/humanabc.htm).

ABC transporters play a role in the transport of drugs (xenobiotics) and drug conjugates. Their role is exemplified by MDR1 (ABCB1 or P-glycoprotein), MRP1 (the multidrug resistance protein 1, ABCC1) and BCRP1 (breast cancer resistance protein 1, ABCG2), whose expression is associated with multidrug resistance (MDR) in cancer cells.[71] Drug resistance results from the ability of the transporters to extrude several classes of anticancer drugs, lowering effective concentrations within the cell. For example, MDR1 is able to cause the greatest resistance to bulky amphipathic drugs such as paclitaxel (taxol) anthracyclines and Vinca alkaloids. SP cells can be isolated from human tumours, and probably contribute significantly to tumour drug-resistance.[72] ABC transporters have emerged as an important field of investigation in the regulation of stem cell biology[73] and, significantly, ABCG2, an ABC transporter found in many stem cells, is upregulated in hypoxic environments, e.g. stem cell niches, mediated by HIF1-α.[74]

3.2. *CSCs in lung cancer*

Evidence for the existence of CSCs in the lung was first described more than 25 years ago in a study by Carney *et al.*[91] In this study, a small population of cells (<1.5%) isolated from the tumours of both adenocarcinoma of the lung and SCLC patients were able to form colonies in a soft agar cloning assay. As with other solid tumours, CD133 expression features predominantly in the search for CSCs in the lung. In NSCLC, one thousand $CD133^+$ cells could form tumours in SCID mice, but 10^4 CD133-negative cells never did.[92] The CD133-positive cells showed enhanced expression of Oct-4 and ABCG2; siRNA knockdown of Oct-4 blocked clonogenicity and enhanced chemosensitivity. Expression of Oct-4 has been claimed to occur in BACs, but the data were unconvincing.[93] In both SCLC and NSCLC a small (<1%) population of CD133-positive cells has been found, with 10^4 of these cells capable of forming tumours in SCID mice with features of the parent tumours.[94] These *in vitro* sphere-forming cells often expressed Oct-4 and Nanog along with CCSP and SP-C. Exploiting the perceived chemoresistance of CSCs, CSCs have been enriched in a NSCLC cell line by treating with the likes of cisplatin and doxorubicin;[95] 5×10^3 drug selected cells regularly formed tumours in SCID mice. These cells expressed CD133, CD117, and the embryonic stem cell markers SSEA-3, TRA1-81 and Oct-4, and had nuclear β-catenin. In the cell line, the SP fraction was 5.2%, but after drug treatment this was increased to 35%. Clearly drug resistance and lung CSCs are heavily entwined. In the A549 NSCLC cell line a large (24%) SP has been found, with enhanced resistance to doxorubicin and methotrexate related to ABCG2 activity.[96] In a number of SCLC cell lines, a subpopulation (1–4%) of urokinase plasminogen activator-positive cells has been found that were more resistant to traditional chemotherapies such as 5-FU, cisplatin and etoposide, seemingly associated with enhanced MDR1 (ABCB1) activity.[97]

3.3. *Implications for treatment*

For tomorrow's oncologist, a variety of druggable targets and strategies related to CSCs present themselves, including:

- Wnt signalling
- Hedgehog signalling

- Notch/Delta signalling
- mTOR
- ABC transporters
- Targeting the stem cell niche
- Small molecule cancer therapeutics (miRNAs, tyrosine kinase inhibitors)
- Gene therapy
- Tissue engineered implants.

Proliferation of CSCs is likely to involve dysregulation of the pathways present in normal stem cell self-renewal such as the Wnt-/β-catenin, PTEN, Notch and Hedgehog pathways, as well as the products of the Bmi1 and other polycomb genes. A detailed account of these pathways in lung development and pathology can be found elsewhere.[98,99] In many tissues key regulators of stem cell renewal appear to be members of the Polycomb group protein family of transcriptional repressors (Bmi1, Rae28, Mel-18). Bmi1 targets genes such as $p16^{INK4A}$ and $p14^{ARF}$ preventing stem cell senescence by respectively maintaining cyclinD/Cdk4 signalling and Mdm2 destruction of p53. Bmi1 is in fact a downstream target of the morphogen sonic hedgehog (Shh) through the latter's activation of the Gli family of latent transcription factors. Shh acts on the receptor complex of Patched (PTCH) and Smoothened (SMO), blocking the restraining influence of PTCH on SMO, resulting in SMO signalling activating the Gli family of transcription factors and so activating target genes like Bmi1. Inhibiting the action of SMO with the anatogonist cyclopamine is a highly effective strategy against some cancers. It has been claimed that almost all SCLCs ubiquitously express Bmi1,[100] and antisense Bmi1 RNA therapy reduces proliferation of A549 cells.[101] The Notch family of receptors is also critical for stem cell self-renewal. Engagement of ligands of the Delta and Jagged families causes cleavage, mediated by the γ-secretase protease complex, of the intracellular portion of Notch and its translocation to the nucleus, where it binds to the transcription factor CSL, changing it from a transcriptional repressor to an activator. The use of γ-secretase inhibitors may have utility in cancers where Notch-signalling is inappropriately activated.[102] As discussed, many pathways activated in human lung cancer converge on mTOR (Fig. 3), with obvious therapeutic possibilities.

Apart from the renewal and proliferation pathways, there are many other potential molecular targets relating to lung CSCs. As many lung tumours have SP fractions, almost certainly enriched for CSCs, targeting ABC transporter activity will be an obvious strategy for overcoming chemoresistance as well as directly eradicating stem cells,[92,95–97] drug resistance being particularly problematic in SCLCs.[103] In NSCLC, a small molecule, YM155, has been used to target the Inhibitor of Apoptotic Protease (IAP) called survivin, rendering the cells more sensitive to radiation induced apoptosis.[104]

Antibody-based targeting of CSCs exploiting the overexpression of the likes of CD133 is another possible approach, while CSCs may reside in areas rich in blood vessels, so-called "vascular niches". So while anti-angiogenic therapy, used to debulk tumours through disruption of their blood supply, is not new, the destruction of the CSC niche adds a new twist to the story. Small molecule therapeutics that target growth factors, growth factor receptors and their kinases, and more specific tyrosine kinase inhibitors, such as imatinib, that target c-Kit-positive cells are also gaining widespread usage. RNA interference will become an increasingly important strategy, and targeting Oct-4 in lung cancer has been shown to increase apoptosis.[105] In lung cancer, adenocarcinoma can be reliably distinguished from SCC based on the expression of a minimal set of 17 genes predicted to be the targets of just three lung-enriched miRNAs (miR-34b/34c/449).[106] This gene signature could be used to predict lung cancer from bronchoscopy biopsies of cigarette smokers. Further therapeutic approaches might well include gene therapy, introducing the likes of prodrug activating enzymes once stem/progenitor cells become better characterised.[107] Finally the prospect of implanting whole tissue implants has become closer to reality with the successful implantation of a donor trachea coated with autologous cells to treat a patient with end-stage bronchomalacia.[108]

4. Conclusion and the Way Forward

This chapter has summarised our current understanding of pulmonary stem cells, noting a diversity of stem/progenitor cells along the length of the conducting airways. A role of BASCs from the BADJ in the histogenesis of

BACs and adenocarcinoma seems compelling. Lung cancer cells are certainly heterogeneous in terms of proliferative activity, clonogenicity and, significantly, chemo- and radioresistance. $CD133^+$ and SP fractions seem to fit our current notions about CSCs, and a number of stem/progenitor-targeted therapeutic approaches seem to be on the horizon for the treatment of this dreadful disease.

References

1. http://info.cancerresearchuk.org/cancerstats/types/lung/mortality/.
2. http://info.cancerresearchuk.org/cancerstats/types/lung/mortality/?a=5441.
3. http://www.cdc.gov/cancer/lung/.
4. Parkin, D.M., Bray, F., Ferlay, J., and Pisani, P. Global cancer statistics. *CA Cancer J Clin* **55**, 43–66 (2002).
5. Borthwick, D.W., Shahbazian, M., Krantz, Q.T., Dorin, J.R., and Randell, S.H. Evidence for stem-cell niches in the tracheal epithelium. *Am J Respir Cell Mol Biol* **24**, 662–670 (2001).
6. Daniely, Y., Liao, G., Dixon, D., Linnoila, R.I., Lori, A., Randell, S.H., *et al.* Critical role of p63 in the development of a normal esophageal and tracheobronchial epithelium. *Am J Physiol Cell Physiol* **287**, C171–C181 (2004).
7. Hong, K.U., Reynolds, S.D., Watkins, S., Fuchs, E., and Stripp, B.R. Basal cells are a multipotent progenitor capable of renewing the bronchial epithelium. *Am J Pathol* **164**, 577–588 (2004).
8. Reynolds, S.D., Giangreco, A., Power, J.H.T., and Stripp, B.R. Neuroepithelial bodies of pulmonary airways serve as a reservoir of progenitor cells capable of epithelial regeneration. *Am J Pathol* **156**, 269–278 (2000).
9. Giangreco, A., Reynolds, S.D., and Stripp, B.R. Terminal bronchioles harbor a unique airway stem cell population that localizes to the bronchoalveolar duct junction. *Am J Pathol* **161**, 173–182 (2002).
10. Giangreco, A., Shen, H., Reynolds, S.D., and Stripp, B.R. Molecular phenotype of airway side population cells. *Am J Physiol Lung Cell Mol Physiol* **286**, L624–L630 (2004).
11. Kim, C.F., Jackson, E.L., Woolfenden, A.E., Lawrence, S., Babar, I., Vogel, S., *et al.* Identification of bronchioalveolar stem cells in normal lung and lung cancer. *Cell* **121**, 823–835 (2005).

12. Giangreco, A., Groot, K.R., and Janes, S.M. Lung cancer and lung stem cells: strange bedfellows? *Am J Respir Crit Care Med* **175**, 547–553 (2007).
13. Jackson, E.L., Willis, N., Mercer, K., Bronson, R.T., Crowley, D., Montoya, R., *et al.* Analysis of lung tumor initiation and progression using conditional expression of oncogenic K-ras. *Genes Dev* **15**, 3243–3248 (2001).
14. Franklin, W.A., Gazdar, A.F., Haney, J., Wistuba, I.I., La Rosa, F.G., Kennedy, T., *et al.* Widely dispersed p53 mutation in respiratory epithelium. A novel mechanism for field carcinogenesis. *J Clin Invest* **100**, 2133–2137 (1997).
15. Meuwissen, R., Linn, S.C., Linnoila, R.I., Zevenhoven, J., Mooi, W.J., and Berns, A. Induction of small cell lung cancer by somatic inactivation of both Trp53 and Rb1 in a conditional mouse model. *Cancer Cell* **4**, 181–189 (2003).
16. Watkins, D.N., Berman, D.M., Burkholder, S.G., Wang, B., Beachy, P.A., and Baylin, S.B. Hedgehog signalling within airway epithelial progenitors and in small-cell lung cancer. *Nature* **422**, 313–317 (2003).
17. Collins, B.J., Kleeberger, W., and Ball, D.W. Notch in lung development and lung cancer. *Semin Cancer Biol* **14**, 357–364 (2004).
18. Dacic, S. Pulmonary Preneoplasia. *Arch Pathol Lab Med* **132**, 1073–1078 (2008).
19. Barth, P.J., Koch, S., Müller, B., Unterstab, F., von Wichert, P., and Moll, R. Proliferation and number of Clara cell 10-kDa protein (CC10)-reactive epithelial cells and basal cells in normal, hyperplastic and metaplastic bronchial mucosa. *Virchows* Arch **437**, 648–655 (2000).
20. Kim, C.F. Paving the road for lung stem cell biology: bronchioalveolar stem cells and other putative distal lung stem cells. *Am J Physiol Lung Cell Mol Physiol* **293**, L1092–L1098 (2007).
21. Politi, K., Zakowski, M.F., Fan, P.D., Schonfeld, E.A., Pao, W., and Varmus, H.E. Lung adenocarcinomas induced in mice by mutant EGF receptors found in human lung cancers respond to a tyrosine kinase inhibitor or to down-regulation of the receptors. *Genes Dev* **20**, 1496–1510 (2006).
22. Dovey, J.S., Zacharek, S.J., Kim, C.F., and Lees, J.A. Bmi1 is critical for lung tumorigenesis and bronchioalveolar stem cell expansion. *Proc Natl Acad Sci USA* **105**, 11857–11862 (2008).
23. Liu, Y., Martinez, L., Ebine, K., and Abe, M.K. Role for mitogen-activated protein kinase p38 alpha in lung epithelial branching morphogenesis. *Dev Biol* **314**, 224–235 (2008).

24. Ventura, J.J., Tenbaum, S., Perdiguero, E., Huth, M., Guerra, C., Barbacid, M., *et al.* p38alpha MAP kinase is essential in lung stem and progenitor cell proliferation and differentiation. *Nat Genet* **39**, 750–758 (2007).
25. Sicinski, P., Zacharek, S., and Kim, C. Duality of p27Kip1 function in tumorigenesis. *Genes and Development* **21**, 1703–1706 (2007).
26. Pei, X.H., Bai, F., Smith, M.D., and Xiong, Y. p18Ink4c collaborates with Men1 to constrain lung stem cell expansion and suppress non-small-cell lung cancers. *Cancer Res* **67**, 3162–3170 (2007).
27. Hui, L., Bakiri, L., Mairhorfer, A., Schweifer, N., Haslinger, C., Kenner, L., *et al.* p38alpha suppresses normal and cancer cell proliferation by antagonizing the JNK-c-Jun pathway. *Nat Genet* **39**, 741–749 (2007).
28. Yang, Y., Iwanaga, K., Gabriela Raso, M., Wislez, M., Hanna, A.E., Wieder, E.D., *et al.* Phosphatidylinositol 3-kinase mediates bronchioalveolar stem cell expansion in mouse models of oncogenic K-ras-induced lung cancer. *PLoS ONE* **3**, e2220 (2008).
29. Yanagi, S., Kishimoto, H., Kawahara, K., Sasaki, T., Sasaki, M., Nishio, M., *et al.* Pten controls lung morphogenesis, bronchioalveolar stem cells, and onset of lung adenocarcinomas in mice. *J Clin Invest* **117**, 2929–2940 (2007).
30. Hill, R. and Wu, H. PTEN, stem cells, and cancer stem cells. *Journal of Biological Chemistry*, in press (2008).
31. Qian, S., Ding, J.Y., Xie, R., An, J.H., Ao, X.J., Zhao, Z.G., *et al.* MicroRNA expression profile of bronchioalveolar stem cells from mouse lung. *Biochem Biophys Res Commun* **377**, 668–673 (2008).
32. Makowski, L. and Hayes, D.N. Role of LKB1 in lung cancer development. *Br J Cancer* **99**, 683–688 (2008).
33. Mori, M., Rao, S.K., Popper, H.H., Cagle, P.T., and Fraire, A.E. Atypical adenomatous hyperplasia of the lung: a probable forerunner in the development of adenocarcinoma of the lung. *Mod Pathol* **14**, 72–84 (2001).
34. Yoshida, Y., Shibata, T., Kokubu, A., Tsuta, K., Matsuno, Y., Kanai, Y., *et al.* Mutations of the epidermal growth factor receptor gene in atypical adenomatous hyperplasia and bronchioloalveolar carcinoma of the lung. *Lung Cancer* **50**, 1–8 (2005).
35. Kanomata, N., Nakahara, R., Oda, T., Aoyagi, Y., Ishii, G., Yokose, T., *et al.* Expression and localization of mRNAs for matrix metalloproteinases and their inhibitors in mixed bronchioloalveolar carcinomas with invasive components. *Mod Pathol* **18**, 828–837 (2005).

60. Sell, S. and Pierce, G.B. Maturation arrest of stem cell differentiation is a common pathway for the cellular origin of teratocarcinomas and epithelial cancers. *Lab Invest* **70**, 6–22 (1994).
61. Berenblum, I. and Shubik, P. The persistence of latent tumour cells induced in the mouse's skin by a single application of 9:10-dimethyl-1:2-benzanthracene. *Br J Cancer* **3**, 384–386 (1949).
62. Alison, M.R. and Islam, S. Attributes of adult stem cells. *J Pathol* **127**, 144–160 (2009).
63. Webb, A., Li, A., and Kaur, P. Location and phenotype of human adult keratinocyte stem cells of the skin. *Differentiation* **72**, 387–395 (2004).
64. Jones, P.H. and Watt, F.M. Separation of human epidermal stem cells from transit amplifying cells on the basis of differences in integrin function and expression. *Cell* **73**, 713–724 (1993).
65. Shackleton, M., Vaillant, F., Simpson, K.J., Stingl, J., Smyth, G.K., Asselin-Labat, M.L., *et al.* Generation of a functional mammary gland from a single stem cell. *Nature* **439**, 84–88 (2006).
66. Stingl, J., Eirew, P., Ricketson, I., Shackleton, M., Vaillant, F., Choi, D., *et al.* Purification and unique properties of mammary epithelial stem cells. *Nature* **439**, 993–997 (2006).
67. Goodell, M.A., Brose, K., Paradis, G., Conner, A., and Mulligan, R. Isolation and functional properties of murine hematopoietic stem cells that are replicating in vivo. *J Exp Med* **183**, 1797–1806 (1996).
68. Zhou, S., Morris, J.J., Barnes, Y., Lan, L., Schuetz, J.D., and Sorrentino, B.P. Bcrp1 gene expression is required for normal numbers of side population stem cells in mice, and confers relative protection to mitoxantrone in hematopoietic cells *in vivo*. *Proc Natl Acad Sci USA* **99**, 12339–12344 (2002).
69. Guo, Y., Lubbert, M., and Engelhardt, M. CD34(-) hematopoietic stem cells: current concepts and controversies. *Stem Cells* **21**, 15–20 (2003).
70. Krishnamurthy, P. and Schuetz, J.D. Role of ABCG2/BCRP in biology and medicine. *Annu Rev Pharmacol Toxicol* **46**, 381–410 (2006).
71. Litman, T., Brangi, M., Hudson, E., Fetsch, P., Abati, A., Ross, D.D., *et al.* The multidrug-resistant phenotype associated with overexpression of the new ABC half-transporter, MXR (ABCG2). *J Cell Sci* **113**, 2011–2021 (2000).
72. Hirschmann-Jax, C., Foster, A.E., Wulf, G.G., Goodell, M.A., and Brenner, M.K. A distinct "side population" of cells in human tumor cells: implications for tumor biology and therapy. *Cell Cycle* **4**, 203–205 (2005).

73. Bunting, K.D. ABC transporters as phenotypic markers and functional regulators of stem cells. *Stem Cells* **20**, 11–20 (2002).
74. Teramura, T., Fukuda, K., Kurashimo, S., Hosoi, Y., Miki, Y., Asada, S., *et al.* Isolation and characterization of side population stem cells in articular synovial tissue. *BMC Musculoskelet Disord* **9**, 86 (2008).
75. Vasiliou, V. and Nebert, D.W. Analysis and update of the human aldehyde dehydrogenase (ALDH) gene family. *Hum Genomics* **2**, 138–143 (2005).
76. Corti, S., Locatelli, F., Papadimitriou, D., Donadoni, C., Salani, S., Del Bo, R., *et al.* Identification of a primitive brain-derived neural stem cell population based on aldehyde dehydrogenase activity. *Stem Cells* **24**, 975–985 (2006).
77. Mirabelli, P., Di Noto, R., Lo Pardo, C., Morabito, P., Abate, G., Gorrese, M., *et al.* Extended flow cytometry characterization of normal bone marrow progenitor cells by simultaneous detection of aldehyde dehydrogenase and early hematopoietic antigens: implication for erythroid differentiation studies. *BMC Physiol* **8**, 13 (2008).
78. Dylla, S.J., Beviglia, L., Park, I.K., Chartier, C., Raval, J., Ngan, L., *et al.* Colorectal cancer stem cells are enriched in xenogeneic tumors following chemotherapy. *PLoS ONE* **3**, e2428 (2008).
79. Park, I.K., Morrison, S.J., and Clarke, M.F. Bmi1, stem cells and senescence regulation. *J Clin Invest* **113**, 175–179 (2004).
80. Molofsky, A.V., Pardal, R., Iwashita, T., Park, I.K., Clarke, M.F., and Morrison, S.J. Bmi-1 dependence distinguishes neural stem cell self-renewal from progenitor proliferation. *Nature* **425**, 962–967 (2003).
81. Molofsky, A.V., He, S., Bydon, M., Morrison, S.J., and Pardal, R. Bmi-1 promotes neural stem cell self-renewal and neural development but not mouse growth and survival by repressing the p16Ink4a and p19Arf senescence pathways. *Genes Dev* **19**, 1432–1437 (2005).
82. Mizrak, D., Brittan, M., and Alison, M.R. CD133: molecule of the moment. *J Pathol* **214**, 3–9 (2008).
83. Alison, M.R., Murphy, G., and Leedham, S. Stem cells and cancer: a deadly mix. *Cell Tissue Res* **331**, 109–124 (2008).
84. Bauer, N., Fonseca, A.V., Florek, M., Freund, D., Jászai, J., Bornhäuser, M., *et al.* New insights into the cell biology of hematopoietic progenitors by studying prominin-1 (CD133). *Cells Tissues Organs* **188**, 127–138 (2008).
85. Burkert, J., Wright, N.A., and Alison, M.R. Stem cells and cancer: an intimate relationship. *J Pathol* **209**, 287–297 (2006).

86. Shmelkov, S.V., Butler, J.M., Hooper, A.T., Hormigo, A., Kushner, J., Milde, T., *et al.* CD133 expression is not restricted to stem cells, and both CD133+ and CD133- metastatic colon cancer cells initiate tumors. *J Clin Invest* **118**, 2111–2120 (2008).
87. Kelly, P.N., Dakic, A., Adams, J.M., Nutt, S.L., and Strasser, A. Tumor growth need not be driven by rare cancer stem cells. *Science* **317**, 337 (2007).
88. Schatton, T., Murphy, G.F., Frank, N.Y., Yamaura, K., Waaga-Gasser, A.M., Gasser, M., *et al.* Identification of cells initiating human melanomas. *Nature* **451**, 345–349 (2008).
89. Quintana, E., Shackleton, M., Sabel, M.S., Fullen, D.R., Johnson, T.M., and Morrison, S.J. Efficient tumour formation by single human melanoma cells. *Nature* **456**, 593–598 (2008).
90. Yoo, M.H. and Hatfield, D.L. The cancer stem cell theory: is it correct? *Mol Cells* **26**, 1–10 (2008).
91. Carney, D.N., Gazdar, A.F., Bunn, P.A., Jr., and Guccion, J.G. Demonstration of the stem cell nature of clonogenic tumor cells from lung cancer patients. *Stem Cells* **1**, 149–164 (1982).
92. Chen, Y.C., Hsu, H.S., Chen, Y.W., Tsai, T.H., How, C.K., Wang, C.Y., *et al.* Oct-4 expression maintained cancer stem-like properties in lung cancer-derived CD133-positive cells. *PLoS ONE* **3**, e2637 (2008).
93. Karoubi, G., Gugger, M., Schmid, R., and Dutly, A. OCT4 expression in human non-small cell lung cancer: implications for therapeutic intervention. *Interact Cardiovasc Thorac Surg* **8**, 393–397 (2009).
94. Eramo, A., Lotti, F., Sette, G., Pilozzi, E., Biffoni, M., Di Virgilio, A., *et al.* Identification and expansion of the tumorigenic lung cancer stem cell population. *Cell Death Differ* **15**, 504–514 (2008).
95. Levina, V., Marrangoni, A.M., DeMarco, R., Gorelik, E., and Lokshin, A.E. Drug-selected human lung cancer stem cells: cytokine network, tumorigenic and metastatic properties. *PLoS ONE* **3(8)**, e3077 (2008).
96. Sung, J.M., Cho, H.J., Yi, H., Lee, C.H., Kim, H.S., Kim, D.K., *et al.* Characterization of a stem cell population in lung cancer A549 cells. *Biochem Biophys Res Commun* **371**, 163–167 (2008).
98. Peacock, C.D. and Watkins, D.N. Cancer stem cells and the ontogeny of lung cancer. *J Clin Oncol* **26**, 2883–2889 (2008).
99. Daniel, V.C., Peacock, C.D., and Watkins, D.N. Developmental signalling pathways in lung cancer. *Respirology* **11**, 234–240 (2006).

100. Koch, L.K., Zhou, H., Ellinger, J., Biermann, K., Höller, T., von Rücker, A., *et al.* Stem cell marker expression in small cell lung carcinoma and developing lung tissue. *Hum Pathol* **39**, 1597–1605 (2008).
101. Yu, Q., Su, B., Liu, D., Liu, B., Fan, Y., Wang, Y., and Meng, X. Antisense RNA-mediated suppression of Bmi-1 gene expression inhibits the proliferation of lung cancer cell line A549. *Oligonucleotides* **17**, 327–335 (2007).
102. Grosveld, G.C. Gamma-secretase inhibitors: notch so bad. *Nat Med* **15**, 20–21 (2009).
103. Subramanian, J. and Govindan, R. Small cell, big problem! Stem cells, root cause? *Clin Lung Cancer* **9**, 252–253 (2008).
104. Iwasa, T., Okamoto, I., Suzuki, M., Nakahara, T., Yamanaka, K., Hatashita, E., *et al.* Radiosensitizing effect of YM155, a novel small-molecule survivin suppressant, in non-small cell lung cancer cell lines. *Clin Cancer Res* **14**, 6496–6504 (2008).
105. Hu, T., Liu, S., Breiter, D.R., Wang, F., Tang, Y., and Sun, S. Octamer 4 small interfering RNA results in cancer stem cell-like cell apoptosis. *Cancer Res* **68**, 6533–6540 (2008).
106. Liang, Y. An expression meta-analysis of predicted microRNA targets identifies a diagnostic signature for lung cancer. *BMC Med Genomics* **1**, 61 (2008).
107. Aboody, K.S., Najbauer, J., and Danks, M.K. Stem and progenitor cell-mediated tumor selective gene therapy. *Gene Ther* **15**, 739–752 (2008).
108. Macchiarini, P., Jungebluth, P., Go, T., Asnaghi, M.A., Rees, L.E., Cogan, T.A., *et al.* Clinical transplantation of a tissue-engineered airway. *Lancet* **372**, 2023–2030 (2008).

Index